Cellular and Molecular Control
of Direct Cell Interactions

NATO ASI Series

Advanced Science Institutes Series

A series presenting the results of activities sponsored by the NATO Science Committee, which aims at the dissemination of advanced scientific and technological knowledge, with a view to strengthening links between scientific communities.

The series is published by an international board of publishers in conjunction with the NATO Scientific Affairs Division

A	**Life Sciences**	Plenum Publishing Corporation
B	**Physics**	New York and London
C	**Mathematical and Physical Sciences**	D. Reidel Publishing Company Dordrecht, Boston, and Lancaster
D	**Behavioral and Social Sciences**	Martinus Nijhoff Publishers
E	**Engineering and Materials Sciences**	The Hague, Boston, and Lancaster
F	**Computer and Systems Sciences**	Springer-Verlag
G	**Ecological Sciences**	Berlin, Heidelberg, New York, and Tokyo

Recent Volumes in this Series

Volume 95—Drugs Affecting Leukotrienes and Other Eicosanoid Pathways
edited by B. Samuelsson, F. Berti, G. C. Folco, and G. P. Velo

Volume 96—Epidemiology and Quantitation of Environmental Risk in Humans
from Radiation and Other Agents
edited by A. Castellani

Volume 97—Interactions between Electromagnetic Fields and Cells
edited by A. Chiabrera, C. Nicolini, and H. P. Schwan

Volume 98—Structure and Function of the Genetic Apparatus
edited by Claudio Nicolini and Paul O. P. Ts'o

Volume 99—Cellular and Molecular Control of Direct Cell Interactions
edited by H.-J. Marthy

Volume 100—Recent Advances in Nervous System Toxicology
edited by Corrado L. Galli, Liugi Manzo, and Peter S. Spencer

Volume 101—Chromosomal Proteins and Gene Expression
edited by Gerald R. Reeck, Graham H. Goodwin,
and Pedro Puigdomènech

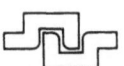

Series A: Life Sciences

Cellular and Molecular Control of Direct Cell Interactions

Edited by

H.-J. Marthy

C.N.R.S.—Pierre and Marie Curie University
Banyuls-sur-Mer, France

Plenum Press
New York and London
Published in cooperation with NATO Scientific Affairs Divison

Proceedings of a NATO Advanced Study Institute on
Cellular and Molecular Control of Direct Cell Interactions in
Developing Systems,
held September 10–22, 1984,
in Banyuls-sur-Mer, France

NATO - ASI - 1984

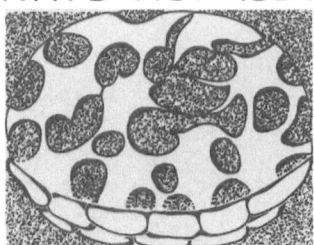

BANYULS/MER - FRANCE

The ASI Logo was created by M.-J. Bodiou (Laboratoire Arago) after
Figure 8 in Chapter 8 by H.-J. Marthy, and shows interconnected cells
in a Cephalopod embryo.

Library of Congress Cataloging in Publication Data

NATO Advanced Study Institute on Cellular and Molecular Control of Direct Cell
Interactions in Developing Systems (1984: Banyuls-sur-Mer, France)
    Cellular and moleculr control of direct cell interactions.

(NATO ASI series. Series A, Life sciences, vol. 99)
    "Proceedings of a NATO Advanced Study Institute on Cellular and Molecular
Control of Direct Cell Interactions in Developing Systems, held September 10–12,
1984, in Banyuls-sur-Mer, France"—Verso of t.p.
    Bibliography: p.
    Includes indexes.
    1. Cell interaction—Congresses. 2. Cellular control mechanisms—Congress-
es. 3. Developmental cytology—Congresses. I. Marthy, H.-J. II. North Atlantic
Treaty Organization. Scientific Affairs Division. III. Title. IV. Series: NATO advanc-
ed science institutes series. Series A, Life sciences; v. 99.
QH604.2.N37   1984   574.3'3   85-28241
ISBN 978-1-4684-5094-1   ISBN 978-1-4684-5092-7   (eBook)
DOI   10.1007/978-1-4684-5092-7

©1985 Plenum Press, New York
Softcover reprint of the hardcover 1st edition 1985

A Division of Plenum Publishing Corporation
233 Spring Street, New York, N.Y. 10013

To my daughter and son Sibylle and Christoph

PREFACE

The NATO Advanced Study Institute on "Cellular and Molecular
Control of Direct Cell Interactions in Developing System" has been
attended by 15 invited main lecturers and 60 participants.
According to its purpose senior scientists, postdoctoral trainees
and graduate students working in areas like biology, biochemistry,
electrophysiology, medicine etc... could discuss their common
interest in the various structural, ultrastructural, molecular and
functional aspects of cell interactions in developing in vivo and
in vitro systems. Whereas the topics of the first week have been
mostly concerned with the general aspects of cell interactions in
embryogenesis (section I and II of this book), the second week has
been mainly devoted to the structures and functions of the direct
cell contact sites at the membrane level as gap junctions,
including electrophysiological aspects, dye coupling and selective
cell-cooperation in some model systems as the neuro-muscular
junctions (section III-V of this book). A multidisciplinary and
stepwise approach, from initial cell contacts in early
embryogenesis up to well defined selective cell cooperation,
appeared to be an efficient means to provide answers to the
question of how cells control, in a dynamic system as given in a
differentiating embryo, their multiple temporary and permanent
interactions so necessary for ordered cell positioning, cell
linking and well established cell-to-cell communication. By
treating in depth numerous biological and molecular aspects on the
formation and functioning of cell-to-cell and cell-to-substrate
interactions in in vivo and in vitro systems, including aspects of
cell adhesion, cell migration and of the extracellular matrix, the
course certainly formed an unique basis, particularly for younger
participants, to get an excellent overview in these fields and thus
a broad basis for research concepts. It was clear indeed from the
very beginning that no clearcut answer would be possible to the
question above. It should be mainly understood to be a "challenge
question" to inspire perspectives when defining, in biological and
molecular terms and increasingly precisely how cells control their
direct interactions by determining the dynamics of their periphery
and the composition of their immediate environment. The comparison
of the results gained in the fields of classical and molecular

biology, physiology and biochemistry as occured at our ASI, provides an excellent means to ensure more accurate interpretations of cell interaction phenomena. This book, which comprizes the contributions of all invited main lecturers and some additional papers of participants, needs therefore to be understood as a sort of "inventory" of the state of actual knowledge and trends in cell interaction research. As the course has shown, it also needs to be, for a larger public, a basis, on which new ideas and new experimental approaches for "answering" the challenge question above may progressively develop.

I take this opportunity to thank the NATO Scientific Affairs Committee for having supported this Advanced Study Institute. The contribution of the "Société Française de Biologie du Développement" also is kindly acknowledged. I wish to thank each of the lecturers and ASI-participants for their effort in making the course a success. I also thank the Direction of the Laboratoire Arago for housing facilities. Important organisational work has been done by Mrs. N. Clara and my wife, U. Marthy. Mr. R. Tait has assisted in many times in text corrections. Finally, the careful and diligent work of Mrs. M. Mutot-Duchêne, who prepared these manuscripts, is greatly acknowledged.

H.-J. Marthy

CONTENTS

CONTENTS

SECTION I

CELL ENCOUNTER : MOLECULAR AND BIOLOGICAL ASPECTS

OF INITIAL CONTACTS

Max M. Burger and Gradimir Misevic

Department of Biochemistry
Biocenter of the University of Basel
Klingelbergstr 70, CH-4056 Basel, Switzerland

" I am fascinated by the detail ... the sum of
many details can shape the contour, can create
the beauty of form. "
Alberto Giacometti
(from a discussion with André Parinaud,
Beyeler Gallery, 1964).

CONTENTS

2. DIMENSION TIME

2.1. Contacts without consequences

2.2. Contacts with temporary consequences

2.3. Permanent consequences

2.4. Evidence that a molecular surface recognition step precedes
     the formation of general intercellular bridges like gap
     junctions

3. CELL-CELL RECOGNITION IN SPONGES

3.1. The aggregation factor has two functional domains

3.2. The polyvalent cell binding domain

3.3. Reconstitution of a polyvalent cell binding domain through
     crosslinking of a small subunit

INTRODUCTION

     The creation of form during embryogenesis is characterized by
mitosis, cell and tissue migration, as well as by stabilization of
the particular configuration of each single cell (shape) and of the
sum of cells (relative positioning). The navigation that cells and
tissues seem to perform during migration, and the recognition of
the final configuration of cellular topography are processes which
will be determined by the encounter between cells.

     We know essentially nothing about the molecular processing
occurring at encounter between cells and very little about their
biology. On the other hand we do know quite a bit about the
biochemistry of cell surfaces of isolated cells, and it is
therefore rather unexpected that not more is known nor has been
done so far to elucidate the processes occurring during
intercellular encounter.

     A rational approach would be to consider separately, the two
different dimensions space and time and thus to analyze first the
structural components that may possibly be involved and then each
of the various steps into which the encounter process can be
subdivided.

     In an addendum we consider the recognition between sponges of
the same species which is mediated via an intercellular macro-
molecule and which can serve as an example of the most simple case
of encounter at the most primitive level of animal cell evolution.

1. DIMENSION SPACE : STRUCTURAL AND MOLECULAR ELEMENTS OF THE CELL
   WHICH MAY BE INVOLVED IN CELLULAR ENCOUNTER

The cellular periphery is built up of several sublayers and it should be considered as an organelle that is involved in the process of cell-cell encounter. At the microscopic level special structures like filopodia, microvilli, perhaps blebs and ruffles, are the first to sense the contact between a cell and its surroundings, as for instance the neighbouring cell. At the molecular level it is obvious that the most peripheral macromolecules will first be in touch with structures on the confronted cell. Then, progressively deeper layers will become involved, from the membrane molecules and substructures down to the cytoskeleton and possibly the cytoplasm, which will become aware and presumably respond to the new contact. It will therefore be necessary to survey and characterize all the structures that are potentially partaking in the encounter.

1.1. <u>Para- and extracellular space</u>

1.1.1. Special cellular protrusions

The earliest contact occurs usually at specialized regions of the cell periphery. Filopodia are the largest protrusions and therefore the most important candidates for the very first contact. They can measure up to 100 μm such as in the secondary mesenchyme cells of sea urchins where they probe the basal lamina reaching through the acellular blastocoel cavity. Microvilli are shorter cellular processes. Ruffles at the tips of migrating cells and blebs are wider cellular extrusions. The extent to which they are directly involved in searching processes is not clear.

Theoretically cellular processes may be formed by alterations in the rigidity of the extracellular and paracellular matrix which may function as holes in a cage. Alternatively changes in the molecular arrangement of the membrane core and most likely specializations in the cytoskeleton may influence formation of cellular processes. One of the key questions is the extent to which the cytoskeleton induces these protrusions and maintains them, and whether they carry molecular arrangements at their tips which can fulfill specialized functions (e.g., clustered paracellular or membrane core molecules).

1.1.2. Extracellular matrix

One may distinguish for operational and practical reasons between an extracellular matrix (ECM) which interdigitates with integral and peripheral glycoproteins of the membrane and which comigrates partially with a moving cell, and the extracellular matrix which never comigrates with cell surface elements. Whatever is pulled along with the glycocalyx of the ECM should be defined as part of the paracellular layer.

The cutoff or gliding zone between two cells is thus somewhere in the extracellular matrix and outside of the paracellular matrix. The shear and ripoff zone when cultured cells are pulled off their artificial substrate or dissociated from each other with protease or EDTA is not necessarily the same as the one where cells glide past each other in the actual tissue. This is a serious obstacle for those who do model analyses of the extracellular matrix which has been secreted onto plastic dishes and which in composition structure and quantity may not reflect the intercellular space but could be an artificial response by cells in culture.

Collagen fibrils (several μm long) are together with proteoglycans the best known space fillers in the extracellular matrix (1). The type of proteoglycans found are primarily chondroitin sulfate, hyaluronic acid and heparan sulfate. They are found in the plastic adhesion sites of fibroblasts, i.e. their cellular footpads. Their composition seems to be dependent on when they are secreted after plating the cells (2). Their size and dimensions are not yet fully known. Even though cartilage proteoglycan aggregates will not be identically organized as those occurring in other tissues it may serve as a reminder of how extracellular matrix proteoglycans in general may look. The backbone is formed by a 1.2 μm long hyaluronic acid molecule to which about 30 molecules (of Mr = 2.5×10^{6}) are attached through small linker glycoproteins, with lectin type bindings (3, 4). Without the linkers the hyaluronate complex is 5 times shorter, with 3 times less proteoglycans attached to the core hyaluronic acid (5). The single proteoglycan molecule which is about 200-300 nm long has about 40-150 chondroitin sulfate chains (Mr = 20,000) and about 30-60 keratan sulfate chains (Mr = 4,000-8,000) attached to a common protein backbone (4, 6-10).

An ever increasing list of large glycoproteins seems to be interwoven among collagens and proteoglycan complexes. Since several of them have binding sites for collagens and proteoglycans and since they also have binding sites on the cell surface, they remain good candidates for the linker elements between extracellular matrix elements and molecules attached to the membrane. The best known examples of such molecules are the fibronectins (60 nm) (11-14), laminin (77 nm in epithelial cells) (15, 16), entactin (endodermal cells), chondronectin (chondrocytes). Although it was felt earlier that fibronectin was absolutely indispensable for the attachment of cells to the substratum and perhaps to other cells, enough data have been accumulated showing that some cells can attach to collagen in the absence of any fibronectin or to collagen devoid of the fibronectin binding fragment (17, 18). There is still some debate as to the exact and unique nature of the cellular receptor for fibronectin, and the carbohydrate moieties of glycolipids and glycoproteins have been suggested. It is possible that more than one type of receptor may be involved (19).

The question of how the extracellular matrix is linked to the paracellular matrix and to the membrane is studied in many laboratories. Eventually we must however solve the problem of how cells migrate ; this means gliding of layers past each other and/or rupture of such linkages. Essentially nothing is known about such processes. It may be more difficult to approach this question in such tightly packed tissues as those in vertebrate organs. In many invertebrates, however, as the sponges for instance, cells (archaeocytes) are migrating even in the adult tissue through vast extracellular spaces filled with collagen and proteoglycan-like material (mesohyle) where the cell mass is less than 10%. Cell migration can be studied there as an almost exclusive cell-extracellular matrix phenomenon.

1.1.3. Glycocalyx

Those molecules that are integrated in the plasma membrane core and that carry carbohydrates on their external portion are the main constituents of the glycocalyx. Whether besides these integral glycoproteins and glycolipids the peripheral glycoproteins, i.e. those that do not need detergents or proteases for extraction but which come off in low salt, can also be counted as part of the glycocalyx, is a matter of definition. They may be part of the paracellular matrix which is carried along when cells move through the extracellular matrix. It is difficult to assign a given species of peripheral glycoproteins to the paracellular or to the immobile extracellular matrix until we find means to determine what is pulled along when cells move in vivo. Most likely there is an interdigitation between the integral glycocalyx, peripheral glycoproteins of the paracellular layer and immobile extracellular matrix, via carbohydrate-carbohydrate interactions, classical protein-protein interactions, as well as lectin-like protein-carbohydrate interactions (13, 20-22).

An ever increasing list of antigens and lectins can be shown on cell surfaces before and after dissociation, many of them integral and some peripheral. Some are glycoproteins and some are glycolipids. They are expected to be involved in contact phenomena since antibodies inhibit or they themselves promote or inhibit aggregation and adhesion. Among them are the slime mold aggregation sites (23), muscle differentiation lectins (24), erythroblast nurse macrophage lectin (25), the cholate extractable uvomorulin which compacts mouse embryos (26, 27), the neural cell adhesion molecule (N-CAM), the neuron-glia cell adhesion molecule (L-CAM) (28-31). The best approach to assess whether glycocalyx molecules are

relevant for contact and attachment to the environment is to have
stable mutations in a given surface molecule. For slime molds that
was fairly easy but such mutants have now been isolated even from
mammalian cells. Thus a mutant which cannot acetylate glucosamine-
6-P and consequently does not build up any GlcNac, GalNac or even
sialic acid in oligosaccharides displays less adhesion to
substrates and particularly fewer focal adhesions (32).Furthermore,
WGA lectin resistant cells could be found which were lacking some
of their peripheral sialic acid thereby exposing galactose. This
surface change correlated with an increase in cell aggregation and
decrease in metastatic capacity (33). Similarly ricin resistant BHK
cells which lost their exposed galactose residues adhered less
(34).

A question that may be with us for a while is whether integral
glycoproteins are always available at early encounter, or whether
their exposition can be modulated by configurational changes or
neighbouring effects with integral and peripheral glycoproteins. If
they were unfolded, they could easily reach points 10 nm and more
above the headgroups of the phospholipids. But do they unfold, even
partially ? Also how much of the carbohydrate portion of the
glycolipids is available to neighbouring cells upon contact ? It is
known that they can serve as receptors for quite a number of
ligands (cholera toxin, tetanus toxin, NK-receptor, etc. (35)), and
so they must be available for protein ligands from outside.

1.1.4. Phospholipid headgroups

This region which in fact belongs already to the membrane core
but is still on the hydrophilic side is rarely involved in direct
physical apposition in encounter or contact between cells. A later
contact phases between myoblasts for instance these headgroups must
act as a barrier for fusion. A rearrangement must occur which
involves these headgroups most probably leading to micelle
formation and then to fusion. Similar processes are expected after
encounter with viruses or for that matter any particulate or even
soluble material leading to endocytosis.

Tight junctions or zonulae occludentes are the closest contact
zones between cells and they appear in the electron microscope to
have partly melted bilipid leaflets. It is however unclear to what
degree the phospholipid headgroups are interdigitated.

Charge repulsion when cells approach each other must primarily
be due to sialic acid of glycoproteins, proteoglycans of the
paracellular layer, acidic amino acids and sialic acid of
glycolipids. To what degree the phospholipid groups are involved
because the outer charges have a considerable degree of physical
flexibility being on movable stalks and exposing thereby the lower
phospholipid head groups is unclear. Liposomal model work will have

to be done with inserted glycocalyx components. Phospholipase treatment of erythrocytes may not be conclusive since alteration of head groups could theoretically alter the position and flexibility of the charge carrying glycoproteins.

1.2. Membrane core

During encounter the relevance of the membrane core may lie in two aspects. First, in the fact that the interacting elements of the glycocalyx are movable at their attachment sites in the lipid core. Thus receptors can move within the plane of the bilayer but they may also move vertically to the plane, but to a very minor degree only. Secondly, paracellular elements will have to communicate directly or indirectly with cytoplasmic elements like the cytoskeleton. This occurs at intermediate junctions between cells as well as at focal junctions with the substratum in vitro and the basal lamina in vivo.

One of the more challenging questions is a detailed molecular tracing of these extracellular and intracellular structural elements through the membrane itself. To what degree transmembranous proteins will guarantee the link, how they interact with their respective partners out- and inside of the plasma membrane and to what degree structural or linker elements are interacting in the membrane itself and are thus linking the two systems is unclear. It should not take too long to resolve this problem since the techniques to obtain answers to such questions (lipophilic crosslinkers, immunelectronmicroscopy, etc.) have improved considerably in the last few years. Ring and spot desmosomes seem to be the prime candidates in epithelial cells primarily because tono- and microfilament networks are seemingly continous through many cells and accumulating in zones where cells are under mechanical stress (36-38). This can be observed for instance in the form of the purse string phenomenon on the lumen side when epithelial tubuli are formed.

1.3. Submembranous filamentous network

Many cells have well developed submembranous network of structural elements as well as a cytoskeleton pervading the cytoplasm down to the organelles and the nucleus. Some cells like the erythrocytes display only the submembranous filamentous network consisting mainly of spectrin (39-41). Similar elements (fodrin, etc.) have been found in other cells as well (42-45). Microfilaments are often found as submembranous structures (fibroblasts, etc.), some of them as aggregates which appear in bundles. Microtubuli are found more in the cytoplasm in general and are most likely not anchored in the membrane at least not directly.

While for the erythrocytes the attachment of spectrin to the membrane is known to be mediated via a peripheral protein (ankyrin) which is linked to the integral membrane glycoprotein (band 3) (46), much less is known for epithelial and fibroblastic cells. Some of the possible candidates for linker proteins also seem to be peripheral proteins ; they are vinculin (47-50), alpha-actinin (51), besides some less characterized proteins (52, 53). Attachment of the cytoskeleton and bundling of microfilaments is not a static but a dynamic phenomenon. Thus for movement, endo- and exocytosis attachment points have to be formed and broken as much as bundles are formed and dissociated. The entrance into and the exit from the membrane of anchoring molecules and the association among them to form bundles may be regulated through modifications in their secondary and tertiary structures. Phosphorylation is a very fashionable model as well as perhaps more exotic modifications like covalent lipid attachment. One may however also consider that changes in the lipid composition as for instance diglyceride formation from phosphatidylinositol may allow the association of a linker protein like Alpha-actinin within the membrane (54), since it has an unexpectedly high and specific affinity to this rare and particular lipid (55).

1.4. Specializations within the cell surface

This book deals with most of the specialized contact zones. In the frame of this article they would have to be dealt with in detail but are therefore only ennumerated. Among them are the tight junctions (zonulae occludentes), ring desmosomes (zonulae adhaerentes), spot desmosomes (maculae adhaerentes), and gap junctions.

The protrusions are of course also specializations but probably not so much of the surface only but equally of the cell itself. They are ennumerated under 1.1.1.

2. DIMENSION TIME

Encounter is most likely not only a process whereby surface molecules determine the outcome alone, but must be a multistep process where several checks and controls are built in guaranteeing the appropriate outcome (56).

In embryonal encounter during morphogenesis a lot of cellular probing seems to be going on as if decisions were made at each step. During gastrulation of sea urchins or synaptogenesis of spinal cord outgrowths on superior cervical ganglion cells, the filopodia or ruffled front probe their environment until they reach their target. It seems that if they miss the target by 5-10 μm they pass by as if chemotactic substances were not being released over

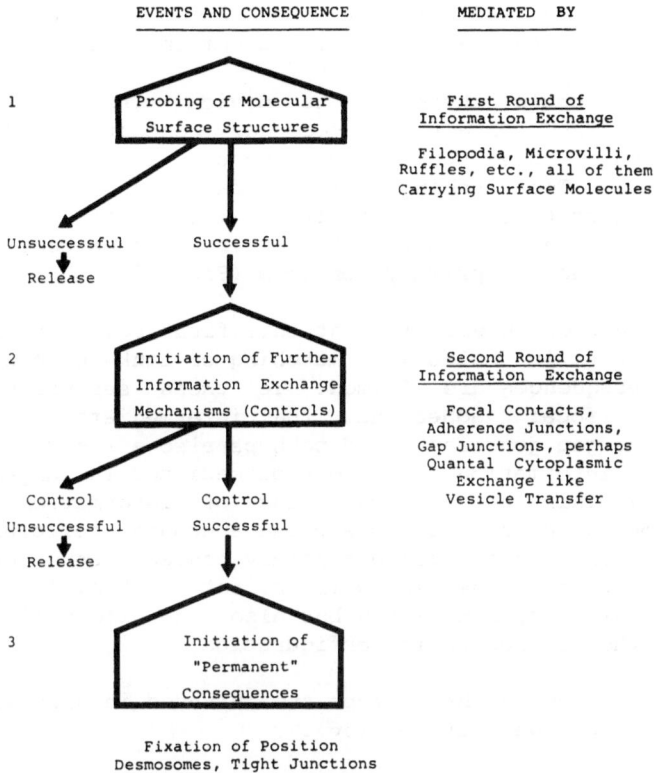

Fig. 1.   An attempt of subdividing the event of cell-cell
          encounter.
          The two recognition and information transfer events
          are considered to be a minimum when a permanent
          consequence should   ensue   from   the   encounter.
          Theoretically the first one could also be mediated
          via special structures that are set   up   upon
          encounter as seen for the second recognition and
          information transfer event. Cells which set up such
          structures (focal contacts, gap junctions, etc.)
          will however first have to make molecular contacts
          in their outer layers as well as their glycocalyx
          before they form contact   structures.   Slightly
          modified from Burger, 1979, and Burger et al., 1980.
          (56, 59).

distances more than 5 μm (57). If the leading front contacts a
nontarget cell surface, it retracts again. If the filopodium
contacts a target cell surface and finds seemingly the "right"

surface structures (58) it attaches and after a while the cell
pulls itself up to the target cell and remains there.

Encounter should therefore be divided into at least two
processes : primary contacts and secondary contacts. We proposed to
assign to each of these two contact phases a "decision" making
process (56). The probing cell surface can either find an
appropriate conatct zone and lock itself in, or it can fail and
continue to probe elsewhere. The latter would be contact without
consequences among the primary contacts (Fig. 1).

If the surface interactions at this first round of information
exchange result in molecular matching or link-ups contacts with
temporary consequences are formed. To these secondary contacts
belong the focal adhesions, the intermediate or adherence
junctions, the gap junctions, and perhaps also a form of quantal
exchange of information which may be mediated in larger packages
like vesicles. This is the second round of information exchange.
Depending on the outcome of this close encounter cellular contacts
may again be ruptured or stable tertiary contact structures which
have permanent consequences will be set up (Fig. 1). This would
include not only differentiation but also the immobilization and
fixation of the morphogenetic configuration.

In the following these steps are surveyed in abbreviated form
and have been reviewed earlier (56).

2.1. Contacts without consequences

Primary fleeting contacts are established while filopodia and
other protrusions are probing neighbouring cells. Molecules in the
paracellular and the glycocalyx layer must be involved. Similarly
cells that are passively moved like lymphocytes during "homing" or
aging erythrocytes will recognize their final destination via
surface molecules. Such initial contacts may not last even long
enough to establish special structures like gap or Abercrombie's
intermediate junctions.

If these contacts do not reveal any affinities because the
proper molecular structures are missing or are not available in
sufficiently exposed positions, in sufficient densities or in
correct patterns, not even temporary contacts will be formed and
signals cannot be exchanged or special structures established.

2.2. Contacts with temporary consequences

For the safety of establishing the correct connections it will
be advantageous for a cell to revise its general position and the
particular primary and fleeting connections established in a second
round of information exchange.

Special structures and mechanisms, some of them not yet known will facilitate this second decision making process. Among them may be the gap junctions, other ways of intercellular communication like the exchange of cytoplasmic elements in vesicular form, as well as the establishment of adhaerens or intermediate junctions and focal contacts to the basal lamina. Some of them, like the adhaerens junctions, can be formed in a few seconds (10-30 s) and could mediate mechanical "information" through the establishment of the appropriate tension lines, that is the cytoskeleton (60). Others like the gap junctions are set up in a short time as well (2-5 minutes) and could mediate small molecular exchanges of the local or general state of the cytoplasm (61). The time range for the exchange of quantal vesicular information is - if it exists at all - unknown. One should not underrate it since exocytosis in neural tissue can occur in the range of milliseconds and endocytosis may not take appreciably longer than the establishment of gap junctions depending on whether the appropriate structures for uptake and uncoating are locally available (62).

After processing the information, a decision is reached as to whether quantity and quality of the information are either appropriate or insufficient and incorrect. The latter will on one hand either lead to the release of the contact protrusion or to the sampling of additional information elsewhere. If the information is appropriate, however, a specific preprogrammed series of responses will follow, which will then belong to the permanent consequences of encounter, as shown in Fig. 1.

2.3. Permanent consequences

Provided the secondary contacts and the second round of information exchange has yielded "satisfactory" results the stage is set for the final responses of encounter. Among them are a series of processes which immobilize the particular cell like the cessation of ruffling, the formation of tight junctions (zonulae occludentes), of spot and ring desmosomes, and last but not least of extracellular matrix, both in between and as a capsule around groups of cells. Shape changes through the final rearrangement of the cytoskeleton will contribute to the appropriate morphology and both biochemical and structural alterations will be initiated as a consequence of differentiation processes set into motion at this point.

2.4. Evidence that a molecular surface recognition step precedes the formation of general intercellular bridges like gap junctions

From the surface topography it seems to be logic to assume, as was done in 1.1., that if surface molecules have a role in

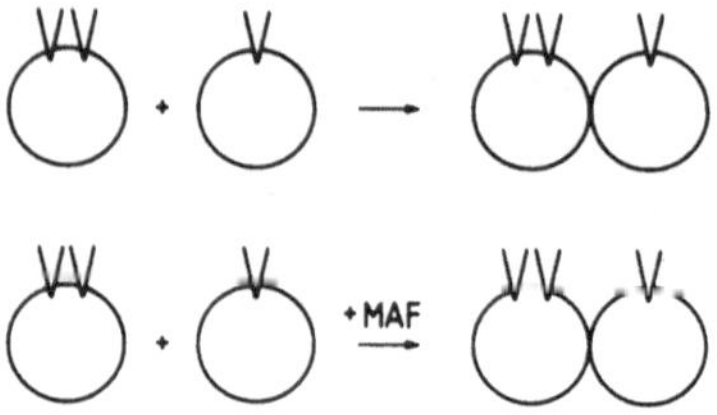

Fig. 2.   Aggregation factor is a prerequisite for inter-
          cellular communication.
          Cells that have been mechanically dissociated and
          still retain their factor at the surface will
          develop a lower membrane resistance between them
          than towards the remainder of their surroundings,
          i.e. they couple (not shown). Cells which are
          stripped of their surface factor will not couple
          even if they are in closest apposition (upper line).
          If aggregation factor is added to the right species
          such cells will couple no later than 30 to 40 min.

encounter they would precede information exchange structures set up
later like gap junctions. Very little evidence is available as to
this question as well as to any particular sequence of events in
encounter. The scheme in Fig. 1 is however supported by studies on
sponge cell interactions, some of which will be discussed in the
next chapter.

Sponge cells of the same species do couple electrically if
brought together under the microscope after dissociating them. If
the recognition and species-specific aggregation mediating
molecules are removed from the surface there is however no coupling
anymore (63). As soon and only if the specific recognition
mediating molecule is added back the cells do they couple again
(see Fig. 2). Sponges, like more and more cell types, are shown to
couple electrically although they do not have classical gap
junctions in freeze-cleavage electronmicrographs. The fact that
they do couple indicates that they must have surface structures
which allow ion and small molecular fluxes like the classical gap
junctions.

Primary contact through communicative surface molecules is
here therefore a prerequisite for the establishment of the
secondary communicative cell-cell channels. A more detailed
analysis with different sponge species and controls will have to be
carried out, and other data in higher animals will be necessary.

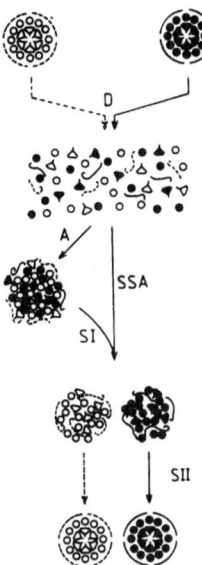

Fig. 3.   Reaggregation of sponge cells.
          Mixtures of cells from two different species undergo
          species-specific   aggregation   (SSA)   followed by
          internal sorting (SII). Unspecific aggregation (A)
          may preceed species-specific sorting out (SI).

3. CELL-CELL RECOGNITION IN SPONGES

     The first multicellular organisms appearing during evolution
had to develop a mechanism for recognizing each other. Since
sponges belong to the most simple and oldest phylum they probably
possess the most archaic cell-cell recognition system.   Studies   on
cell-cell recognition in sponges may thus bring some insights into
cellular recognition in general. A practical advantage when working
with sponges is that dissociation of cells can be carried out
without the use of proteolytic enzymes. This then enables us to
study   the   intact   surface molecules which are involved in
recognition processes.

     In some species of marine sponges dissociated cells will
undergo immediately species-specific reaggregation when incubated
together without first forming mixed aggregates (Fig. 3) (64-67).
During the last two decades efforts have been made to isolate
molecules which mediate species-selective recognition leading to
species-selective reaggregation of some of these sponges (68-71).
In the marine sponge Microciona prolifera a minimum of three

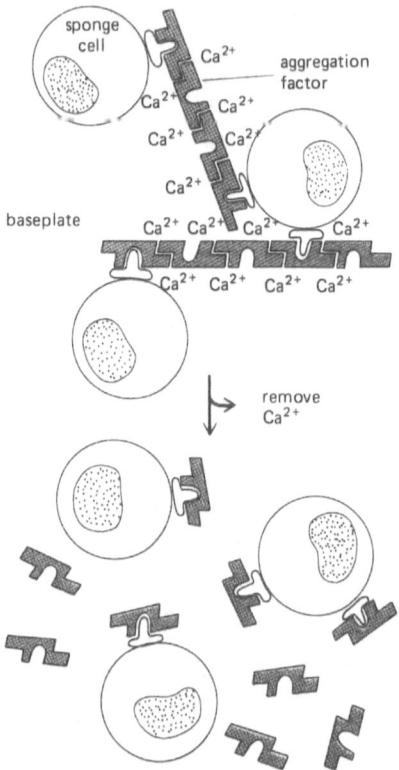

Fig. 4.   Requirement for the three components for the spe-
          cies-specific reaggregation of sponge cells.
          Sponge cells dissociated in Ca^{2+} / Mg^{2+} free
          seawater (CMF-SW) release aggregation factor (AF)
          that is not directly attached to the cell surface
          receptor called baseplate. Extensive washing of
          cells allows baseplate bound AF to dissociate (not
          shown). Both AF and Ca^{2+} must be added for
          reaggregation. Treatment of AF-free cells with low
          salt releases a receptor or baseplate (BP) which
          interacts with the AF. Hypotonically treated cells
          lacking BP do not aggregate in response to AF, but
          can be restored to do so by preincubation of cells
          with baseplate (70). From Molecular Biology of the
          Cell, eds. Alberts B., Bray D., Lewis J., Raff M.,
          Roberts K. and Watson J.D.

components are found to be necessary for aggregation :

1) a proteoglycan-like molecule, named aggregation factor
2) cell surface receptor(s) and
3) Ca^{2+} ions (Fig. 4)

3.1. The aggregation factor has two functional domains

Microciona prolifera aggregation factor (MAF) is a proteoglycan-like molecule of Mr = 2 x 10^7 containing 50-60% carbohydrate and 40-50% protein by weight (68-72). In electron micrographs MAF appears as a fibrous complex molecule with 15-16 arms extending like rays from a circular backbone (72). The MAF arms are 1100 Å long and the circular backbone has a diameter of 800 Å. In the presence of EDTA, MAF dissociates only after several weeks into the circular backbone, arms and arm fragments (72). By use of urea and EDTA at an elevated temperature (80°C), however, dissociation could be achieved in a few hours (73).

Jumblatt et al. (1980) have demonstrated that MAF mediates species-specific reaggregation of Microciona prolifera cells via two functional domains, a cell binding domain and an MAF-MAF interaction domain. Binding of MAF to cells via the cell binding domain does not require the physiological 10 mM Ca^{2+}, contrary to the MAF-MAF polymerization which occurs in the presence of 10 mM Ca^{2+} (74, 75). The MAF-MAF domain is also irreversibly inactivated by EDTA treatment or extensive periodate oxidation (74), but neither of these treatments had any effect on the MAF cell binding activity.

Although the species-specific aggregation is mediated via the two functional domains of MAF, it is not essential that both domains should possess species-specific interactions. Jumblatt et al. (1980) have shown that the cell binding domain of MAF is species-specific, but it is still not clear to which extent the MAF-MAF Ca^{2+} dependent interactions might not also be species-specific. As suggested by Burkart and Burger (1981) species-specific interactions of MAF with itself would be based on different charge spacings within the different aggregation factor molecules from different species (21). This would permit a good fit of negative charges only between two aggregation factor molecules from the same species. In addition, specificity could also arise from chemical differences in the aggregation factor-aggregation factor binding domain.

3.2. The polyvalent cell binding domain

In an attempt to isolate the cell binding domain, MAF was dissociated and then treated with proteolytic enzymes. The resulting fragments were purified and the presence of cell binding activity was tested (Fig. 5). Interestingly all proteolytic fragments regardless of size were bound only to homotypic cells (73), indicating that each of them has an intact cell binding domain (Table 1). After prolonged digestion of MAF 60% of its mass was converted into the four cell binding fragments (Fig. 5). It was calculated from the recovery of the material in the fragments that

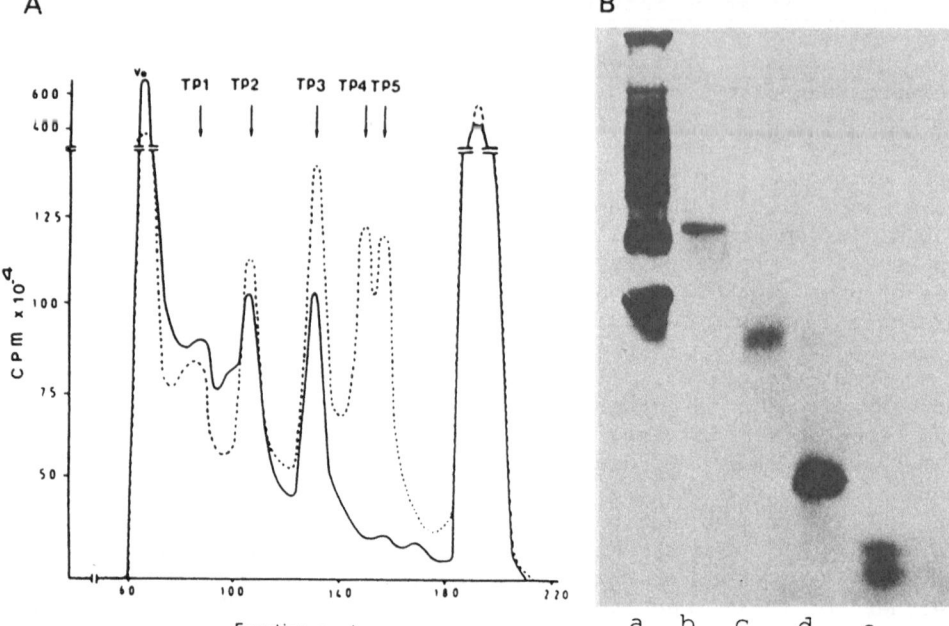

Fig. 5. A) Gel filtration on Sephacryl S-200 of trypsin
treated dissociated ^{125}I-MAF.
0.05 mg/ml of ^{125}I-MAF was treated with trypsin
(final concentration 2 mg/ml) for a period of 2 h
(———) and 24 h (---). At the end of the
incubation time 0.5 mg of MAF was applied to a
Sephacryl S-200 column. The column was eluted with
0.25 M NH_4HCO_3 and 1 ml fractions were collected,
counted in a Packard Gamma spectrometer and analyzed
for protein, neutral hexoses and uronic acid. Peak
fractions TP4 and TP5 were usually pooled because
they could only be occasionally partially resolved.

B) SDS electrophoresis on a 7.5-20% linear gradient
polyacrylamide gel, and autoradiography of ^{125}I
labelled MAF tryptic fragments.
a) 5.0 µg void volume fraction 22 x 10^4 cpm/µg
b) 2.0 µg TP1,   7.2 x 10^4 cpm/µg
c) 2.0 µg TP2,   7.2 x 10^4 cpm/µg
d) 2.9 µg TP3,   6.3 x 10^4 cpm/µg
e) 1.5 µg TP4, 14.6 x 10^4 cpm/µg
After staining and drying the gel was autoradio-
graphed on a NS-2T Kodak X-ray film.

these are present to the extent of several hundred copies in the intact MAF. This finding clearly indicated that the cell binding domain of MAF is as suggested polyvalent (76, 77). It has been discussed that specificity of cell adhesion molecules usually arises via single high affinity interactions (23, 28-31, 76, 77) ; alternatively here in sponges, a multiple number of interaction sites on MAF seems to mediate specificity of aggregation.

The experiments which indicated polyvalency of the cell binding domain still left open the question of whether the cell binding domain was composed of one or several types of binding sites. Several experiments were performed to answer this question. In the first approach the affinity of binding of the fragments to homotypic cells was measured. The apparent association constants were found to decrease proportionately with the decrease of the fragments' molecular weight. Conversely, the number of binding sites per cell increased as the fragment size decreased (Table 2). The simplest explanation for such a correlation would be that the higher affinity of the larger fragments is a product of the higher number of copies of a low affinity cell binding site. If this

Table 1. Binding specificity of ^{125}I-MAF and the MAF fragments.
Binding of radioiodinated MAF or fragments to CMF M. prolifera, M. fusca and C. celata cells was carried out in CMFT-SW under standard assay conditions for 20 min. at 22°C (73). After incubation cells were layered on and centrifuged through 0.1% bovine serum albumin, 10% sucrose in CMFT-SW. Supernatants were aspirated and the pellet counted on a Packard Gamma spectrometer. The number of bound molecules / cell was calculated from the specific radioactivity and Avogadro's number.

MAF frag- ments	Number of molecules applied / assay	% bound to		
		M. prolifera	C. celata	M. fusca
	$\times\ 10^{-12}$			
MAF	0.1	33	2.2	2.1
TP1	11.4	31	8.1	7.9
TP2	17.8	29.4	7.3	6.8
TP3	16.1	31	7.8	6.7
TP4	20.0	45	12	11.1

Table 2. Characteristics of ^{125}I-MAF tryptic fragments binding
to homotypic cells.
K_a values, B_{max} values + S.E. (from four different
experiments), were determined in the same way as
described in the legend of Table 1. Starting
amounts/assay volume were: MAF, 1.6×10^{-4} nmol, TP1,
1.89×10^{-2} nmol, TP2, 2.95×10^{-2} nmol, TP3, 2.67×10^{-2} nmol, TP4, 3.3×10^{-2} nmol.

Peak fractions after trypsin	Mr	K_a (x 10^{-6})	B_{max}
	x 10^{-3}	M	x 10^{-4}
TP1	124	39 + 4.0	100 + 10
TP2	70	21 + 2.1	150 + 15
TP3	27	7.1 + 1.8	270 + 68
TP4	10	3.4 + 1.1	350 + 45

explanation is true one would expect that the fragments and MAF all
have a similar amino acid and sugar composition. Table 3 shows that
a high degree of similarity does indeed exist for the major amino
acids and sugars, except for the lower fucose content of the
smallest fragment. This, together with the finding that larger
fragments could be partially converted into the smaller ones by
further proteolysis, supports even more the hypothesis of the
polyvalence of a single type of cell binding unit.

3.3. Reconstitution of a polyvalent cell binding domain through crosslinking of a small subunit

If this hypothesis of polyvalency is correct the cell binding
unit should be present in the smallest glycoprotein fragment of Mr
= 10,000 of which there are more than 1,000 copies in MAF. It is,
however, still not clear whether both carbohydrate and protein
portions, or either one of them is the actual carrier of the cell
binding activity. Since we have proposed that the 10,000 dalton
glycoprotein fragment (TP4) carries a cell binding site of low
affinity which is present in more than 1,000 copies in MAF, it
should be possible to reconstitute the high binding affinity equal
to that of MAF by crosslinking this fragment into a large
artificial polymer. Crosslinking of the 10,000 dalton fragment was
only possible by use of diepoxy butane in combination with
glutaraldehyde and by periodate oxidation followed by
glutaraldehyde. The other bifunctional crosslinking reagents

Table 3. Amino acid and carbohydrate composition of MAF and its tryptic fragments.

Amino acid analysis was performed on a Durrum D-500 amino acid analyzer. Values present the average from four different values on two different preparations. Maximal error was less than 5% of each value. Trp and Lys were not determined.

Carbohydrate composition: values represent the average of three determinations, and maximal error was less than 5% of each value.

% of total amino acid present in

	MAF	Vo	TP1	TP2	TP3	TP4
Asx	15.6	16.4	14.0	13.7	13.9	13.2
Thr	10.0	9.9	10.0	10.2	8.5	7.1
Ser	5.7	6.8	8.6	9.7	10.2	12.0
Glu	11.5	12.9	13.0	12.6	10.4	11.6
Pro	7.7	6.4	5.9	4.1	5.2	3.5
Gly	6.8	10.3	9.8	13.1	13.2	13.4
Ala	7.1	7.4	7.0	6.7	8.8	7.6
Val	7.0	6.8	7.9	7.5	5.8	5.3
Met	2.2	2.5	2.1	1.1	2.3	3.3
Ile	4.4	3.8	4.5	3.9	4.1	4.7
Leu	7.2	7.1	6.8	8.0	7.1	7.3
Tyr	3.4	2.3	3.0	2.7	3.3	3.1
Phe	4.5	3.4	4.5	3.1	2.4	3.2
His	3.1	0.7	0.5	0.6	0.5	0.8
Lys	0.7	0.7	0.8	1.1	3.2	2.0
Arg	3.1	2.6	1.9	1.9	1.0	1.9

% of total carbohydrate

	MAF	Vo	TP1	TP2	TP3	TP4
Fuc	16,9	20,5	16,1	10,5	11,6	12,1
Man	5,0	5,4	7,6	10,5	9,4	9,4
Gal	14,5	16,3	18,3	18,7	14,9	21,8
GalNac	1,5	2,2	0	0	0	0
GlcNac	12,0	14,0	16,1	19,2	17,4	19,8
uronic acid	16,6	18,0	18,0	18,6	18,6	22,7

specific for amino or amino and carboxyl groups were inefficient due to the low number of amino groups present in the fragment. The polymers with Mr > 1.5 x 10^7, which is in the range of the intact MAF, were separated by gel filtration on a BIO-GEL A - 1.5 m

Table 4. Binding characteristics of reconstituted TP4 polymers
and of MAF to homotypic cells.
Binding was assayed as described in Table 1. The
number of bound molecules / cell was calculated from
the specific radioactivity and Avogadro's number.
TP4EG5 is a polymer of TP4 crosslinked with diepoxy
butane and glutaraldehyde. TP4PG5 is a polymer of TP4
crosslinked with glutaraldehyde after periodate
oxidation.

ligand	concentration of ligand	% bound to fixed cells of		
		M. prolifera	H. oculata	C. celata
	nMolar			
TP4	83.00	45.0	7.6	9.8
TP4EG5	0.014	34.2	3.9	3.2
TP4PG5	0.015	30.0	2.5	2.9
MAF	0.014	35.9	3.6	2.1

Table 5. Binding characteristics of large TP4 polymers and
of MAF to homotypic cells.
K_a values out of B_{max} values were determined as
described in Table 1. Maximal error determined from
three experiments is less than 12% of each value.

ligand	Mr	K_a	B_m
	$\times 10^{-3}$	$\times 10^{-6} M$	$\times 10^{-4}$
TP4	10	3.4	350.0
TP4EG5	15,000	10,000	0.1
TP4PG5	15,000	8,000	0.24
MAF	20,000	45,000	0.12

column. The cell binding specificity of the reconstituted polymer
was preserved, showing that the crosslinking did not destroy the
cell binding domain (Table 4). When affinity to homotypic cells was
measured it was found that the apparent association constant was
indeed in the same range as that of intact MAF (Table 5).

Since it was possible to reconstitute the high affinity of the cell binding domain by crosslinking a cell binding fragment of low affinity, we propose that the high affinity of binding is based on a single type of multiple weak interactions. The polyvalent interaction of MAF with the cell surface is thus one of the steps necessary for the specificity of MAF aggregation promoting activity.

This work was supported by the Swiss National Foundation for Scientific Research, Grant 3.269.82 as well as the Ministry of the City and Canton of Basle.

REFERENCES

1. P. Bornstein and H. Sage, Annu. Rev. Biochem., 49 : 957-1003 (1980).
2. M.W. Lark and L.A. Culp, J. Biol. Chem., 259 : 6773-6782 (1984).
3. V.C. Hascall and D. Heinegård, J. Biol. Chem., 249 :4232-4241 (1974).
4. V.C. Hascall and S.W. Sajdera, J. Biol. Chem., 245 :4920-4930 (1970).
5. J.A. Buckwalter, L.C. Rosenberg and L.H. Tang, J. Biol. Chem., 259 : 5361-5363 (1984).
6. L. Rosenberg, W. Hellmann and A.K. Kleinschmidt, J. Biol. Chem., 250 : 1877-1883 (1975).
7. V.C. Hascall, J. Supramol. Struct., 7 : 101-120 (1977).
8. D. Heinegård and I. Axelsson, J. Biol. Chem., 252 : 1971-1979 (1977).
9. V.C. Hascall and J.H. Kimura, Meth. Enzym., 82 : 769-800 (1982).
10. U. Lindahl and M. Höök, Annu. Rev. Biochem., 47 :385-417 (1978).
11. S. Stenman and A. Vaheri, J. Exp. Med., 147 : 1054-1064 (1978).
12. K.M. Yamada and K. Olden, Nature, 275 : 179-184 (1978).
13. R.O. Hynes, In: "Cell Biology of Extracellular Matrix", ed. Hya E.D., (Plenum, New York), pp. 295-334 (1982).
14. K.M. Yamada, In: "The Glycoconjugates", ed. Horowitz M. (Academic Press, New York), Vol. 3, pp. 331-362 (1982).
15. J. Engel, E. Odermatt, A. Engel, J.A. Madri, H. Furthmayr, H. Rohde and R. Timpl, J. Molec. Biol., 150 : 97-120 (1981).
16. R. Timpl, H. Rohde, R.P. Gehron, S.I. Rennard, J.M. Foidart and G.R. Martin, J. Biol. Chem., 254 : 9933-9937 (1979).
17. K. Rubin, S. Johnsson, I. Pettersson, C. Ocklind, B. Oebrink and M. Höök, Biochem. Biophys. Res. Commun., 91 : 86-94 (1979).

18. F. Grinnel and M.H. Bennett, J. Cell Sci., 48 : 19-34 (1981).
19. R.C. Hughes, S.D.J. Pena and P. Vischer, In: "Cell Adhesion and
      Motility", eds. Curtis A.S.G. and Pitts J.D., (Cambridge
      University Press, Cambridge), pp. 329-356 (1979).
20. D.A. Rees, "Polysaccharide Shapes" (Chapman and Hall, London),
      pp. 62-73 (1977).
21. W. Burkart and M.M. Burger, J. Supramol. Struct. Cellular
      Biochem., 16 : 179-192 (1981).
22. L.A. Fransson, L.C. Carlstedt, L. Cöster and A. Malmström, In:
      "Cell Function and Differentiation", eds. Akoyunoglou G.,
      Evangelopoulos A.E., Georgatsos J., Palaiologos G.,
      Trakatellis A. and Tsiganos C.P., (Alan R. Liss, New York),
      Part A, FEBS Vol. 64, pp. 435-445 (1982).
23. K. Müller and G. Gerisch, Nature, 274 : 445-449 (1978).
24. S.H. Barondes, In: "Cell Adhesion and Motility", eds. Curtis
      A.S.G. and Pitts J.D., (Cambridge University Press,
      Cambridge), pp. 309-324 (1980).
25. F.G. Harrison and C.J. Chesterton, Nature, 286 : 502-504
      (1980).
26. F. Hyafil, C. Morello, C. Babinet and F. Jacob, Cell, 21 :
      927-934 (1980).
27. F. Hyafil, C. Babinet and F. Jacob, Cell, 26 : 447-454 (1981).
28. G.M. Edelman, Science, 219 : 450-457 (1983).
29. G.M. Edelman, Proc. Natl. Acad. Sci. USA, 81 : 1460-1464
      (1984).
30. W.J. Gallin, G.M. Edelman and B.A. Cunningham, Proc. Natl.
      Acad. Sci. USA, 80 : 1038-1042 (1983).
31. J.P. Thiery, A. Delouvée, W. Gallin, B.A. Cunningham and G.M.
      Edelman, Dev. Biol., 102 : 61-78 (1984).
32. J. Pouyssegur, M. Willingham and I. Pastan, Proc Natl. Acad.
      Sci. USA, 74 : 243-247 (1977).
33. J. Finne, T.W. Tao and M.M. Burger, Cancer Res., 40 : 2580-2587
      (1980).
34. P. Vischer and R.C. Hughes, Eur. J. Biochem., 117 : 275-284
      (1981).
35. S.A. Hakomori, Annu. Rev. Biochem., 50 : 733-764 (1981).
36. P. Drochmans, C. Freudenstein, L. Wanson, T.W. Laurent, J.
      Keenan, R. Stadler, R. Leloup and W.W. Franke, J. Cell
      Biol., 79 : 427-443 (1978).
37. H. Müller and W.W. Franke, J. Cell Biol., 163 : 647-671
      (1983).
38. D. Billig, A. Nicol, R. McGinty, P. Cowin, J. Morgan and D.
      Garrod, J. Cell Sci., 57 : 51-71 (1982).
39. V.T. Marchesi, Science, 159 : 203-205 (1968).
40. T.L. Steck, J. Cell Biol., 62 : 1-19 (1974).
41. J.F. Hainfeld and T.L. Steck, J. Supramol. Struct., 6 : 301-311
      (1977).
42. J. Levine and M. Willard, J. Cell Biol., 90 : 631-643 (1981).
43. K. Burridge, T. Kelly and P. Mangeat, J. Cell Biol., 95 :
      478-486 (1982).

44. J.R. Glenney, P. Glenney and K. Weber, Proc. Natl. Acad. Sci. USA, 79 : 4002-4005 (1982).
45. E. Lazarides and J. Nelson, Cell, 31 : 505-508 (1982).
46. J.M. Tyler, W.R. Hargreaves and D. Branton, Proc. Natl. Acad. Sci. USA, 76 : 5192-5196 (1979).
47. B. Geiger, Cell, 18 : 193-205 (1979).
48. K. Burridge and J.R. Ferameisco, Cell, 19 : 587-595 (1980).
49. B. Geiger, K.T. Tokuyasu, A.H. Dutton and S.J. Singer, Proc. Natl. Acad. Sci. USA, 77 : 4127-4131 (1980).
50. J.A. Wilkins and S. Lin, Cell, 28 : 83-90 (1982).
51. E. Lazarides and K. Burridge, Cell, 6 : 289-298 (1975).
52. P. Maher and S.J. Singer, Cell Motility, 3 : 419-429 (1983).
53. B. Oesch and W. Birchmeier, Cell, 31 : 671-679 (1982).
54. R.K. Meyer, H. Schindler and M.M. Burger, Proc. Natl. Acad. Sci. USA, 79 : 4280-4284 (1982).
55. P. Burn, A. Rotman, R.K. Meyer and M.M. Burger, Nature, (in press).
56. M.M. Burger, In: "The Role of Intercellular Signals : Navigation, Encounter, Outcome", ed. Nicholls J.G., (Dahlem Konferenzen, Berlin), pp. 119-134 (1979).
57. K.H. Pfenninger and R.P.Rees, In: "Neuronal Recognition" ed. Barondes S.H., (Plenum Press, New York), pp. 131-178 (1976).
58. M. Spiegel and M.M. Burger, Exp. cell Res., 193 : 377-382 (1982).
59. M.M. Burger, G.N. Misevic, J. Jumblatt and F. Mir-Lechaire, In: Adv. Physiol. Sci. : Physiology of Non-excitable Cells, ed. Salanki J. (Pergamon Press, Oxford), Vol. 3, pp. 233-240 (1980).
60. J.E.M. Heaysman and S.M. Pegrum, Exp. Cell Res., 78 : 71-78 (1973).
61. J.D. Sheridan, In: "Intercellular Junctions and Synapses : Receptors and Recognition", eds. Feldman J. Gilula N.B. and Pitts J.D., (Chapman and Hall, London), Series B, Vol. 2, pp. 39-59 (1978).
62. J.E. Heuser, T.S. Reese, M.J. Dennis, Y. Jan, L. Jan and L. Evans, J. Cell Biol., 81 : 275-300 (1979).
63. W.R. Loewenstein, Dev. Biol., 15 : 503-520 (1967).
64. M.V. Wilson, J. Exp. Zool., 5 : 245-258 (1907).
65. P.S. Galtsoff, Biol. Bull., 57 : 250-260 (1929).
66. T. Humphreys, Dev. Biol., 8 : 27-47 (1963).
67. D.R. McClay, Biol. Bull., 141 : 319-330 (1971).
68. P. Heukart, S. Humphreys and T. Humphreys, Biochemistry, 12 : 3045-3050 (1973).
69. C.B. Cauldwell, P. Henkart and T. Humphreys, Biochemistry, 12 : 3051-3055 (1973).
70. G. Weinbaum and M.M. Burger, Nature, 244 : 510-512 (1973).
71. W.E.G. Müller and R.K. Zahn, Exp. Cell Res., 80 : 95-104 (1973).

72. S. Humphreys, T. Humphreys and Y. Sono, J. Supramol. Struct., 7
      : 339-351 (1977).
73. G.N. Misevic, J. Jumblatt and M.M. Burger, J. Biol. Chem., 257
      : 6931-6936 (1982).
74. J. Jumblatt, V. Schlup and M.M. Burger, Biochemistry, 19 :
      1038-1042 (1980).
75. D.J. Rice and T. Humphreys, J. Biol. Chem., 258 : 6394-6399
      (1983).
76. M.M. Burger and J. Jumblatt, In: "Cell and Tissue
      Interactions", eds. Lash J. and Burger M.M., (Raven Press,
      New York), pp. 155-172 (1977).
77. G.N. Misevic and M.M. Burger, In: "Embryonic Development" eds.
      Burger M.M. and Weber R., (Alan R. Liss, New York), part B,
      pp. 193-209 (1982).

CELLULAR AND MOLECULAR CONTROL OF

CELL INTERACTIONS IN EMBRYOGENESIS

Adam S.G. Curtis

Department of Cell Biology
University of Glasgow
Glasgow, G12 8QQ, UK.

INTRODUCTION

One of the main interests of developmental biologists over the last fifty years has been to achieve an understanding of the mechanisms whereby cells are placed and held in their correct positions in the embryo. Thus a great deal of interest has been devoted to developing an understanding of morphogenetic movements. In turn the consideration of these mechanisms raises questions about the involvement of components of the plasmalemma and of the cytoskeleton in such events. Another way of looking at this is to state that we are now investigating the molecular biology of cell interactions and thus in turn the aspects of cell behaviour which control morphogenesis.

It is perhaps surprising that we have not yet managed to carry out a comprehensive survey of the situation at the cellular level. However it is worthwhile making a bare list of the main mechanisms which have been proposed as being responsible for morphogenetic movement.

At the level of individual cells.

Mobilisation of cells followed by directional or random movement finally terminated by arrest of movement.

Mobilisation.
Polarisation of microfilaments, (18)
Decrease in adhesion
Chemokinesis (which may result from change of adhesion)

Cell surface change to aid insertion of microfilaments

Directionality.
  Chemotaxis
  Topographical effects (e.g., contact guidance, (17)
  Availability of   space, a consequence of contact
  inhibition, (1)
  Imposition of microfilament orientation, (see also
  under topographical effects)
  Persistence of movement in absence of cues.

Arrest of cell movement.

  Rise in cell adhesion
  Lack   of available space into which spreading or
  movement might occur, a   consequence of contact
  inhibition of movement.
  Damage to locomotory system of the cell
  Loss of orientation of movement leading perhaps to
  oscillations around a point.

If we analyse this list in terms of the cues which a cell may
receive from its surroundings we obtain the following three sets.

  Changes in adhesion. These may be direct, e.g., the
  presentation of a very adhesive surface to a cell, or
  more indirect, e.g., the action of a substance on a
  cell which leads to a change in the synthesis of
  adhesive subtances for the cell surface.

  Diffusible signals, leading mainly to positive or
  negative chemotaxis.

  Topographical cues, see (17). These probably represent
  adaptations   of   the microfilament system and its
  insertions on the plasmalemma to the shape of the
  underlying substrate. Thus it was found (18) that
  fibroblasts could not move over convex discontinuities
  and that their microfilaments were unable to bend
  sufficiently to allow a cell to have an   intact
  continuous microfilament system on a convex surface
  with the system crossing the discontinuity.

In essence the cues are either ones of adhesion or of
microfilament orientation and persistence or ones in which we
suspect that microfilament organisation is important, e.g., contact
inhibition,   see (2). Microfilamentous and other cytoskeletal
structures are dealt with in this conference by the contributions
from Franke, Garrod and Reggio, though important data on the rate
of turnover of such structures or their response to changes in the

external environment of the cell are still largely lacking.

We know little about the extent of and nature of the interactions between the microfilamentous system and a cellular substrate because nearly all our information on microfilaments and cell contacts has come from studies on the adhesion of cells to glass substrates (see (28). In part this lack of information comes from the fact that interference reflection microscopy cannot be used to study cell-cell contacts (see later for a discussion of this). We ought to be able to answer questions such as :

1. Does the microfilament system align in response to substrate topography ?

2. If the answer to the first question is yes, what types of and scale of topography is required ?

3. Is the alignment of materials secreted by the cell to form the extracellular matrix related to the alignment of the cells, or to that of the microfilament system ?

4. Do cells orient the substrate when this is possible, either by doing mechanical work on the substrate, as suggested by Garrod in this symposium or by oriented deposition or oriented lysis ? Do cells perform a type of grapho-epitaxy on the surface (see (31) for definitions and (39) for a remarkable example of calcite deposition) in relation to oriented substrates ? In other words do cells respond to substrate topography by orienting the deposition of products they secrete ? It is interesting to note the tracts of oriented fibronectin reported by (40) and others in embryos, and to speculate whether they arose by grapho-epitaxy by oriented movements of surrounding cells or by mechanical stressing of large regions of the embryo.

Thus it should be clear that the two main areas we need to examine in order to understand positioning are adhesion mechanisms and microfilaments dynamics. I intend to devote most of this paper to the first topic.

CELL ADHESION

At first sight it is surprising that we are still very unsure about the nature of the mechanisms of cell adhesion.

In part the problem results from the fact that there are perhaps as many as nine different morphological types of adhesion (see Garrod, this symposium), though it is possible that these may

actually be reduced to three basic types, the contact which
includes the zonula adherens, the macula adherens and the terminal
bar type structures, the zonula occludens and the gap junction. In
part the problem arises from the fact that very many molecules
possess adhesive proportios to a greater or lesser degree and it is
probable that it is all too easy to set up systems which produce
effective adhesions of test objects such as cells but which do not
actually resemble the adhesions of living cells.

The gap junction and the zonula occludens are ones in which
components of the plasmalemma of each cell actually make contact.
All the other types of adhesion, considered from the morphological
point of view, are ones in which an electron-lucent gap is seen
between the plasmalemmae. The nature of this gap is perhaps crucial
to an understanding of cell adhesion. It is filled with adhesive
molecules or are we seeing a secondary minimum type of adhesion ?

This question needs explanation and this is best done in terms
of the models of adhesion derived from colloid chemistry (3).
Adhesive molecules might extend from one plasmalemma to the other
having some more or less specific bonding to a component or
components of the plasmalemma. This is definable in colloid
chemical terms as a bridging system. Another type of system is that
known as a secondary minimum adhesion (3, 12) in which the surfaces
settle in a relatively weak adhesion at a distance determined by a
balance between the short range electro-static forces of repulsion
and the electrodynamic forces of attraction, which are long range
in nature. Experimental evidence has been published which
establishes that secondary minimum effects can occur with a
separation between the plasmalemmae of about 6 to 15 nm, see (20,
11). If the secondary minimum explanation is correct cell adhesion
of the zonula adherens, macula adherens and gap junctions types
should be fairly weak and, if the possibility of the simultaneous
operation of bridging systems can be excluded, the gap should not
be crossed by bridging molecules. The zonula occludens and gap
junction type systems can be identified with the primary minimum
type of adhesion (12), in which electrostatic forces of repulsion
are too weak to prevent the electrodynamic forces bringing the
surfaces into molecular contact, both by their strength and by
their morphology. The other model of adhesion arising from colloid
chemistry is that of the sterically stabilised system (25) in which
molecules in between the adhering bodies modify, shield or extend
adhesive interactions arising primarily from the surfaces
themselves. If biological adhesions are sterically stabilised we
would expect to find a gap between plasmalemmae containing a
greater or lesser amount of some hydrogel.

So it appears that colloid chemistry provides several models
which might be relevant to cell adhesion but that there is no very

clear diagnostic feature which would allow the acceptance or
dismissal of a particular explanation. A second problem arises when
we consider the reliability of the evidence for the nature of the
gap seen between plasmalemmae in so many electron micrographs of
cell contacts. Unfortunately though we have rigourous experimental
proofs of the adequacy of electron micrography for the study of
objects such as mitochondria or plasmalemmae no such work has been
completed for the study of the study of cell contacts.

METHODS FOR STUDYING CELL CONTACTS

The status of work on the morphology of cell contacts is
reminiscent of that on the microtrabecular system (27), because the
reliability of the electron microscopic image has not been
corroborated by the application of other methods to live cells. Of
course the lack of research in this area suggests opportunities. If
we had the answer to the problem we would have a better
understanding of the mechanism of cell adhesion. Obviously we need
procedures for investigating cell contact morphology in living
cells.

Interference reflection microscopy

Interference reflection microscopy was introduced in 1964 (11)
though not used greatly until ten years later. The technique can
provide a measurement of the thickness of a gap between a cell and
the glass substrate to which it adheres. Unfortunately the
technique does not provide good measurements for gap thicknesses
less than 15 nm and is confined to cell-glass contacts.
(Cell-polystyrene contacts are almost invisible by this method of
microscopy because of the refractive index of the polystyrene.) The
technique depends upon the fact that light reflected from one
refractive index discontinuity will interfere with that reflected
from a second such discontinuity nearby on the ray path. Thus cells
grown on glass can provide an image of the dimensions of the
cell-substrate gap as an interference image. The best published
description of the physics of the situation is that of Gingell
(21). The technique suffers from a lack of lateral resolution and
from its inability to distinguish gaps of say 5 nm from those of 12
nm. The full potential of the method has not been exploited because
it should be possible to use it to measure the refractive index of
the gap material and hence estimate the concentration of proteins
etc in that region. The refractive index values usually assumed in
attempts to quantitate the method have the interesting property
that very small changes in the values, in particular in that of the
gap, result in large changes in the contrast of image that should
be seen.

Evanescent wave microscopy

This technique (38) depends upon the fact that light, when totally internally reflected at an interface, penetrates as an evanescent wave for up to about 100 nm into the other medium. If any object is there present to scatter or fluoresce in this wave, the light from that object will penetrate through the other medium. Thus if light is totally internally reflected at a glass interface on which a cell is growing in water any fluorescent molecule in the cell surface or in the gap between the cell and substrate will be induced to fluoresce, provided that it lies within 100 nm of the glass surface. Thus, in pinciple, the dimensions of any gap between cell and substrate which is accessible to a fluorochrome molecule can be detected. It might with difficulty be possible to measure the thickness of the gap from the fluorescence intensity or with more ease by measuring the size limit for permeation of fluorochromes into the gap.

Fluorescence energy transfer microscopy

I should like to introduce the idea of using this type of microscopy, which is applicable to cell-cell contacts as well as to cell-substrate situations. Fluorescence energy transfer spectroscopy (33) has already been used to measure the distance between two sites on a molecule in situations where each site can be labelled with a fluorochrome. The only other requirement is that one fluorochrome is excited by the wavelength of light emitted by the other.

The attractions of this technique are manifold :
(i) the lateral resolution of the observation is not limited by the numerical aperture of the objective lens but by the intensity of excitation of the first fluorochrome, the level at which the type of molecule bearing the fluorochrome has been labelled and by the proximity to, the density of the second fluorochrome and the sensitivity of the detection system. Thus if appropriate fluorochromes are used and sensitive detection it should be possible to detect a close approach of molecules from two separate surfaces to each other, even when very few molecules are involved.

(ii) since two different plasmalemmae may be labelled, each with a different fluorochrome, the system can be applied to cell to cell contacts as well as to cell-substrate contacts (where the substrate might be labelled with a second fluorochrome).

(iii) the labelling levels of each fluorochrome can be assessed by conventional fluorescence microscopy to provide calibration of the system, while the transfer of energy provides

(iv) the measure of the approach of the two surfaces. The method is

particularly sensitive in the region 0.5 to 5 nm, a range in which
other techniques fail.

(v) the fluorescent probes can be attached to different molecular
types which may lie at different levels in the membrane and this
provides an internal consistency check as well as a means of
mapping the position of the various components of the membrane.

    Dr. R. Mohan and I have been developing this technique.
Preliminary results on contacts between BHK cells show that the
distance between the glycoproteins on one cell surface and the
lipids on another cell surface is such that transfer is almost
undetectable, in other words these two components probably lie more
than 4 nm apart.

PERMEATION TECHNIQUES

    If the structure between two adhering cells is a gap
containing few or no macromolecules the gap should be open to
penetration by appropriately small particles. The rate of
permeation could be treated as an example of filter-bed flow (24)
to provide a measure of the void volume and void viscosity. Thus
the rate of permeation of a tracer particle such as ferritin into
the contacts of a tissue should provide some information on the
properties of such structures. Slow permeation would argue that the
gap was narrow or/and filled with a viscous material. An early
experiment of this type (7) was carried out using a ferritin
tracer, and though the geometrical details of the gaps have to be
assumed, the rate of permeation was so high that it seems likely
that few macromolecules bridged the gap. This type of experiment
cries out for repetition.

CELL SLIDING AND CELL MOVEMENT

    If cell bonding to substrates is very strong or if it cannot
be turned-over in some manner by the cell, then cells should be
unable to move. Passive movement of the cells under conditions
which they cannot turnover their bonds would imply that cell
adhesion was relatively weak. Obviously cooling cells stops their
movement but this may also inhibit turnover of the bonding system.
However, is there evidence that cells whose motile systems have
been inhibited will slide under gravity ? If there is, the
conclusion must be that for those cells adhesion is a weak system
which probably does not require direct bonding. It has already been
shown (34) that cells will slide under gravity under conditions in
which their adhesiveness is diminished by chelation of calcium and
in which they are non-motile. Recently we have confirmed this for
BHK fibroblasts in a low ionic strength medium.

These observations imply that either adhesive bonding molecules are relatively few in number and easily broken and reformed as the cell moves, or that bonding is by some process such as secondary minimum adhesion which does not require direct chemical links. However these experiments were carried out at reduced temperature and this fact revives a long-lasting controversy. The problem is : Can cells form normal types of adhesion at metabolically unfavourable temperatures ? There has been much comment in the literature on this matter, (12, 37). If a summary is made of those experiments in which a requirement for adhesion of a raised temperature (say 20 to 37 C) was demonstrated it turns out (12) that trypsinised cells require a temperature at or close to the body temperature of the animal from which they come in order to reform adhesions, but that cells which have not been subjected to protease treatment e.g., lymphocytes, (or which have been allowed an appreciable time to recover from such treatment) adhere well at low temperature. This point needs further experimental test for if it can be shown that cells do not require a period of metabolic activity to establish adhesions it is less likely that a bridging mechanism, requiring enzyme activity, occurs.

PLASMALEMMAL MOLECULES AND CELL ADHESION

This section appears as a very brief precis of results from this field. It should become clear to the reader that a problem arises as to whether the molecules reported as being involved in cell adhesion act directly as the adhesive molecules (adherons, 30) or in a more indirect manner, for example simply determining the display and conformation of other molecular species, which are in fact the real adhesive structures.

Evidence for the involvement of plasmalemmal lipids in adhesion was reported (14), of plasmalemmal proteins by very many authors of whom those who have worked on fibronectin (41), laminin (23) and on cell-adhesion molecules (CAMs) (29) are perhaps the most noteworthy. Some of the proteins are glycoproteins but it is still unclear as to whether the glycosidic parts of these molecules play an important role in adhesion. Relatively few reports have suggested that glycolipids are of importance though it is worth referring to one (36).

MOLECULAR REQUIREMENTS FOR CELL ADHESION

It is easy to carry out experiments in which various substances are added to systems in which cells are in the process of forming adhesions and then to show a change in adhesion. It is

also easy to extract substances from tissues which are being
dispersed (dissociated) into single cells and to show that cell
adhesion is increased when the substance is returned to the system.
It is, however, much more difficult to demonstrate that any of
these substances are the direct molecular agent or agents which
stick the cells to each other or to a non-living substrate.

First, it should be remarked in this context, that any
molecule, whose removal from, or return to, a cellular system
affects adhesion, may be acting indirectly. By the term
'indirectly' I mean that the molecule may be acting as a supporting
or modifying molecule for the actual adhesive species. Similarly,
the inhibition of cell adhesion with an antibody directed against a
specific sequence on cell-surface associated antigen may be nothing
more than the result of attaching long hydrophilic non-adhesive
molecules to the surface, see later.

Second, these molecules may be acting relatively
non-specifically as components of a colloid interaction system,
such as an electrodynamic-electrostatic force balance system (e.g.,
DLVO system, (3, 12) which produces cell adhesion, either in the
primary minimum (less than 1 nm gap) or in the secondary minimum
(5-15 nm gap). Increase in the surface charge density on the cell
surface or a decline in the hydrophobicity would act to alter
adhesion.

Third, the adhesion produced by the addition of such molecules
to the system may be of a type which differs from that which occurs
in life.

It is worth remarking at this point that most large molecules
are either clearly adhesive or clearly anti-adhesive in many
situations and that such molecules tend to adsorb to many surfaces.
Thus alpha-1-antitrypsin (15) adsorbs to plain polystyrene and
though rendering the surface wettable inhibits cell adhesion.

TEST SYSTEMS FOR BRIDGING MOLECULES IN ADHESION

If we believe that a molecular bridging agent is involved in
cell adhesion there are a number of tests that should be made of
our hypothesis :

Measurement of the number of binding sites for the bridging
agent on cell surfaces should be made as well as measurement of the
affinity of the receptor for the bridging molecule. If the molecule
is believed to be bivalent, the 'double site' affinity should be
measured, as this will give a measure of the strength of an
adhesive interaction for given concentrations of the bridging
agent.

Thus, if we hold the view that adhesion is effected by a very specific molecular mechanism we would expect that the number of binding sites would be low and the affinity high. A small number of binding sites would be required to correlate with the rather low strength of cell adhesion obtained in physical measurements (12). Alternatively, we might find that the affinity constant was of the fairly low value of, say, 1000 ; in which case it might be reasonable to expect that there would be relatively large numbers of sites. We must also consider cross-reactivity with other proteins etc. of fairly similar sequence.

These considerations relate to the concentrations of competitive molecules which must be used in competitive inhibition experiments, which might identify specific molecules involved in adhesion. For instance, hepta- and other small peptides from fibronectin have been used (26) as inhibitors of cell adhesion to plastic. The affinity constant of the sequence most likely to represent the cell binding part of the molecule is 3000 and hence high concentrations of these agents were required for inhibition. Such concentrations begin to approach the level at which other rather less specific cellular effects may occur as well. Some workers e.g., (5) have used observations that high concentrations of certain sugars, in the range from 50 to 200 millimolar, inhibit cell adhesion to argue that cell adhesion involves glycosidic or other glyco-molecules. The argument seems fallacious since these concentrations have many rather non-specific effects, though of course, other better evidence does point in the same direction for some cell interactions, e.g. (36).

Fab methods have become widely used for the demonstration of the presence of more or less specific molecules involved in cell adhesion. Gerisch (19) was the first person to introduce this technique. The principle of the test is to prepare an antibody against a putative adhesion molecule located in the cell surface and then to show that the Fab (univalent) form of this antibody blocks cell adhesion. Obviously a control antigen is required which is present in similar or greater surface density than the 'adhesive' antigen. Fab to this control antigen should have no effect on adhesion. Such an experiment was carried out for Dictyostelium adhesion (6) but in another paper it was also shown that this control antigen lay deeper in the membrane than the putative adhesion antigen. This difference in location may be sufficient to explain the difference in effect on adhesion.

Since Fabs are soluble molecules of appreciable size, their binding to a surface receptor (antigen) by their one reactive site may then coat the surface with that is in effect a non-adhesive covering. If this occurs we should be much more doubtful about the validity of the Fab test in inhibiting cell adhesion as one which identifies cell surface molecules involved in adhesion. It may only identify cell surface molecules present in sufficient number and in location appropriate to bind enough Fab to inhibit adhesion. Recently I tested this possibility (13) by incorporating an artificial antigen into a cell surface and showing that Fabs against this antigen inhibited adhesion.

SPECIFICITY IN CELL ADHESION

It is clear that some types of cell adhesion, for instance mating types in certain yeasts (9), are highly specific. The evidence for such specificity in many other systems is lacking, even for sponges (see Burger and Misevic, in this volume and (8)). Steinberg (32) made the important observation that the sorting-out patterns in aggregates of chick embryonic tissue show evidence of a graded property involved in sorting-out which is most simply interpreted as a single quantitative property varying between cell types. Though many of the pairs of cell types that were combined for sorting-out experiments in Steinberg's work do not meet in normal animals other do and in any event the results agree perfectly with earlier ones on germ layers in amphibian embryos (34). Steinberg's experiments have two important consequences. First, it seems likely that sorting-out in aggregates is a realistic model for morphogenesis. Second, the fact that a graded property appears to be responsible for the sorting-out implies that, if adhesion is linked to this process, then the important feature about adhesion is not any specificity that may be found but its quantitative value. In the last analysis cell movements in morphogenesis appear to be related to the exchange of one set of neighbours for another and this suggests that adhesive differences of a quantitative type may be of importance.

This conclusion is consistent with the finding that cells adhere well to a very wide range of types of material. However this set of results has led in its turn to variations in interpretation.

Cells secrete extracellular materials and some components of such materials may be adhesive. Thus the cell may modify any surface to which such materials can adsorb so that its adhesiveness is altered. Experimental test of such a happening is difficult, but see the ingenious work by Culp (10). Obviously the experiment cannot be done in the presence of serum or plasma because these media contain fibronectin and other adhesion-promoting proteins (see Vaheri, this volume). However it has been argued (22) that fibronectin or other nectins are synthesised and exported by many cell types as the main and essential adhesive molecule. Nevertheless there are reasons to question the importance of fibronectin as an adhesive molecule.

It has been found (16) that BHK fibroblasts will adhere to and spread on highly hydroxylated polystyrene even after all exogenous and endogenous fibronectin has been removed from the system. This result disproved the claim that fibronectin is an essential molecule for fibroblast adhesion. Indeed with the sole possibility that the cycloheximide-poisoned cells in these experiments carry a pool of a long-lasting and slowly secreted adhesive molecule it is hard to envisage the existence of any secreted material being used by the cells for adhesion to such surfaces.

MODIFICATION OF SURFACE BY CELLS

This type of investigation is still in its infancy. Culp (10) has used the approach of stripping cells off surfaces and identifying the copmponents left on the surface. This type of experiment has several difficulties. First, the cells appear to leave behind various cytoplasmic components such as actin which suggests that the fracture takes place at least some of the time at the wrong level. Second, the cells may leak proteins which adsorb to the surface at the moment of detachment. Subtle attempts have been made by Aplin and Hughes (4) to label by photoaffinity methods only those proteins in the correct position for adhesion molecules. Perhaps this technique might be refined by using evanescent wave methods (see above) to ensure that the photolabelling took place only in the right zone close to the substrate. Such work is in any event closely linked to that on the nature of the gap (see above). In any event it is clear that cells may be able to modify other surfaces by adsorbing their own products onto them. One particularly interesting example for the immunologists is afforded by the realisation that histocompatibility antigen fragments of both Class I and II types can be bound by cells that do not normally bear them and that this has effects on adhesion (Gallagher, in press).

It is also worth considering the possibility that various cell surface enzymes could affect cell adhesion. Proteases released at the cell surface might modify the adhesion of surrounding cells. Phospholipases might possibly have a similar effect. The enzymic system which generates superoxide anions, particularly well developed in leucocytes, can probably modify the adhesions of various types of cells : it certainly acts powerfully on platelet and leucocyte adhesion.

CONCLUSION

Thus there are two main schools of thought about the role of adhesion in morphogenesis. First, some believe that we shall identify specific molecules directly involved in adhesion which control positioning of cells. These may be laid down as 'tracks' to lead cells to the correct sites, or they may provide adhesive traps which fix some cell type in a given final site. If this is true we shall then have to seek out the mechanisms which specify the sites of synthesis or adsorption of these molecules. The other approach holds that many different types of molecule are more or less adhesive and that the adhesiveness of a cell represents a summation of all these events. In turn adhesiveness and motility interact, perhaps with topographic reaction (17), to specify cell behaviour which in turn locates cells with respect to one another.

The first approach would lead us to expect that there might be quite a lot of cell behaviour mutants : few, if any, have yet been found. Alternatively protagonists of this school would suggest that precise antibodies will be found which will have significant and direct effects on cell adhesion (see above).

The alternative approach, namely that adhesion is effectively non-specific and quantitative in effect leads to the expectation that many different molecules, both on the cell surface, and more indirectly those beneath, will make a contribution to adhesion but that no one of them will appear to be uniquely important.

REFERENCES

1. M. Abercrombie, In vitro, 6 : 128-142 (1970).
2. M. Abercrombie, Proc. Roy. Soc. Lond., B 207 : 129-147 (1980).
3. A.W. Adamson, Physical chemistry of surfaces. 4th Edition.
        Wiley-Interscience, New-York (1982).
4. J.D. Aplin and R.C. Hughes, J. Cell Sci., 50 : 89-103 (1981).
5. S.H. Barondes and S.D. Rosen, in: 'Surface Membrane Receptors',
        R.A. Bradshaw, W.A. Frazier, R.C. Merrell, D.I. Gottlieb and
        R.A. Hogue-Angeletti Eds., Plenum Press, New-York, pp. 39-55
        (1976).
6. H. Beug, F.E. Katz, A. Stein and G. Gerisch, Proc. Nat. Acad.
        Sci. USA., 70 : 3150-3154 (1973).
7. M.W. Brightman, J. Cell Biol., 26 : 99-123 (1965).
8. M.M. Burger, W. Burkart, G. Weinbaum and J. Jumblatt, Symp.
        Soc. Exp. Biol., 32 : 25-50 (1978).
9. M. Crandall, Symp. Soc. Exp. Biol., 32 : 105-120 (1978)
10. L.A. Culp, J. Cell Biol., 63 : 71-83 (1974).
11. A.S.G. Curtis, J. Cell Biol., 20 : 199-215 (1964).
12. A.S.G. Curtis, Prog. Biophys. and Mol. Biol., 27 : 315-386
        (1973).
13. A.S.G. Curtis, in press.
14. A.S.G. Curtis, C. Chandler and N. Picton, J. Cell Sci., 18 :
        375-384 (1975).
15. A.S.G. Curtis and J.V. Forrester, J. Cell Sci., in press.

16. A.S.G. Curtis, J.V. Forrester, C. McInnes and F. Lawrie, J. Cell Biol., 97 : 1500-1506 (1983).
17. A.S.G. Curtis and M. Varde, J. Natl. Cancer Inst., 33 : 15-26 (1964).
18. G.A. Dunn and J.P. Heath, Exp. Cell Res., 101 : 1-14 (1976).
19. G. Gerisch, Curr. Topics Dev. Biol., 14 : 243-270 (1980).
20. D. Gingell and J.A. Fornes, Biophys. J., 16 : 1131-1153 (1977).
21. D. Gingell and I. Todd, J. Cell Sci., 41 : 135-149 (1980).
22. F. Grinnell, Intl. Rev. Cytol., 53 : 65-144 (1978).
23. H.K. Kleinman, R.J. Klebe and G.R. Martin, J. Cell Biol., 88 : 473-485 (1981).
24. L.D. Landau and E.M. Lifshitz, Fluid Mechanics. Pergamon Press, Oxford. (1959).
25. D.H. Napper, J. Coll. Interfac. Sci., 32 : 106-114 (1970).
26. M.D. Pierschbacher and E. Ruoslahti, Nature, 309 : 30-34 (1984).
27. K.R. Porter, M. Beckerle and W. McNiven, in: 'Modern Cell Biology' vol. 2, ed. B. Satir, A.R. Liss, New-York, pp. 259-302 (1983).
28. D.A. Rees, R.A. Badley and A. Woods, in: 'Cell Adhesion and Motility', ed. A.S.G. Curtis and J.D. Pitts, Cambridge Univ. Press, Cambridge, pp. 389-408 (1980).
29. U. Rutishauser, Nature, 310 : 549-554 (1984).
30. D. Schubert, M. LaCorbiere, F.G. Klier and C. Birdwell, Cold Spring Harbor Symp. Quant. Biol., 48 : 539-549 (1983).
31. H.I. Smith and D.C. Flanders, Appl. Phys. letters, 32 : 349-350 (1978).
32. M.S. Steinberg, in: 'Cellular Membranes in Development', ed. M. Locke, Academic Press, New-York, pp. 321-366 (1964).
33. L. Stryer, Ann. Rev. Biochem., 47 : 819-846 (1978).
34. P.L. Townes and J. Holtfreter, J. Exp. Zool., 128 : 53-120 (1955).
35. P.H. Weigel, J. Cell Biol., 87 : 855-861 (1980).
36. P.H. Weigel, E. Schmell, Y.C. Lee and S. Roseman, J. Biol. Chem., 253 : 330-333 (1978).
37. L. Weiss and D.L. Kapes, Exp. Cell Res., 41 : 601-608 (1966).
38. R.M. Weisz, K. Balakrishnan, B.A. Smith and H.M. McConnell, J. Biol. Chem., 257 : 6440-6446 (1981).
39. A. Williams, Calc. Tissue Res., 6 : 11-19 (1970).
40. C.C. Wylie and J. Heasman, Phil. Trans. Roy. Soc. Lond., B 299 : 177-183 (1982).
41. K.M. Yamada, J. Cell Biol., 78 : 520-541 (1978).

SECTION II

THE ADHESIONS OF EPITHELIAL CELLS

D.R. Garrod

Cancer Research Campaign
Medical Oncology Unit, CF99
Southampton General Hospital
Southampton, SO9 4XY

INTRODUCTION

Epithelia are tissues which line or cover the cavities and surfaces of the bodies of animals. They consist of single or stratified sheets of cells which are applied to be underlying tissue on one face, the basal surface, and have a free edge adjacent to a luminal space on the other face, the apical surface. All epithelia thus have a structural apico-basal polarity which is always of great functional importance. For example, in absorptive epithelia such as that of the small intestine, absorption takes place at the apical surface of the cells and the absorbed material is transported basally. In protective epithelia such as the epidermis, the apical surface must be strong in order to resist friction and pressure, while the basal region of the epithelium must generate new cells to replace the stressed apical tissue.

It is the purpose of this review, not to discuss the polarised function themselves, but to consider a fundamental property of epithelial cells by means of which the polar organisation is maintained. That property is cell adhesion. In relation to their polarity, epithelia must have three adhesive properties as follows: (i) at the basal surface they must adhere to the underlying tissue; (ii) within the epithelium there must be strong cohesion in order to maintain the epithelial integrity ; (iii) at the apical surface adhesiveness must be low in order to prevent occlusion of the lumen. In simple epithelia which are only one cell thick, all of these properties must be vested in every cell, which means that each cell must have a polarised organisation of its own and especially a polarly organised set of adhesive properties. In

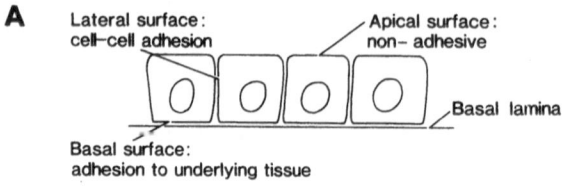

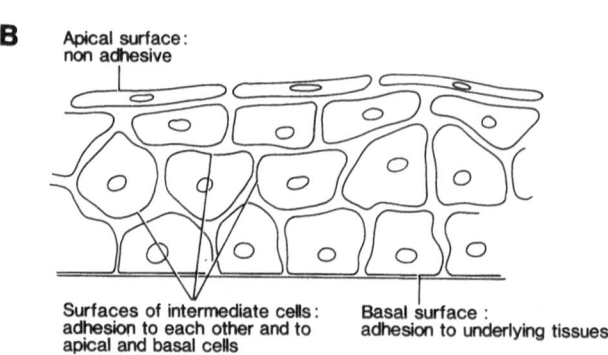

Fig. 1. Diagram showing the adhesive properties of cells in simple (A) and complex or stratified (B) epithelia.

complex or stratified epithelia only cells in the basal layer need to adhere to the underlying tissue and only the apical cells need to have non-adhesive upper surfaces. The cells in between need to adhere to each other, to the apical surfaces of the basal cells and the basal surfaces of the apical cells. In a stratified epithelium therefore there is need for a division of labour or differentiation in properties between cells in different layers (Fig. 1).

Such polarised organisation of adhesive properties requires a certain complexity of adhesion mechanisms. We have pointed out previously that no cell type should be expected to have only one single adhesion mechanism (34,36). In the case of a highly organised cell such as a cell in a simple epithelium there are a minimum of four, and possibly a maximum of eight or ten, different molecular mechanisms. Each contributes towards the overall adhesiveness of the cell and subserves a different specialized function. The diagram shown in Fig. 2 indicates these adhesion mechanisms for a simple, non-stratified epithelial cell. This figure shows ten different adhesion mechanisms, five involved in cell-cell adhesion and five in cell-substratum or cell-basal lamina adhesion. It is likely that all of these mechanisms, and possibly more, may be present, simultaneously, in many epithelial cells.

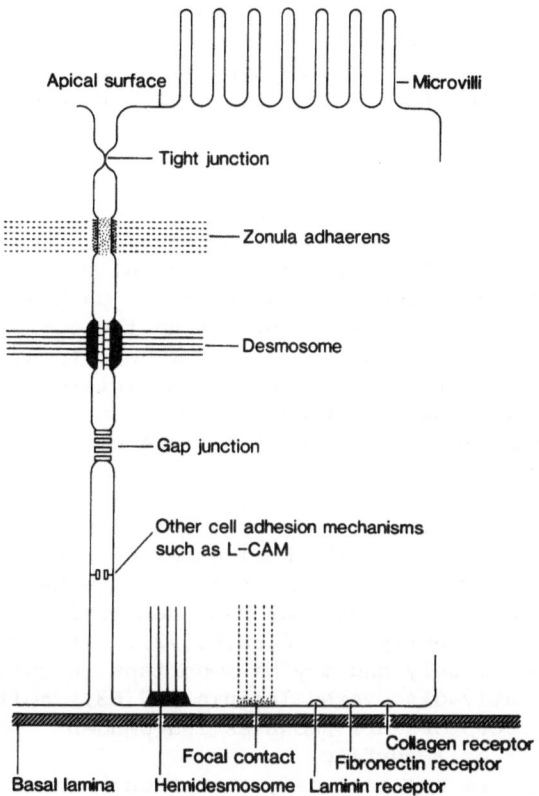

Fig. 2. Diagram showing the adhesion mechanisms of a "typical" simple epithelial cell.

      The   cell-cell   adhesions   consist   of   four   types   of
ultrastructurally   identifiable   junctions, plus one mechanism, the
liver cell adhesion molecule, L-CAM (27), which has   been   detected
at   the   molecular level and which has been shown in this figure in
the region where, ultrastructurally, there appears   to   be   contact
between   the   apposed plasma membranes of the cells but where is no
junctional specialization. It should be be stressed that there   may
be   several such 'molecular-only' mechanisms, i.e. mechanisms which
have   not   been   localised   to   particular   junctions. Thus Ogou,
Yoshida-Noro   and   Takeichi   (78)   demonstrated   that   a monoclonal
antibody   raised   against   teratocarcinoma   cells   also   inhibited
adhesion   of   hepatocytes   and precipitated molecules of 124 kd and
100 kd fom the latter. The former is similar in molecular weight to
the   liver   cell adhesion molecule, L-CAM (27) while the latter may
be similar to the 105 kd glycoprotein demonstrated   in   hepatocytes
and known as Cell-CAM 105 (76). In mammary epithelial cells Damsky,

Richa, knudsen and Buck (21) have denoted a putative adhesion molecule as Cell-CAM 120/80 since a monoclonal antibody which disrupts cell-cell interactions recognised glycoproteins with these molecular weights, the 120 kd from cell extracts and the 80 kd from conditioned medium.

Of the junctions, two, the desmosome and the Zonula adhaerens, are believed to have primarily adhesive functions. It should be stressed that the precise functions of these junctions are not understood : they will be discussed more fully below. The others, the tight or occluding junction and the gap junction, have other primary roles in occluding and controlling the permeability of the extracellular space, and intercellular communication, respectively (see 41). Since their functions depend upon intercellular molecular contact and binding, however, they must contribute a component to intercellular adhesion (see 36 for references to the adhesive function of these junctions).

The cell-substratum (basal lamina) adhesion characteristically shows one type of junction, the hemidesmosome, which is present in many types of epithelia. The focal contact which has been studied mainly in fibroblasts (47,85) has been demonstrated in epithelial cells relatively recently (5,46,68,68,73). It probably has a transitory existence only and may be important in relation to wound healing. The recently-discovered laminin (67,83), collagen (72,88), and fibronectin receptors are shown as independent 'molecular-only' sites on the basal cell surface.
Fibronectin receptors are included, even though there is no direct evidence for them in epithelial cells, because Mattey and Garrod (68) have shown that anti-fibronectin IgG inhibits adhesion of corneal epithelial cells to fibronectin coated substrata. It is possible however, that these receptors for basal lamina and matrix components could be associated with either of the junctional structures. Here again it should be stressed that other 'molecular-only' mechanisms may be present, so that the diagram in Fig. 2 probably represents a minimum number of mechanisms.

To conclude this introduction I wish to stress that the formation of epithelia during embryonic development, their maintenance during adult life and their repair during wound healing, depends on the integrated functioning of all these mechanisms. This is emphasised in disease conditions such as cancer (epithelial cancer, or carcinomas, constitute 90% of human cancer), pemphigus and pemphigoid where the lesions and abnormal cell behaviour may result from abnormalities, disorganisation or breakdown of the adhesion mechanisms.

It should also be pointed out that because there is a variety of different epithelia which are specialized in different ways, different adhesion mechanisms will predominate in different

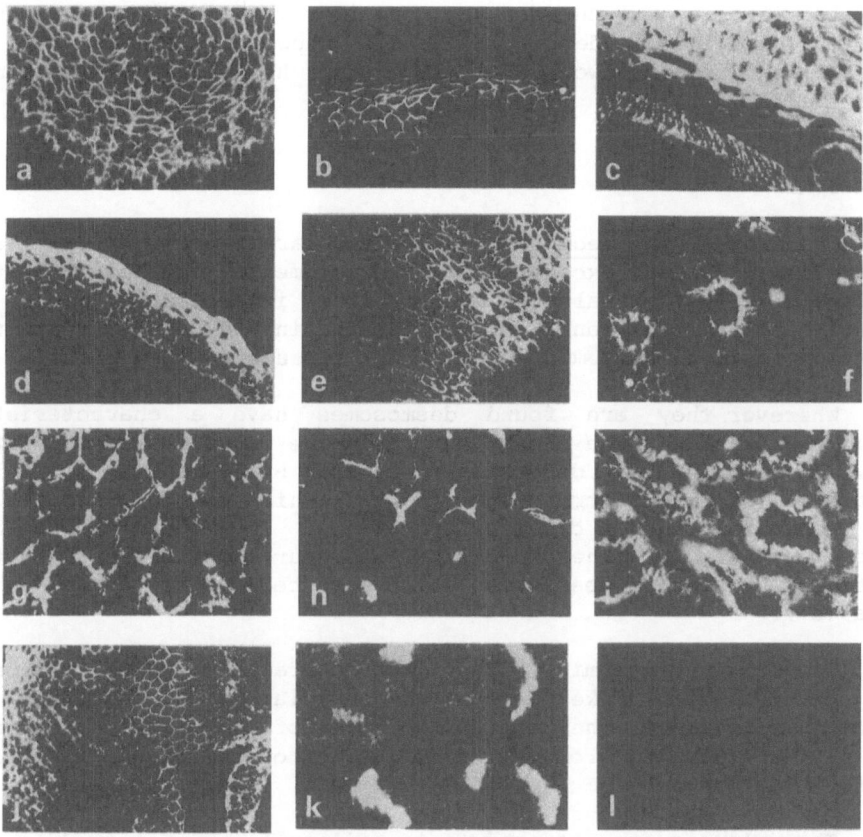

Fig. 3. Photographs showing the distribution of desmosomal antigens (desmoplakins) and, therefore, of desmosomes in different tissues. The tissues are as follows: (a) bovine nasal epithelium, (b) human skin, (c) frog skin, (d) chick embryo cornea, (e) human oesophagus, (f) guinea pig stomach, (g) rat liver, (h) axolotl liver, (i) human kidney, (j) human breast epithelium, (k) bovine heart, (l) rat skeletal muscle which has no desmosomes. (e), (i) and (j) were provided by Mr. John Parry. Other photographs are reproduced, considerably reduced in size, from (16) and (17). Scale bar (same for all photographs) = 20 μm.

situations. For example, desmosomes are more numerous in epidermis and other stratified epithelia such as the oesophagus, bladder and cervix, whereas the zonula adhaerens is probably more highly developed in simple epithelia.

It is now my purpose to consider the adhesion mechanisms outlined here in more detail. I shall, however, concentrate on desmosomes since the work of my laboratory has been much concerned with these.

DESMOSOMES

Desmosomes or <u>maculae adhaerentes</u> are present in all epithelial tissues except one, the pigmented epithelium of the retina (23). They are also present in the intercalated discs of heart muscle. The panel of photographs in Fig. 3 illustrate the distribution of desmosomes in different tissues.

Wherever they are found desmosomes have a characteristic structure (Fig. 4). The plasma membranes are parallel and separated by 25-35 nm in the desmosomal region. Half way between the membranes is an electron dense mid-line which may be connected to the plasma membrane by cross bridges (84). Such connections are also suggested by the knob-like structures which appear on the surface membranes of desmosomes which have been split with 9.5 M urea (31).

On the cytoplasmic side of the membranes are dense plaques, the features which make desmosomes immediately recognisable in electronmicrographs. They are of the order of 15-20 nm in thickness and may be separated from the inner leaflet of the plasma membrane by few nanometers.

The plaques are attached to tonofilaments which extend to the interior of the cell. These filaments are usually composed of cytokeratin (48) although they may be desmin in the heart (58) and vimentin in the meninges (60). A more detailed consideration of the ultrastructure of desmosomes is given by Garrod and Cowin (35).

Analysis of the molecular composition of desmosomes has been made possible by the development of techniques for their isolation (24,90,33). A modification of these procedures has been developed by Gorbsky and Steinberg (44), enabling preparation of desmosomal cores, structures which consist of parallel plasma membranes with mostly plaques and little cytokeratin.

A polyacrylamide gel showing the protein composition of desmosomal cores is presented in Fig. 5. The postulated locations of the components in the desmosome are indicated by arrows pointed to Fig. 4. A summary of desmosomal components, their names and some of their properties are given in Table 1.

I shall not repeat information presented in Table 1, but wish to discuss one particular aspect with which we have been involved.

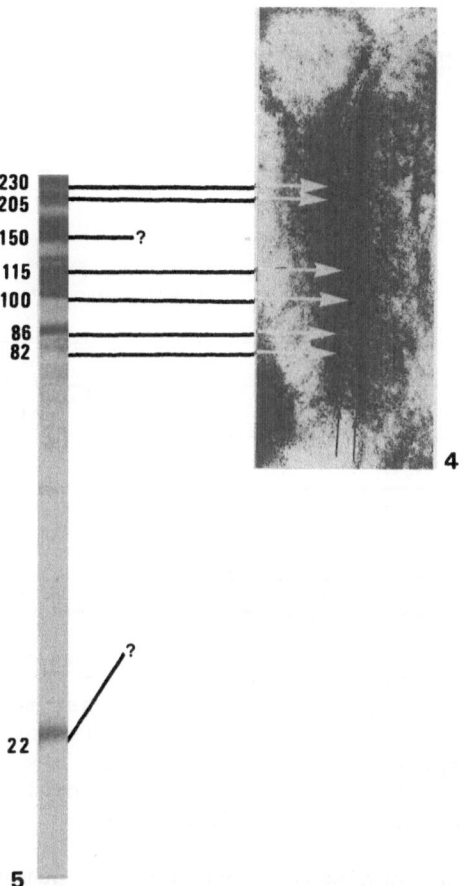

Fig. 4.   An electron micrograph of a desmosome from bovine
          nasal epithelium. The plasma membranes of the
          adhering cells are indicated by the small black
          arrows at the bottom of the photograph. The white
          arrows indicate the postulated locations of the
          protein and glycoprotein components shown in Fig. 5.

Fig. 5.   10% polyacrylamide gel of desmosomal cores prepared
          from bovine nasal epithelium by the method of
          Gorbsky and Steinberg (44). The numbers on the left
          hand side indicate the molecular weights of the
          components (x10^{-3}). In this preparation there are
          two additional bands, below the 205 kd protein and
          between the 115 kd and 100 kd bands. These are
          believed to be breakdown products of other bands and
          do not appear in all preparations. For further
          details see Table 1.

Table 1. Summary of desmosomal components.

Number (Franke et al.)	Molecular Weight ($\times10^3$) (Franke et al.)	Molecular Weight ($\times10^3$) (Gorbsky and Steinberg)	Protein or Glycoprotein	Name	Postulated Location	Isoelectric Point
1	250	230	P	Desmoplakin I	Cytoplasmic plaque	7.2-6.8
2	215	205	P	Desmoplakin II		7.2-6.8
3 group	175 - 164	150 (triplet)	GP		?	acidic
4 a	130	115 (doublet)	GP	Desmocollins	Intercellular	4.85 4.93
4 b	115	100	GP			
5	83	86	P		Cytoplasmic plaque	6.3-7.0
6	75	82	P		Cytoplasmic plaque	basic
		22	GP		Intercellular	

Data from : Gorbsky and Steinberg (1981)
            Franke et al. (1981)
            Franke et al. (1982) Mueller and Franke (1983)

The horizontal lines group the components in accordance with their immunological and biochemical relationships as determined by monoclonal antibodies (Gorbsky, Cohen and Steinberg, 1983), polyclonal antibodies (Cowin and Garrod, 1983), lectin binding (Gorbsky et al., 1983) and peptide mapping (Mueller and Franke, 1983).

This is the question of which components are involved in cell-cell adhesion. Gorbsky and Steinberg (44) found that the bands labelled 150, 115 and 100 in Fig. 5 are glycosylated and are enriched in desmosomal core preparations compared with the non-glycosylated components. They concluded that these glycoproteins may function as the intercellular adhesive of the desmosome.

Using polyclonal antibodies raised in guinea pigs against desmosomal components isolated from bovine nasal epithelium (16), we have conducted a study of the distribution of desmosomal components in Madin-Darby Bovine Kidney (MDBK) cells (18). It has been our policy in previous publications to refer to desmosomal components by their molecular weights, adopting those reported by Gorbsky and Steinberg (44) (see Table 1). We found that only antibodies to the 115 kd and 100 kd components bound to the surfaces of living cells. Binding of antibody to the 150 kd glycoprotein was no greater than binding of antibodies to cytokeratin and the desmosoplakins. Furthermore, if cells were incubated in univalent Fab' fragments derived from anti-115/100, desmosome formation appeared to be inhibited.
Fab' from control antibodies, including anti-150 were without effect on desmosome formation (Fig. 6). (It should be noted that all antibodies, polyclonal or monoclonal, raised against the 115 kd glycoprotein react also with the 100 kd glycoprotein and vice versa – see Table 1.)

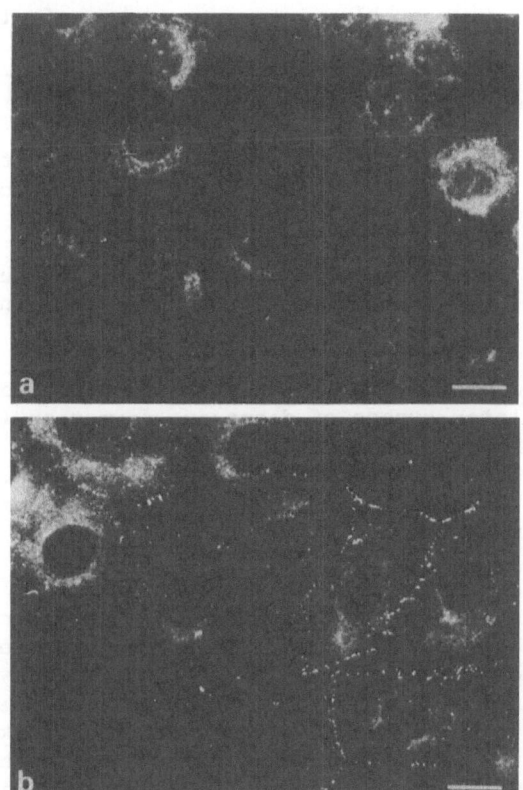

Fig. 6.   Inhibition of desmosome formation in MDBK cells by
           anti-desmocollin Fab'. Cells were harvested and
           replated (a) in anti-115/100 (desmocollin) Fab' and
           (b) in anti-150 Fab'. After 24 hours of culture the
           cells were fixed in methanol and stained for the
           presence of desmosomal plaques with anti-desmoplakin
           antibody. Desmosomal plaques are present in (b) but
           not in (a). Reduced from (18) from which further
           details may be obtained. Scale bar = 20 µm.

    From these results we concluded that the 115 kd and 100 kd
glycoproteins are located on the cell surface and are involved in
desmosomal adhesion. We named these glycoproteins desmocollins I
and II, respectively, to denote their adhesive role (18).

    It is important to note that while our results have firmly
demonstrated that the desmocollins are present on the cell surface,
they do not prove unequivocally that the 150 kd glycoprotein(s) is
entirely cytoplasmic. Rather, they simply show that we were unable
to detect the 150 kd glycoprotein on the surfaces of living cells.

There are several possible interpretations of this result, as follows :

(a)   The 150 kd glycoprotein is cytoplasmic.

(b)   Part of the 150 kd glycoprotein is on the surface, but our antibodies recognise only that part which is located cytoplasmically or within the membrane.

(c)   Part of the 150 kd glycoprotein is on the surface but is masked by other components, thus being inaccessible to antibody binding.

(d)   The 150 kd glycoprotein is inserted into the desmosome only during the later stages of desmosome formation (see below) and is inaccessible to antibody binding because the extracellular material of the desmosome is impenetrable to the antibody.
        Either (b), (c) or (d) would suggest that the 150 kd glycoprotein may be a transmembrane component.

(e)   A final possibility which should be mentioned is that the 150 kd glycoprotein is a precursor of the desmocollins and is partially de-glycosylated when inserted into the membrane. In relation to this possibility, Cohen, Gorbsky and Steinberg (13) found that monoclonal antibodies raised against desmosomal cores reacted either with the 150 kd glycoprotein or against the desmocollins but not against both. It could be argued, however that these antibodies may be directed against the carbohydrate portions of the molecule, and therefore would simply detect differences in glycosylation. Cohen et al. (13) also reported differences in lectin binding between 150 kd and the desmocollins. We have recently found that some polyclonal guinea pig antibodies raised against desmocollins react also with the 150 kd glycoprotein (98). This may indicate similarities between the molecules. However, we have so far never encountered the reverse situation : antibodies raised against the 150 kd glycoproteins have always proved antigen-specific.

        There is still much to be learned about relationships between these molecules and the possibility that other components such as the 22 kd glycoprotein (see Table 1) are involved in desmosomal adhesion.

EVOLUTIONARY CONSERVATION OF DESMOSOMAL COMPONENTS
AND DESMOSOMAL ADHESION MECHANISMS

        Desmosomes are widespread throughout the epithelial tissues of different vertebrate animals and desmosome-like structures are also

found in many invertebrate species (65,66,71,80,91). Wherever they
occur they have ultrastructural features similar, though not
necessarily identical, to those outlined above. We have used our
antibodies against desmosomal components to study the distribution
of those components among different vertebrate animals (16,17).
As our first criterion of desmosomal comparison we used fluorescent
antibody staining. Our main results were as follow :

1)   The epidermis of all vertebrates from man to the frog
     (including a lizard and a bird) stain equally brightly for all
     the desmosomal components to which we have antibodies (i.e.
     all components shown in Table 1 except the 22 kd glyco-
     protein).

2)   In the epidermis of axolotl and trout, staining for desmosomal
     plaque constituents is present, but staining for desmosomal
     glycoproteins is greatly reduced or absent.

3)   In non-epidermal tissues of mammalian species as well as
     chick, lizard and frog, staining for the desmocollins is less
     intense than in the epidermis, while staining for the
     desmosomal plaque constituents as well as for the 150 kd
     glycoprotein is undiminished.

These results suggest that the components of desmosomes are to
a large extent conserved between the different tissues of
vertebrate animals. At the same time they may be interpreted to
indicate that there may be some chemical differences between the
desmocollins of epidermal and non-epidermal tissues. The plaque
constituents appear to be more highly conserved in terms of
antigenicity than the desmocollins. Observation of desmoplakins
between human and bovine tissues has also been demonstrated by
Franke et al. (32) and Mueller and Franke (74) on the basis of
fluorescent antibody staining and peptide mapping. Suhrbier and
Garrod (98) have conducted a comparison of desmosomal components in
human, bovine, canine, chick and frog cells by gel electrophoresis
and immunoblotting. The main conclusions are that the plaque
constituent have identical molecular weights in all these species,
and there are some minor variations in the molecular weights of the
glycoprotein which may be due to differences in glycosylation.

Our results suggest that there may be more differences in the
composition of the desmocollins in desmosomes from different
sources. It is therefore crucial to ask whether these differences
are functionally important in relation to cell adhesion. In order
to test this we have made combinations between five different kinds
of cells from different animal species and different tissue origin
(70). The cells used were HeLa cells (human cervical epithelium),
MDBK cells, MDCK cells (similar to MDBK but derived from canine
kidney), chick embryonic corneal epithelial cells (Ccep) and adult

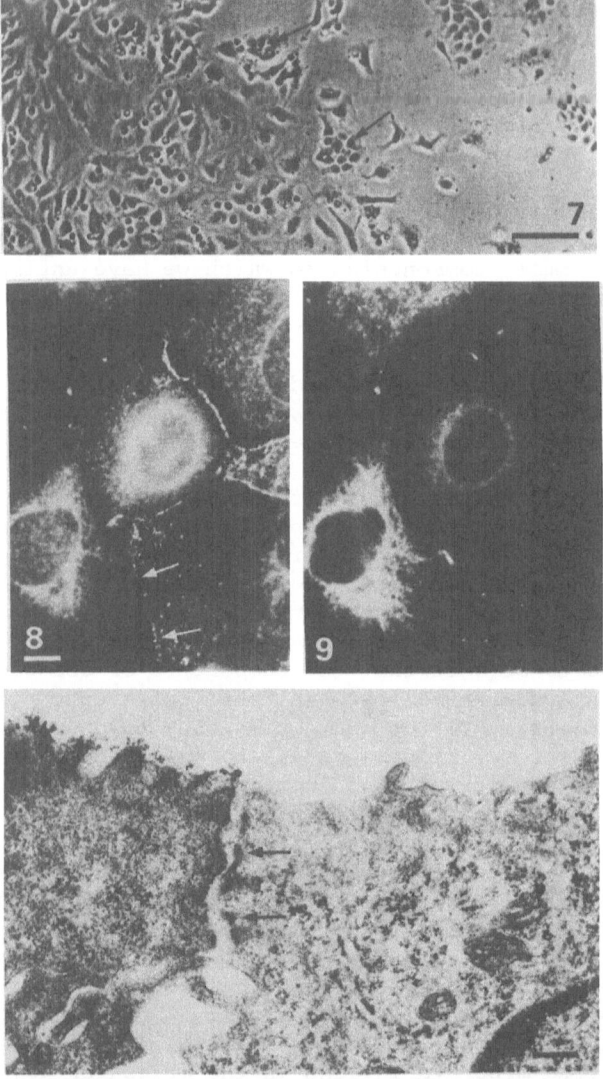

Fig. 7.   Mixed culture of HeLa and <u>Rana pipiens</u> corneal
          epithelial cells (Fcep) (arrows). Scale bar = 200 μm

Fig. 8.   Mixed culture of HeLa and Fcep cells stained with
          anti-desmoplakin staining (arrows) at HeLa/Fcep
          contacts. Rhodamine fluorescence. Scale bar = 20 μm

Fig. 9.   Same field as Fig. 8 stained with anti-keratin
          monoclonal antibody LE61 which stains HeLa but not
          Fcep cells. Fluorescein fluorescence.

Rana pipiens (frog) corneal epithelial cells (Fcep). Desmosome formation between the various cell types was assessed by fluorescent antibody staining and confirmed by electron microscopy. Identification of the cell types was by morphology both at the light and electron microscope levels, and also by staining with the anti-keratin monoclonal antibody, LE61, which has specificity for simple epithelia (64). In our experiments it stained HeLa and MDBK cells, and thus aided the distinction between these cells in combinations with non-staining cells.

The result of our study was very straightforward. Desmosome formation occurred between all of these cell types (Fig. 7, 8, 9 and 10). We have therefore greatly extended the work of Overton (81) who demonstrated desmosome formation between embryonic chick corneal and mouse epidermal cells. We have also demonstrated for the first time desmosome formation between cells of ectodermal and non-ectodermal origin. This is important because distinct differences in fluorescent staining with anti-desmocollin antibodies were found between epidermal and non-epidermal cells (17 and see above), and it has been demonstrated that desmosomes from different sources show different susceptibility to trypsin, EDTA and deoxycholate (6).

The formation of desmosomes between human cells and frog cells, as well as between epidermal and non-epidermal cells, suggests that the mechanism of intercellular binding is highly conserved. Presumably, mutual desmosome formation is initiated by a recognition event between the adhesive components of desmosomes on opposed cell surfaces (see below). Our results suggest therefore that compatibility or molecular complementarity exists between the desmosomes of frogs and men. The adhesion sites have not changed significantly since frogs and men had a common ancestor. Furthermore, even if further analysis confirms the existence of slight chemical differences between desmocollins from different sources, these differences are not important in adhesion and therefore may be presumed to be in regions of the molecules other than the adhesion recognition sites.

Whatever the precise function of desmosomal adhesion may be (see below), we may conclude that it is sufficiently vital in forming or maintaining epithelial organisations, that there is

---

Fig. 10. Electron micrograph showing desmosomes (arrows) at contact between HeLa and Fcep cell. The latter is on the left and is identified by the fuzzy material on the upper surface. Scale bar = 0.2 µm. For further details of Figs. 7 to 10 see (70).

strong evolutionary pressure against changing it. Desmosomal adhesion is a vital component in the organisation of vertebrate tissue structure and has therefore been conserved. If one views desmosome formation in this way, there is no surprise in the lack of specificity or selectivity of desmosome formation. The desmosome is a mechanism which has evolved to join epithelial cells together. Wherever it occurs it is composed of essentially the same molecular constituents and employs the same cell-cell adhesion-recognition mechanism. Eventually it will be of interest to extend studies such as those reviewed above to determine whether the desmosomes of fish and invertebrates are also similar.

It should be stressed, however, that there is one obvious difference between the desmosome of different tissues and that is the pattern in which the desmosomes are arranged (see Fig.3). When the precise function of desmosomes is understood, we may expect the different tissue patterns to be of functional significance. We must try to understand how these different patterns or arrangements are formed and what controls their formation.

MODULATION OF DESMOCOLLINS
DURING DEVELOPMENT OF EPITHELIAL POLARITY IN MDBK CELLS

MDBK cells are an established line of simple epithelial type from bovine kidney. In vivo they were presumably components of the kidney tubules. Once they reach confluent density in culture they form a monolayer which has many of the features of a true simple epithelium (see Fig. 2). In particular the cells become polarised having a basal surface which adheres to the culture substratum, a lateral surface which forms adhesions with other cells via zonulae adhaerentes and desmosomes and an apical surface which is non-adhesive. The lateral surfaces also develop tight junctions which occlude the intercellular space and give rise to a high trans-epithelial resistance. The upper surface which is exposed to the medium and free from contact possesses microvilli.

In confluent MDBK cells, desmosomes are concentrated principally in a sub-apical band between the lateral surfaces of adjacent cells. This can be shown by fluorescent antibody staining (Fig. 11) or by electron microscopy (32). It is probable that each fluorescent spot in Fig. 2 corresponds to a desmosomal plaque, though this has not been directly demonstrated by fluorescent antibody staining followed by electron microscopy on the same cell.

When such cells are treated with EGTA their lateral adhesions are broken and their desmosomal plaques internalised (59). The desmocollins, however, remain at the cell surface for hours after EGTA addition (18), while treatment of the cells with trypsin and

EDTA completely removes all trace of desmocollin staining from the surface (18). Such desmocollin-free cells can be seeded onto a tissue culture substratum and the reappearance of desmocollins follows, together with the development of epithelial polarity. (It should be noted that if the cells are treated with trypsin in the presence of calcium, desmocollin staining is not removed (Mattey, unpublished observations.)

The first reappearance of desmocollin staining in these cells is at about 6 hours after seeding. At 24 hours desmocollin staining is distributed evenly over the entire upper surface of the cells (Fig. 12) and there is slight concentration in regions of intercellular contact. Staining with univalent Fab' antibody fragments reveals an essentially monodispense distribution on the upper cell surface, whereas staining with IgG gives patching, but not capping (18).

Subsequently there occurs a modulation or redistribution of desmocollin staining which appears to be associated with the development of epithelial polarity. Thus, after about 3 days in culture, desmocollin staining is much reduced on the upper cell surface, and that which is present being patchily distributed over the monolayer and concentrated towards the regions of intercellular contact (Fig. 13). By six or seven days, desmocollin staining is entirely absent from most of the monolayer surface, being detectable only adjacent to free edges (Fig. 14). Treatment of such a monolayer with EGTA, however, causes separation of cell-cell contacts and the reappearance of bright desmocollin staining (Fig. 15).

Our interpretation of these results (18) is as follows. As epithelial polarity develops the adhesive molecules known as desmocollins are progressively removed from the upper cell surface and expressed only on the lateral (and possibly basal) surfaces. We have not demonstrated that the desmocollins on lateral surfaces are present solely and exclusively in desmosomes. However, the progressive confinement of adhesion molecules to the lateral cell surface and removal from the luminal surface, is a rather satisfying demonstration of the acquisition of the polarised adhesive properties outlined in Fig. 2. Similar modulation of a 35 kd surface component of unknown function in MDCK cells has been described by Herzlinger and Ojakian (53).

The situation with regard to desmosomal plaque components is less clear. In early cells (up to 24 hours) there is sometimes punctate staining for desmosomal plaque constituents within the cytoplasm (18), but we have not yet carried out experiment to determine whether this represents precursors of future desmosomes or remnants of desmosomes broken down and internalised during trypsinisation.

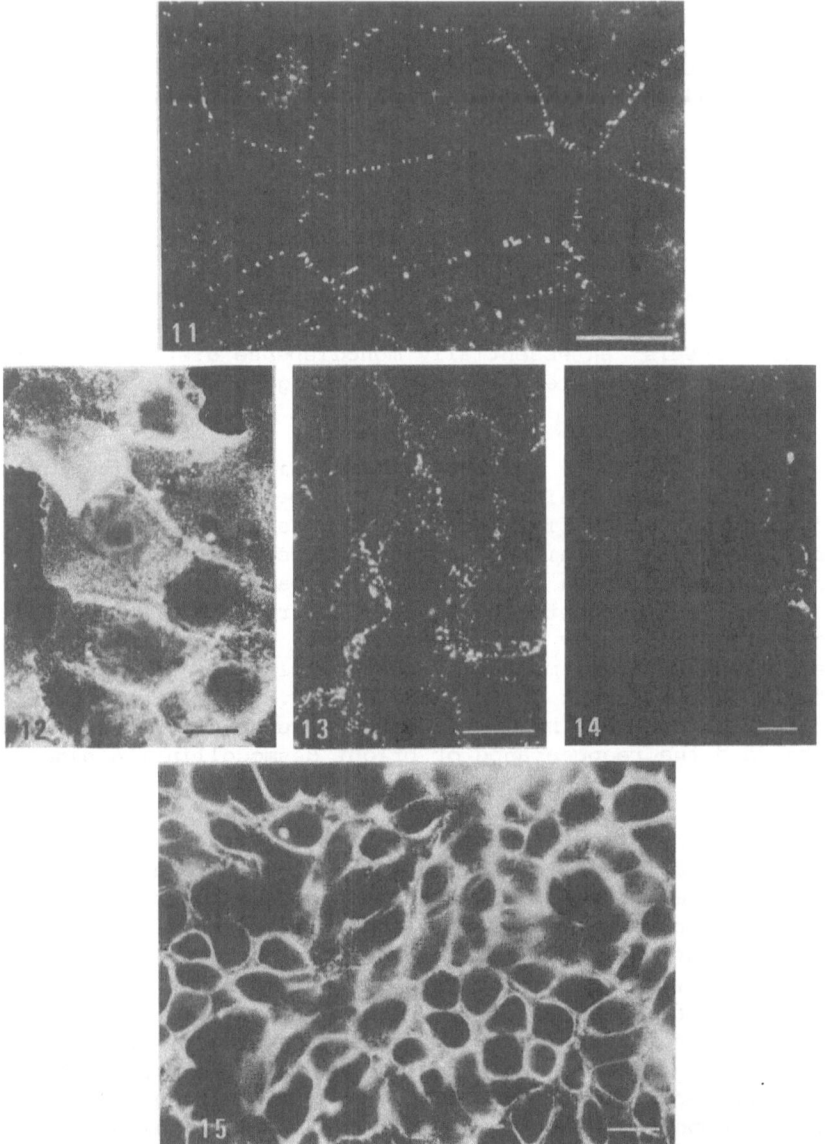

Fig. 11. Desmosomes between MDBK cells after confluent culture
          for several days revealed by fluorescent staining
          with anti-desmoplakin antibody. (Scale bar for all
          figures 20 μm).

Fig. 12. Anti-desmocollin staining of living MDBK cells after
          24 hrs. in culture.

At any stage from about 18 hours onwards, punctate staining for desmosomal plaque constituents may be found in regions of intercellular contact. It seems likely that this may be taken as desmosome formation, suggesting that this occurs progressively through the period of development of epithelial polarity. Two points should be stressed. Firstly, the staining patterns obtained with antibodies to the 150 kd glycoprotein have always been identical to those obtained with anti-desmoplakin. Secondly, there is never any general diffuse staining of the cytoplasm for these components as is found with keratinocytes (see below).

Not too much emphasis should be placed on the timing of the modulation events described above. We believe that they are essentially related to the development of epithelial polarity rather than to time in culture, and the precise timing may vary with culture conditions such as seeding density, medium composition, etc...

An in vivo example of the redistribution of desmosomal proteins which seems not unlike that which we have described for MDBK cells occurs in the development of the rat mammary gland (26). During lumen formation from immature end buds, the proteins microvillin (2) and p80 (7) move to the apical poles of the luminal cells, in contact with lumen, while the desmoplakins go to the basal surface in contact with the myoepithelial cells.

CALCIUM-INDUCED MODULATION OF DESMOSOMAL COMPONENTS IN HUMAN KERATINOCYTES

In contrast to the formation of desmosomes and modulation of desmosomal components by MDBK cells, which takes place slowly over a period of hours or days, calcium-included desmosome formation by cultured keratinocytes occurs rapidly. Keratinocytes from a variety of mammals may be grown successfully in culture, where they form multilayers which resemble the intact epidermis (45,87). If the cells are maintained in low calcium medium (0.1 mM) they do not stratify but remain as a monolayer (52). Raising the calcium

Fig. 13. As Fig. 12 after 5 days in culture.

Fig. 14. As Figures 12 and 13 after 7 days in culture

Fig. 15. Anti-desmocollin staining of comparable culture to that shown in Fig. 14 after treatment for 15 minutes with 1 mM EGTA in culture medium. All figures reproduced, with size reduction, from (18).

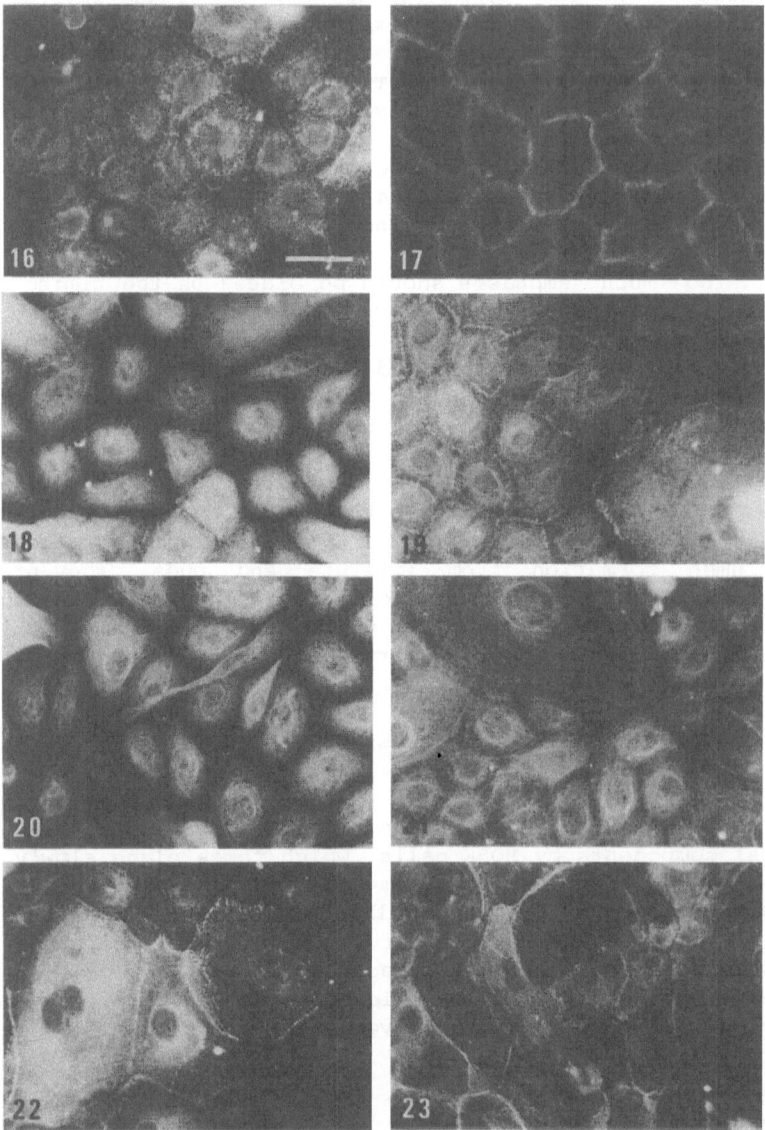

Fig. 16. Human keratinocytes in LCM stained with anti-
desmocollin antibody. Scale bar = 40 μm. All figures
in plate at approximately same magnification.

Fig. 17. As figure 16 but 15 minutes after raising calcium
concentration. Note peripheral concentration of
desmocollins.

concentration in the culture to 1.8-2 mM causes the onset of stratification and differentiation. The first signs of stratification appear about nine hours after raising the calcium concentration and the process continues for days or weeks.

A much more rapid event following increase in calcium concentration is the redistribution or modulation of desmosomal components, and desmosome formation. Electron microscope studies of mouse (49) and human (Mattey unpublished) keratinocytes have shown that when maintained in low calcium medium (LCM) they have very few desmosomes. Indeed the membranes of adjacent cells may be quite widely separated. On raising the calcium concentration desmosome formation occurs extremely rapidly. Hennings and Holbrook (49) report fairly extensive desmosome formation within 1 hour. Our own results with human keratinocytes show an essentially similar picture.

Using fluorescent staining with our antibodies against desmosomal components, we have undertaken a study of the modulation or distribution of desmosomal components during calcium-induced desmosome formation in human keratinocytes (102).

Keratinocytes in LCM have the desmocollins distributed evenly over their surfaces, while the desmoplakins and the 150 kd glycoprotein appear to be evenly and diffusely distributed throughout the cytoplasm (Fig. 16 and 18). In these cells the

---

Fig. 18. Human keratinocytes in LCM stained with anti-150 antibody.

Fig. 19. As Fig. 18 but 120 minutes after raising calcium concentration. Note peripheral and almost fibrillar pattern of staining.

Fig. 20. Human keratinocytes in LCM stained with guinea pig anti-bovine keratin polyclonal antibody.

Fig. 21. As Fig. 20 but 120 minutes after raising calcium concentration. Note establishement of "bridges" between adjacent cells.

Fig. 22. Human keratinocytes 9 hrs. after raising calcium concentration. Stained with anti-desmoplakin antibody. Note punctate staining in regions of overlap.

Fig. 23. As Fig. 22 but 24 hrs. after raising calcium concentration. Overlap is more extensive.

cytokeratin is concentrated as a basketwork around the cell nucleus and does not extend to the cell periphery. (In a few instances one does find desmoplakin staining concentrated towards the periphery where there is apposition between cell membranes. It is therefore probably not true to say that keratinocytes in LCM possess no desmosomes, but they certainly possess very few.)

When the calcium concentration is increased, an extremely rapid and dramatic redistribution of desmosomal components takes place (Fig. 17). After 15 minutes there is already a concentration of both surface and cytoplasmic components towards the regions of cell contact (Fig. 18 and 19). (It seems probable that the first thing which happens when calcium is raised is that the cells spread to establish contact.) Over a period of about 2-3 hours this concentration is progressively increased and there is a concomitant decrease in the diffuse staining for desmocollins on the cell surface and for plaque constituents in the cytoplasm. Furthermore, the keratin cytoskeleton becomes reorganised so that fine strands may be seen extending towards the cell periphery and being opposed by corresponding strands in the adjacent cell (Fig. 20 and 21). It is as though a link has been established between the cytoskeletons of adjacent cells, the links being mediated by desmosomes.

The redistribution of desmosomal components thus corresponds in timing with the formation of desmosomes seen by electron microscopy. The pattern of desmosomal staining at the cell periphery, particularly that for the desmoplakins and the 150 kd glycoprotein, is punctate. It may be that each punctum or spot represents a desmosome though this has not been formally proved. Later on, when stratification is underway, very clear punctate staining is observed (Fig. 22 and 23). It is reasonable to assume that in each case a spot represents a desmosome between the upper and lower surfaces of stratified cells.

The picture which emerges is that the keratinocyte in LCM is provided with all the necessary components for desmosome formation. When the calcium level is raised a trigger is activated, resulting in rapid redistribution of these components and desmosome formation. This obviously provides a superb situation for the study of desmosomal assembly. Such studies are being undertaken.

DESMOSOME FORMATION AND ITS CONTROL

The desmosome is a fairly complex organelle involving a number of different molecular interactions. These may be listed as follows:

   (a) intercellular recognition and binding between adhesive mo-
       lecules (desmocollins) on apposed cells,

    (b) transmembrane linkage between extracellular adhesive com-
ponents and plaque constituents,

    (c) Linkage between plaque and cytoskeletal intermediate fila-
ments, usually cytokeratin,

    (d) some type of lateral binding between components which acts
at right angles to the other linkages to unite the
desmosome into a single structure. This may be
extracellular, intramembrane or intracellular, or all
three.

The fact that desmosomes remain intact with virtually
undisturbed structure when isolated (90), and are extremely
insoluble, attests to the strengths of these interactions. The
observation that treatment of isolated desmosomes with metrizamide
to produce desmosomal cores removes the plaque material and causes
disorganisation of the intercellular structure (44) suggests that
the plaques are important in maintaining the arrangement of the
intercellular binding material.

There have been several ultrastructural descriptions of the
sequence of events involved in desmosome formation. An excellent
description of desmosome formation during embryonic development in
Fundulus was given by Lentz and Trinkaus (66). Overton (81)
described desmosome formation in Ccep cells which had been
dissociated with trypsin, finding the first signs of desmosome
formation about 3 hours. After the commencement of reaggregation.
Dembitzer et al. (22) used a human cervical cell line which had
been trypsin dissociated, allowed to recover from trypsin damage
and centrifuged together. Under these conditions, complete
desmosome formation required 60-90 minutes, the first event being
an accumulation of electron dense material between the cell
membranes. This timing is of the same order as or slightly slower
than that found with calcium-induced desmosome formation in
keratinocytes (49,102,83). Formation of other types of junctions
also occurs within seconds or minutes (review of Garrod (34) for
references).

We are now in a position to begin to consider the molecular
events in desmosome formation. If we consider a keratinocyte in LCM
or a MDBK cell isolated in culture 24 hours (say) after
trypsinization we find that there are no desmosomes and that the
desmocollins are evenly distributed over the cell membrane. At this
stage the desmocollins are not attached to the cytoskeleton because
they can be removed from the cell by treatment with Triton X100
(Suhrbier, unpublished). They are also freely mobile within the
membrane because they can be patched by antibody (18). An even
distribution of adhesion molecules is a logical stage for a cell
which has no adhesions : any part of its surface is then available
for adhesion formation should it come into contact with another
cell.

I believe that the initial event or trigger for desmosome formation is recognition and binding between desmocollin (and possibly other) molecules on the surfaces of cells which come into contact. This is followed by a lateral recruitment or association of other molecules by a process of patching resulting in the type of 'rafting' of surface molecules referred to by Rees et al. (85) in the formation of focal contacts. A good analogy may be to think of the process as the operation of a type of two-dimensional zip fastener* in which the slide travels outwards in all directions from a central initiation point causing interlocking of more teeth, the desmocollins. The possibility that there are cross bridges between the desmosomal midline and the plasma membrane, which alternate in opposite sides of the midline (84) in fact reminds one very much of a zip fastener which has interlocking teeth. Interlocking teeth may also be suggested by the knobs present on desmosomes split by 9.5 M urea (31). Rather than a catch travelling outwards, however, we should probably think of desmocollins moving towards the initiation point and clicking into place. The analogy with a zip fastener should perhaps be reserved as a structural one rather than an operational one.

The necessity for patching or rafting is probably to be explained in terms of cooperativity of binding between adhesion molecules, which increases the strength of the adhesion. Here again, the analogy with a zip fastener is useful : binding between two teeth of a zip is almost non-existent, but when all are engaged a strong binding results.

I believe binding-recognition to be the initiating event because desmosomes do not form at free surfaces. They are organelles requiring the participation of two cells which possess the necessary components for desmosome formation. Where such recognition occurs, even if it is between a human cell and a frog cell, desmosomes will be formed.

The assembly of plaques on the cytoplasmic side of the desmosomal membrane and the attachment of tonofilaments to the plaques are believed to occur as a consequence of the recognition event between cells. Thus the keratinocyte in LCM possesses all the components necessary for desmosome formation, in a diffuse distribution. When the calcium concentration is raised and cell-cell recognition occurs the cytoplasmic components of the

---

* The phrase 'two-dimensional zip fastener' was first suggested to me by Sir Michael Stoker to describe something which I was clumsily trying to demonstrate by interlocking my fingers.

Low (Ca²⁺)

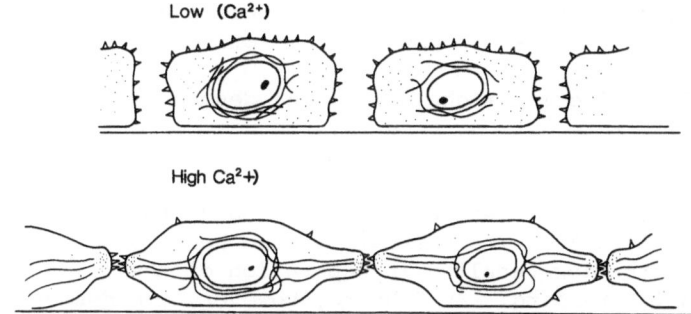

High Ca²⁺)

Fig. 24. Diagram   showing   redistribution of desmosomal   components   and   keratin   cytoskeleton   in   keratinocytes
induced   by   calcium   switching.   Desmocollins   are
represented   by 'teeth' on cell surface, cytoplasmic
desmosomal components by dots in the   cytoplasm   and
cytokeratin filaments by black lines in cytoplasm.

desmosome   and   the   cytokeratin filaments undergo a redistribution
and assembly process which is   topologically   associated   with   the
sites   of   recognition   on   the   cell   surface. This results in the
establishment of symetrical organelles, desmosomes,   which   provide
links   between the cytoskeletal networks of adjacent cells, thereby
uniting the monolayer into an integral   whole   :   the   cytoskeletal
network   of each cell becomes linked to that of every other via the
desmosomal connections (Fig.24).

At present it is not   clear   how   calcium   triggers   desmosome
formation.   The   observation   that   calcium   protects   desmocollins
against degradation by trypsin suggests   that   there   is   a   direct
interaction   between   calcium   and   the   desmocollins. It is possible
that such an interaction causes a   configurational   change   in   the
desmocollin   molecule   and   that   this   change is the trigger which
promotes desmosome assembly. This   would   imply   that   desmocollins
have   similar   properties   to   other   calcium   dependent   adhesion
molecules   (27,78).   It   is   not   known   whether   calcium   has   a
cytoplasmic   role   in   desmosome formation as well as a role in the
cell surface. The work of Hennings, Holbrook and Yuspa   (50)   using
calcium   channel blockers and ionophores would seem to suggest that
the primary role of calcium may be on the cell surface. If this   is
so,   calcium   concentration   would   seem   a   rather   unsatisfactory
mechanism for the cell to   regulate   its   own   desmosome   formation
because control of the extracellular concentration of an individual
ion would probably be beyond an individual cell. (The concentration
of   calcium   in   extracellular   fluid   is   of the order of 2mM, the
concentration which promotes desmosome formation   by   keratinocytes
in   vitro.)   Hennings,   Holbrook and Yuspa (51) have suggested that
intracellular   potassium   concentration   may   be   important   in

regulating keratinocyte differentiation but we have no information
relating to intracellular potassium and desmosome formation.

The effect of calcium chelation on desmosome stability has
also provided interesting results. The desmosomes of MDBK cells
seems to remain perpetually susceptible to EGTA treatment (18,58).
Keratinocytic desmosomes seem to acquire resistance to calcium
removal, however. Watt et al. (102) found that treatment of
cultured keratinocytes with EDTA within two hours after raising the
calcium concentration resulted in separation of the cells and loss
of desmosomal staining. At 3 hours, however, the desmosomes were
not broken down by EDTA : the cells detached from the substratum
but did not separate from eachother. (Overnight culture of 3 hours
keratinocytes in LCM eventually resulted in loss of desmosomes
without detachment from the substratum.) We interpret this result
to indicate the occurrence of some type of stabilizing event in
formation of keratinocyte desmosomes, which does not occur in MDBK
cells. This result may accord with the difference in sensitivity to
EDTA between desmosomes in simple epithelia and those of stratified
squamous and many glandular epithelia (6).

Another way in which desmosome formation may be controlled is
to regulate the synthesis of one or more desmosomal components.
Presumably desmosomes would be unable to form unless all their
components were present in the cell. It may be that synthesis of
components controls desmosome formation in development, that is to
say there is developmental regulation of desmosomal components. We
have no direct evidence for this at present but have done some
preliminary experiments with an in vitro differentiation system of
human teratocarcinoma cells (kindly provided by Dr. C.F. Graham).
These small rather rounded cells if attached to a gelatin
substratum and treated with retinoic acid differentiate into
flattened epithelial cells over a period of about 7 days (Fig. 28).
The precursor or stem cells have no desmosomes as demonstrated by
fluorescent antibody staining and electron microscopy (E.M. by Dr.
D.L. Mattey), and the keratin cytoskeleton is extremely poorly
developed (Fig. 25 and 27). The differenciated cells, however, show
a well developed, well spread cytokeratin network and extensive
staining for desmosomes along contact regions (Fig. 26 and 27).
Desmosome formation has been confirmed by electron microscopy. The
study of desmosome formation during retinoic acid-induced
differentiation in these cells may further the understanding of
control of desmosome formation.

DESMOSOMES IN TISSUE ORGANISATION

It must be stated, that we do not fully understand the role of
desmosomes in tissue organisation. Because they are arranged in
characteristic patterns in different tissues we assume that their

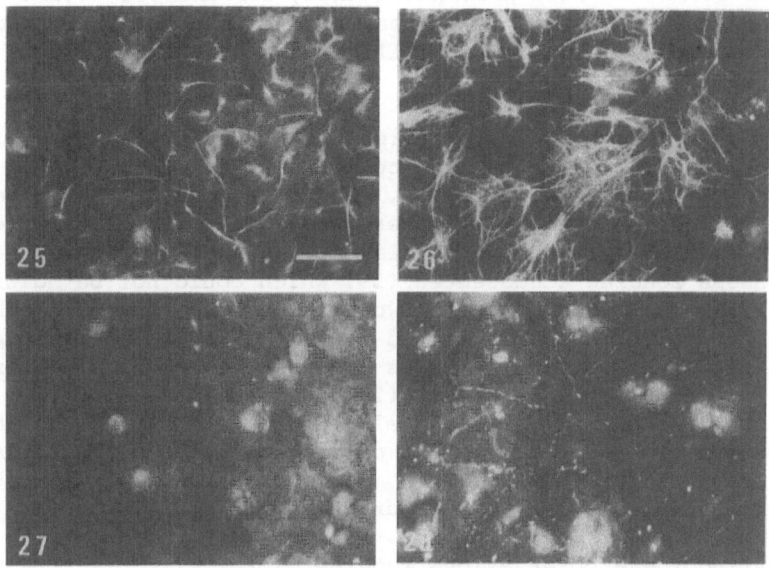

Fig. 25. Human teratocarcinoma stem cells stained with anti-
         keratin polyclonal antibody. Scale bar = 40 μm. All
         figures at same magnification.

Fig. 26. As Fig. 25 but after cells were treated for 7 days
         with   retinoic   acid.   Note   establishment   of
         "bridges".

Fig. 27. Human keratocarcinoma stem cells stained with anti-
         desmoplakin   antibody showing absence of desmo-
         somes.

Fig. 28. As Fig. 27 but after cells were treated for 7 days
         with retinoic   acid. Note staining of desmosomal
         plaques along cell boundaries.

adhesive function is fundamental to the organisation of those
tissues. It follows that the formation of desmosomes should play
some role in the development of tissue organisation   during
embryonic development. Where epithelial replacement continues
during adult life, such as in epidermis and the epithelial lining
of the intestine, desmosome formation presumably continues to be an
important process. We do not yet know, however, what would be the

consequences for development or for tissue organisation if desmosome formation were prevented. What, for example, would be the consequences for the mammalian embryo if the formation of desmosomes between trophoblast cells at the blastocyst stage (25) were to be inhibited ?

It is my view that the principal function of the desmosome may be at the tissue rather than the cellular level. This view is based on three considerations. Firstly, when desmosome formation between MDBK cells was inhibited (18), there was remarkably little effect on the morphology of the monolayer. This was presumably because MDBK possess zonulae adhaerentes (59) which could not be disrupted by anti-desmocollin Fab' and would therefore maintain adhesion between the cells. Secondly, and related to the first point, retinal pigmented epithelial cells possess only zonulae adhaerentes (Fig. 29) and no desmosomes (23). Zonulae adhaerentes would therefore seem sufficient alone to maintain epithelial organisation. Thirdly, disruption of the keratin cytoskeleton by micro-injection of anti-keratin monoclonal antibody caused no observable change in cell behaviour (62).

Putting these observations together I think it is reasonable to argue that the principal function of the desmosome is to link together the keratin cytoskeletons of adjacent cells providing a cytoskeletal continuum throughout the tissue. These components are therefore of little significance at the level of the individual cell. When the components of structure contributed by individual cells are linked together, however, one obtains a framework which supports the whole tissue. Where hemidesmosomes are present in the basal cells of an epithelium the filamentous network may extend beyond the epithelium itself since, in some cases at least, hemidesmosomes are directly linked through the basement membrane to the collagen fibres of the underlying connective tissue by anchoring filaments (28 and see below). The number and pattern of desmosomes in a tissue may therefore be determined in relation to the structural requirements of the whole epithelium providing links in the cytoskeletal scaffolding necessary to brace the epithelium against stress.

Whether inhibition of desmosome formation causes tissue disruption or not may be expected to depend on the relative abundance of different adhesion mechanisms in the tissue concerned. It is possible that inhibition of desmosome formation in keratinocytes, for example, where desmosomes appear to be the predominant adhesion mechanism, may profoundly influence the adhesion of the cells. In this context it should be pointed out that the so-called 'closed' and 'open' morphologies of different clones of human breast epithelial cells are correlated with presence or absence, respectively, of staining for desmosomes (96).

There is a little evidence that desmosome formation may play a role in controlling cell position in experimental situations. Overton (81) found that when combinations of Ccep cells and embryonic mouse epidermal cells were combined in culture, they sorted out to adopt a sphere-within-a-sphere arrangement with the internal position being adopted by the cell type which possessed the greater number of desmosomes per length of cell membrane. A similar result was obtained by Wiseman and Strickler (104) in sorting out of chick embryonic heart cells of different ages. Nicol and Garrod (75), studying sorting-out of chick embryonic tissues in monolayer, concluded that cell types which possess desmosomes adopted the internal situation with respect to types which do not. An explanation of these phenomena may be sought in terms of the differential adhesion hypothesis (DAH) of Steinberg (92,93,94,95). Put in its simplest terms the explanation would be that in mixed aggregates cells which adhere more strongly to each other sort out internally with regard to those which adhere less strongly. Cells which possess many desmosomes are more adhesive than those with less, and cells which possess desmosomes are more adhesive than those which do not. This illustrates how the presence of desmosomes in a tissue may play a role in determining the positioning of that tissue during embryonic development.

Of great importance is the problem of the possible role of junctional breakdown, in the invasive and metastatic behaviour of malignant cells. Junctional breakdown may release cells from their tissue locations, enabling them to spread to neighbouring tissues or to other parts of the body. If this view is correct, and the evidence is far from convincing (103), it is extremely important to understand desmosome formation in order to understand breakdown. Desmosome breakdown may also be important in pemphigus, a serious blistering disease of the skin. Blistering occurs as a consequence of splitting within the epidermis due to loss of adhesion between epidermal cells (29). At present there is controversy about whether anti-desmosomal antibodies are found in the serum of pemphigus patients (43,56). It is possible that autoantibodies arise as a consequence of the disease rather than act as causative agents. This would be consistent with irregular occurrence of antibodies among pemphigus patients.

Desmosome breakdown and reformation may also be important in the organisation of normal tissues where cellular repositioning occurs. Thus in the epidermis there is continual division of cells in the germinative layer and a continual trafficking of cells from the basal layer outwards. We do not know whether these events are accompanied by breakdown and reformation of desmosomes or, alternatively, repositioning of desmosomes in the cell membrane. Franke et al. (33) show an example of a cell dividing in culture in which the desmosomes would appear to be intact. An experiment of Klymkowsky et al. (62) in which microinjection of anti-cytokeratin antibody into a cell caused breakdown of its cytokeratin filaments, and bundling of the cytokeratin filaments in an adjacent cell in contact with the first cell, suggests that desmosomes are mobile within the cell membrane. Another in vitro observation may suggest, however, that desmosome breakdown and reformation is involved in stratification of epidermal cells. In cultured keratinocytes, fluorescent antibody staining for the protein involucrin a marker of differentiation in the epidermis, detects, within a monolayer shortly after raising the calcium concentration, those cells which are destined to leave the monolayer and become incorporated into the upper layers (101). Watt et al. (102) found that involucrin-positive cells in such monolayers also stained diffusely for desmoplakins and the 150 kd desmosomal glycoprotein. These cells also showed reduced peripheral staining for desmosomal components. This may indicate that desmosomal components are disassembled in these cells, and that they may have reduced numbers of desmosomes. This provides circumstantial evidence that desmosome breakdown and stratification may go together. The desmosomes would presumably reform in a cell as stratification was completed.

ZONULA ADHERENS

The other major type of adhesive junction of epithelial cells is the zonula adherens, also known by the rather unsatisfying name of "intermediate junctions". Confusion has existed in the past between the various type of junctions, but the distinction between them now seems to be quite clear and based on several criteria, ultrastructural, compositional and immunological.

The classical zonula adherens is found between simple epithelia cells, for example, in the gut where it forms the second element of the terminal bar (30). It is characterised by an intercellular space of about 20 nm occupied by homogeneous, amorphous material of low density, by parallel opposed plasma membranes over distances of 0.2 to 0.5 μm, by a conspicuous band of dense material located on the cytoplasmic face of the membrane and by association with microfilaments (Fig. 2) (for electron micrographs see 15). The zonula adherens may extend entirely around the apico-lateral surface of a cell where the cell adheres to another (Fig. 29).

Ultrastructural criteria are not necessarily entirely reliable for distinguishing between the various types of adherens junctions, but biochemical and immunological distinctions can now be drawn between them. Thus the classical zonula adherens is associated with cytoplasmic microfilaments which are composed of actin (99). Futhermore, alpha-actinin (19,40), vinculin (37,38,39) and tromyosin (37) have all been found in association with the epithelial zonula adherens (Fig. 29). It has recently been shown

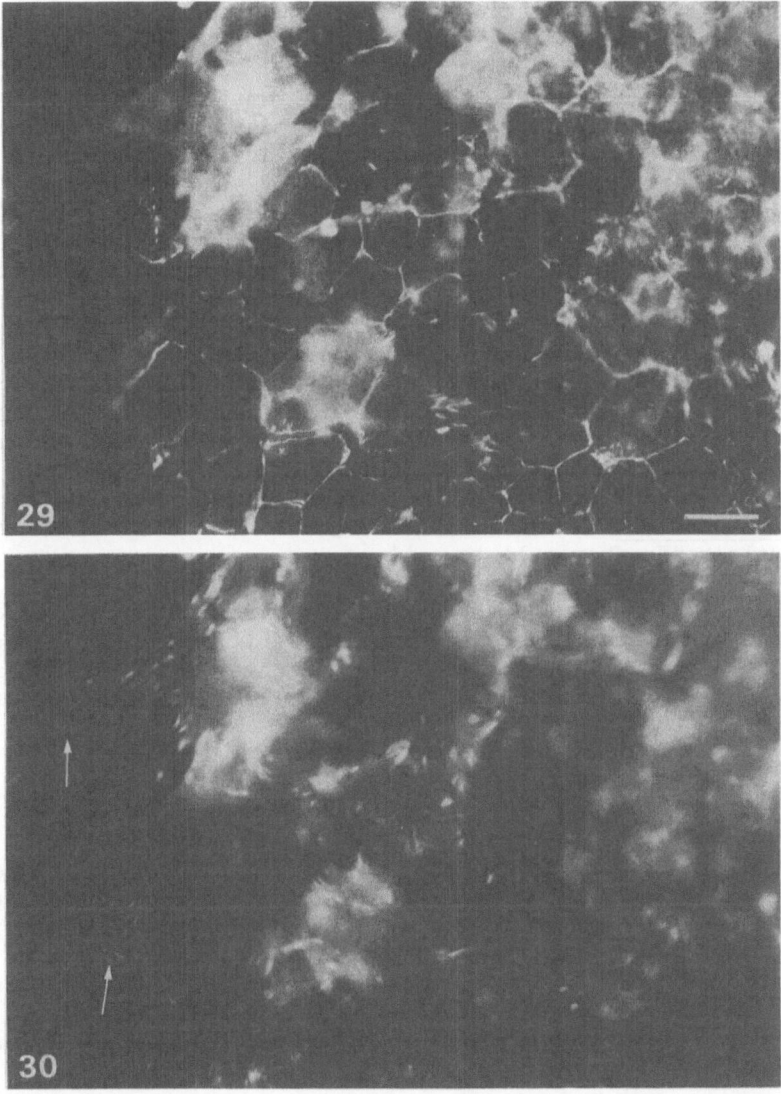

Fig. 29. Chick embryonic pigmented retinal epithelial cells
stained with rabbit anti-chick vinculin antibody.
Microscope focused to show staining of Zonulae
adhaerentes between apico-lateral cell contacts.
Scale bar = 20 μm.

Fig. 30. Same field as Fig. 29 with microscope focused to
show vinculin staining of focal contacts (arrows) on
the sub- stratum.

that the protein talin undergoes association with vinculin (11). These features clearly distinguish these junctions from the macula adherens or true desmosome.

The zonula adherens belongs to a family of junctions which have similar biochemical characteristics, especially the insertion of actin filaments and the association of vinculin and alpha-actinin with the cytoplasmic face of the membrane. These related structures include the fascia adhaerens in the intercalated disc of heart and the distal tubule of the nephron (see 80), the dense bodies of smooth muscle and the focal contacts which cells form with the substratum in tissue culture (38,40) (Fig. 30). The association of these structures with actin suggests that they are part of an adhesive-contractile system involved in cell motility. The contractile properties of the terminal web of intestinal epithelial cells have indeed been elegantly demonstrated by Burgess (10). The contractile role of an apical band of microfilaments has also been demonstrated by the work of Baker and Schroeder (3) and Karfunkel (57) on neural tube closure during embryonic development. The zonula adherens is the adhesive junction which links these contractile bands from cell to cell.

Fasciae adherentes have been isolated from intercalated discs of rat and mouse heart by Colaco and Evans (14). These preparations contained glycoproteins of molecular weights 180 kd, 88 kd and 38 kd which may be candidates for adhesive molecules of these junctions.

HEMIDESMOSOMES

These are organelles which occur at the base of an epithelium where it makes contact with the basal lamina and by means of which epithelial cells adhere to the underlying substratum or matrix. I am not aware of a systematic survey of the occurrence of hemidesmosomes. They are particularly prevalent in the epidermis, however, where they occur at the epidermal-dermal junction (8). The remarks below refer particularly to epidermal hemidesmosomes of the skin or cornea.

Hemidesmosomes , like desmosomes, have a prominent, electron-dense cytoplasmic plaque from which tonofilaments, presumably cytokeratin in nature, course into the cytoplasm (61). They should be distinguished from the half desmosomes which are created through the breakdown of intercellular adhesion when desmosomes are trypsinised (79). On the extracellular face of the plasma membrane hemidesmosomes abut the lamina lucida of the basal lamina (Fig. 30). A number of authors have reported fine filaments extending from the plasma membrane of the hemidesmosome into the lamina rara (63). Also, on the dermal side of the lamina densa, opposite the

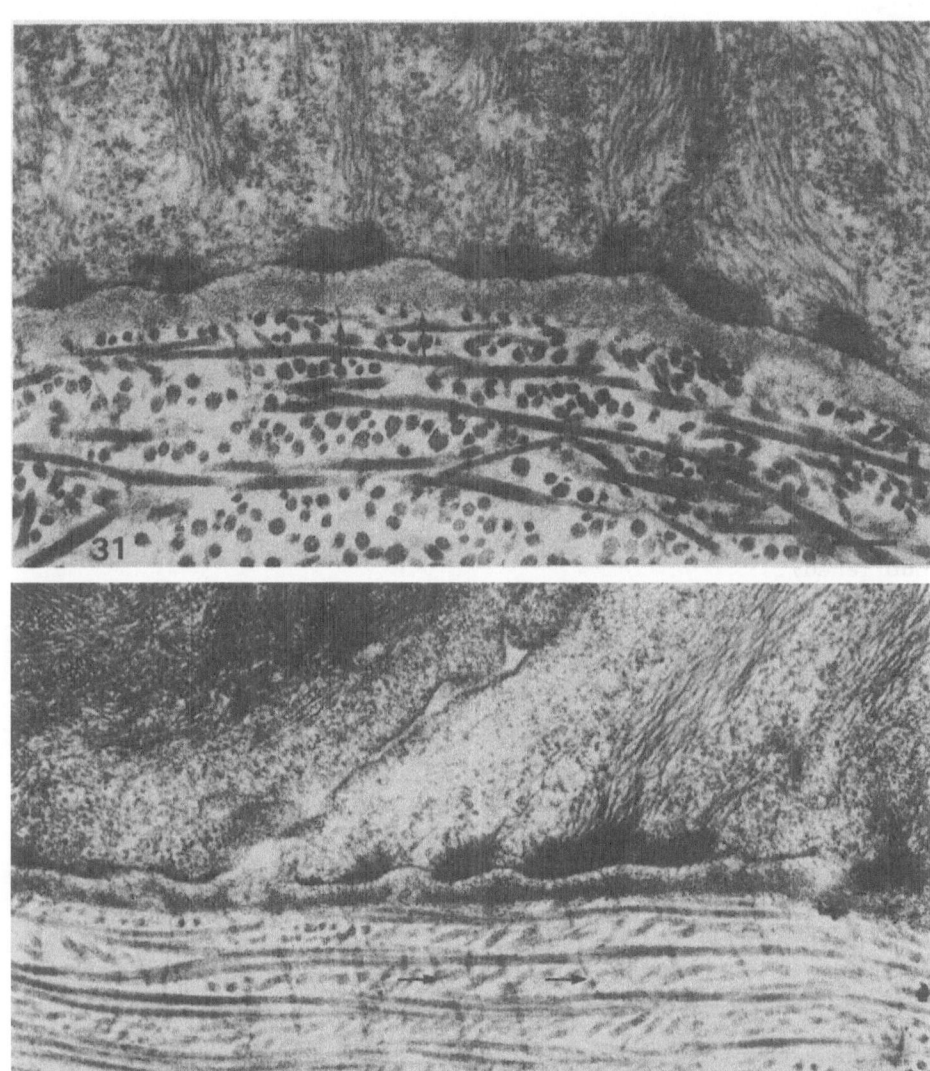

Fig. 31. Electron-micrograph showing hemidesmosomes at the epidermal-dermal junction of the axolotl. Note the anchoring filaments (arrows) which traverse the entire basal lamina. Scale bar = 0.2 µm.

Fig. 32. Electron-micrograph showing anchoring fibres (arrows) of the axolotl. Scale bar = 0.2 µm.

hemidesmosomes are thicker filaments known as anchoring fibrils (55).

These are frequently banded by cross-striations though the pattern and periodicity of banding are distinct from that found in collagen fibres (82). Gipson et al. (42) have reported that when rabbit cornea epithelium is removed from underlying stroma with dispase and then replaced, hemidesmosomes form opposite the anchoring fibrils of the stroma, implying some functional connection between the two. We have recently shown (28) that there is in fact a direct physical connection between the dermal anchoring fibrils and hemidesmosomes in the skin of adult amphibians. In both the axolotl (Ambystoma mexicanum) and the frog (Rana pipiens) electron microscopy shows that anchoring filaments, about 12 nm in thickness, pass directly from the plasma membrane of the hemidesmosome, through the entire thickness of the basal lamina, where may unite to form anchoring fibrils (Fig. 31). The latter mesh with the collagen fibres of the dermis (Fig. 32). We have preliminary evidence that a similar arrangement is present in the epithelial-matrix junctions of some mammalian tissues including the epidermis and the oesophagus (Ellison and Garrod, unpublished).

This gives an entirely new concept of the structural relationship between the epidermis and the dermis. Instead of an effective discontinuity between the two structures as presented by the basal lamina, there is in fact a direct filamentous continuity between the desmosome-linked, cytokeratin scaffolding of the epidermis and the collagen fibres of the detmis. The link is provided by the hemidesmosome-anchoring filament-anchoring fibril system (Fig. 33) (28).

Gipson et al. (42) suggested that hemidesmosome formation occurs only on substrata containing anchoring fibrils. Furthermore they showed some species specificity in that rabbit corneal epithelium formed more hemidesmosomes on rabbit corneal basement membrane than on corneal basement membranes from other species. In the light of our results this may imply the presence of a receptor on the hemidesmosomal membrane for anchoring filament components, anchoring filaments being the extensions anchoring fibrils within the basal lamina.

There are other reports, however, that hemidesmosomes can form on simple substrata in the absence of anchoring fibrils. Christophers and Wolf (12) showed that retinoic acid treated epidermis from adult guinea pigs ears formed hemidesmosomes on a substratum of tissue culture plastic (coated of course with layer of protein adsorbed from the medium). Billig et al. (5) showed that embryonic chick corneal epithelium formed hemidesmosomes on a substratum of adsorbed gelatin. Mattey, (unpublished) has extended

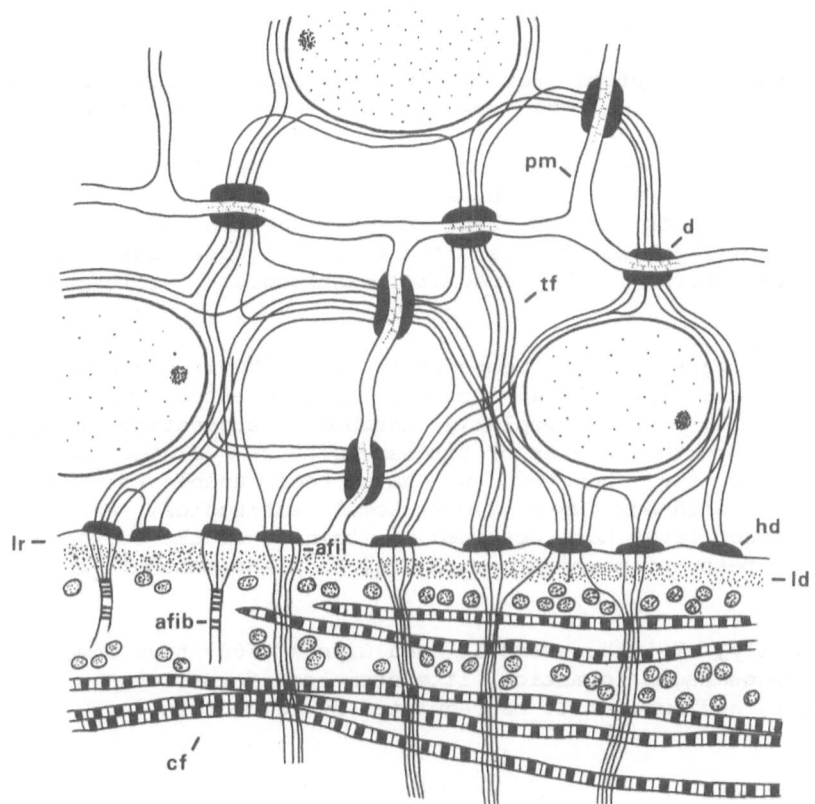

Fig. 33. Diagram showing suggested filamentous continuity between anchoring fibril-collagen system of the dermis and cytokeratin-desmosome system of the epidermis. This continuity is mediated by the hemidesmosomes and anchoring filaments.
afil = anchoring filament, afib = anchoring fibril, cf = collagen fibre, d = desmosome, hd = hemidesmosome, ld = lamina densa, lr = lamina rara, pm = plasma membrane, tf = tonofilaments.
Reproduced from (28).

these observations to show that the latter tissue will form hemidesmosomes on lens capsule or a substratum of types I and III collagen derived from rat skin. This would seem to indicate either that hemidesmosome formation does not require a specific substratum (any type of collagen may be adequate), or that epidermal cells can, in some circumstances, secrete the matrix components necessary for hemidesmosome formation. These components may also be the constituents anchoring filaments and fibrils. At present the only suggestion as to the nature of the latter is that they may consist

of type VII collagen (4). It is also interesting to speculate on
what may be the relationship between hemidesmosomes and
cell-surface receptors for matrix components such as laminin
(67,83). These receptors are probably involved in epithelial
cell-matrix adhesion and so they could be located in the plaque
membrane of the hemidesmosome. However, the laminin receptor,
otherwise known as connexin, appears to bind to actin (9) whereas
one might expect it to bind to cytokeratin if it were a
hemidesmosomal component. The ultrastructural localisation of these
matrix receptors with specific antibodies will be most
interesting.

With regard to the composition of the hemidesmosomal plaque,
Franke et al. (32) have reported immunoelectron microscopical
staining with anti-desmoplakin antibody, suggesting that these
components are the same as or similar to those of the desmosomal
plaques. So far, however, we have been unable to repeat this
observation with our own anti-desmosomal antibodies, and at the
fluorescence level have obtained no staining of the hemidesmosomal
region or basement membrane with any anti-desmosomal antibodies
(16,17).

Finally, Trinkaus-Randall and Gipson (100) have demonstrated
that hemidesmosome formation, like desmosome formation, is calcium
dependent, and appears to be regulated by calmodulin.

FOCAL CONTACTS

Focal contacts are a region on the lower surfaces of cells in
culture where the cell membrane is separated from the substratum by
10-15 nm : they appear dark by interference reflection microscopy
(1,20,46,47,54). They are located at the peripheral ends of actin
microfilament bundles or stress fibres (Fig. 34 and 35) (47) and
show positive discrete staining for the cytoplasmic proteins
α-actinin and vinculin (39,40,89) (Fig. 29).

Focal contacts are present in both fibroblasts (see above
references) and epithelial cells (5,46,68,69,73). Their probable in
vivo counterparts are regions at the ends of microfilament bundles
where cells make contact with their substratum, the extracellular
matrix. In epithelial cells they represent a second cell-substratum
adhesion mechanism, in addition to the hemidesmosome system. What
is their function in epithelial cells ?

In their cytoplasmic associations with actin, α-actinin and
vinculin, they are homologous with a variety of intercellular and
intracellular junctions and structures. These are the zonula
adherens of epithelial cells, the fascia adherens of heart muscle,

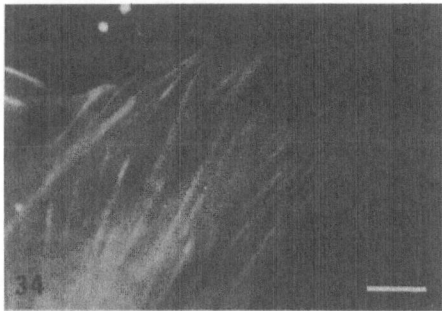 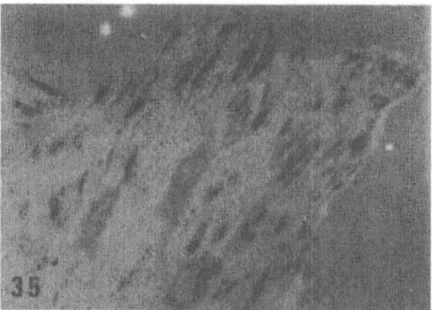

Fig. 34. Part of HeLa cell stained with fluorescent rabbit anti-chicken actin antibody showing microfilament bundles. Scale bar = 10 μm.

Fig. 35. Same field as Fig. 34 showing focal contacts. Note that these are located at the ends of microfilament bundles. Reproduced from (73).

the Z-bands of skeletal muscle and the dense bodies of smooth muscle (37). The essential feature of all these structures is an association with the contractile mechanisms of cells, which generate cell motility. In cultured embryonic corneal epithelial cells, focal contacts occur largely, if not exclusively, in cells at the free edges of epithelial islands. Furthermore, although they are not directly associated with fibronectin on the substratum (68,73 see also Morgan and Garrod, 1984), their presence depends upon the existence of an extracellular matrix. Disruption of that matrix by glycoaminoglycanases or collagenase results in diminution in number or absence of focal contacts (68,69). Finally, the peripheral cells of epithelial islands clearly exert force on the substtratum because they are able to reorganise the extracellular matrix into fibrils (69).

These observations suggest that the focal contacts of epithelial cells may be an in vitro manifestation of wound healing activity. They are clearly located at free edges, and those edges can clearly exert a tractile force on the substratum. It seems probable that focal contacts or their in vivo equivalents are not present in the normal, undamaged epithelium, but are formed at the free edges of wounded epithelia to bring about closure of the wound. They represent an alternative mechanism for interaction with the matrix and have a specific function. The assembly of microfilaments extending towards the membrane of lung cell process in amphibian skin cells during wound healing has been described by Represh and Oberpriller (86). I believe that we have observed the in vitro counterpart of this activity in studying focal contacts of corneal epithelial cells.

The observations of Sugrue and Hay (96) that soluble laminin or collagen in the medium suppress the basal blebbing of corneal epithelial cells have shown that the presence of these components of the basal lamina promote the stabilisation of the basal membrane. This may suggest how the balance between different matrix adhesion mechanisms is controlled. The presence of basement membrane components may contribute to the predominance of the stable adhesion of the normal epithelium, over the active motile mechanism of wound healing.

The molecular mechanism of focal contact interaction with the matrix or substratum is open to some doubt. There is controversy about whether fibronectin is a component of the focal contacts of fibroblasts (see 73 for references). We believe that fibronectin, although a crucial component of the extracellular matrix, may not be directly involved in focal contact formation in corneal epithelial cells (68). Morgan and Garrod (73) provided evidence that fibronectin was not involved in focal contact formation by HeLa cells.

Oesch and Birchmeier (77) have reported the association of a 60 kd antigen with the extracellular face of focal contacts of chick embryonic fibroblasts. Woods et al. (105) show an association between heperan sulphate proteoglycan and focal contacts in fibroblasts. These observations may be useful clues to the molecular basis of focal contact formation.

CONCLUSION : THE FUNCTIONS OF ADHESION MECHANISMS AND DIFFERENCES
               BETWEEN EPITHELIA

The fact that epithelial cells possess a variety of different intercellular adhesion mechanisms means that each must have a different function. These functions have been alluded to throughout the preceding text but I now wish to draw them together in order to stress the variety of different adhesive functions required by epithelial cells. The suggested functions are set out in Table 2.

An analytical consideration of the adhesive behaviour of cells has led to the belief that individual cell adhesion mechanisms are shared by a variety of different cell types (36). While it is not true to say that all cell types possess all adhesion mechanisms (for example fibroblasts and pigmented retinal epithelial cells do not possess desmosomes) it is likely that cell-type specific adhesion mechanisms, once believed to be extremely important in development, are probably rare among animal tissue cells (34). This being so, wherein lie the differences in adhesive properties between different types of epithelial cells ? Our studies on desmosomes lead to the realisation that three things may vary

Table 2. Suggested functions of epithelial cell adhesions.

Type of Adhesion	Function
Desmosome	Link between keratin or other intermediate filament systems of individual cells - coupling between cytoskeletal scaffolding of tissue resistance to stress.
Zonula adherens	Link between contractile apparatus of individual cells - control of epithelial shape.
"Molecular only" cohesion mechanisms	Adhesion between non-junctional membranes of individual cells.
Hemidesmosome	Adhesion to underlying connective tissue stroma via anchoring filaments and anchoring fibrils - link between stroma and cytoskeletal scaffolding of epithelium.
Focal contact	In vitro manifestation of terminal adhesions of microfilament bundles involved in wound healing.
"Molecular only" adhesion mechanisms	Adhesion to basal lamina.

between the adhesion of different cell types :

1) The distribution of adhesive junctions.
2) The number of adhesive junctions.
3) The stability of adhesive junctions.

Thus keratinocytes possess large numbers of EDTA-resistant desmosomes distributed all over the cell surface (102), whereas cells of simple epithelia possess EDTA-sensitive desmosomes between their lateral surfaces (18). It is the balance of those adhesive properties set out in Table 2, required for normal functioning of a particular epithelium, that dictates what number, distribution and stability of adhesion mechanisms are necessary. The differences in adhesive properties of different epithelial cells are determined in terms of these three parameters rather than by possession of different cell-type specific adhesion mechanisms.

ACKNOWLEDGEMENTS

    I thank Drs. Derek Mattey and Geoff Shellswell for critical reading of the manuscript. Work in our laboratory is supported by the Cancer Research Campaign and the Medical Research Council.

REFERENCES

1.   M. Abercrombie and G.A. Dunn, Exp. Cell Res., 92 : 57–62
     (1975).
2.   A. Allen, R. Dulbecco, P. Syka and M. Bowman, Proc. Natl.
     Acad. Sci. USA (in press) (1984).
3.   P.C. Baker and T.E. Schroeder, Dev. Biol., 15 : 432 (1967).
4.   H. Bentz, N.P. Morris, L.W. Murray, L.Y. Sakai, D.W. Hollister
     and R.E. Burgeson, Proc. Natl. Acad. Sci. USA, 80 : 3168–3172
     (1983).
5.   D. Billig, A. Nicol, R. McGinty, P. Cowin, J. Morgan and D.
     Garrod, J. Cell Sci., 57 : 51–71 (1982).
6.   J.Z. Borysenko and J.P. Revel, J. Anat., 137 : 403–422
     (1973).
7.   A. Bretscher, J. Cell Biol., 97 : 425–432 (1983).
8.   R.A. Briggaman and C.E. Wheeler, J. invest. Dermatol., 65 :
     71–84 (1975).
9.   S.S. Brown, H.L. Malinoff and M.S. Wicha, Proc. Natl. Acad.
     Sci. USA, 80 : 5927–5930 (1983).
10. D.R. Burgess, J. Cell Biol., 95 : 853–863 (1982).
11. K. Burridge and P. Mangeat, Nature, 308 : 744–746 (1984).
12. E. Christophers and H.H. Wolf, Nature, 256 : 209–210 (1975).
13. S.M. Cohen, G. Gorbsky and M.S. Steinberg, J. Biol. Chem., 90
     : 243–248 (1983).
14. C.A.L.S. Colaco and W.H. Evans, J. Cell Sci., 52 : 313–325
     (1981).
15. C.A.L.S. Colaco and W.H. Evans, Electron Microscopy of
     Proteins, 4 : 332–363 (1983).
16. P. Cowin and D.R. Garrod, Nature, 302 : 148–150 (1983).
17. P. Cowin, D.L. Mattey and D.R. Garrod, J. Cell Sci., 66 :
     119–132 (1984 a).
18. P. Cowin, D.L. Mattey and D.R. Garrod, J. Cell Sci., (in
     press) (1984 b).
19. S.W. Craig and J.V. Pardo, J. Cell Biol., 80 : 203–210
     (1979).
20. A.S.G. Curtis, J. Cell Biol., 20 : 199–215 (1964).
21. C.H. Damsky, J. Richa, D. Solter, K. Knudsen and C.A. Buck,
     Cell, 34 : 455–466 (1983).
22. H.M. Dembitzer, F. Herz, A. Schermer, R.C. Wooley and L.G.
     Koss, J. Cell Biol., 85 : 695–702 (1980).
23. R.J. Doherty, J.G. Edwards, D.R. Garrod and D.L. Mattey, J.
     Cell Sci., (in press) (1984).
24. P. Drochmans, C. Freudenstein, J.C.Wanson, L. Laurent, T.W.
     Keenan, J. Stadler, R. Leloup and W. Franke, J. Cell Biol., 80
     : 231–247 (1978).
25. T. Ducibella, D.F. Albertini, E. Anderson and J.D. Biggers,
     Dev. Biol., 45 : 231–250 (1975).
26. R. Dulbecco, W.R. Allen and M. Bowman, Proc. Natl. Acad. Sci.
     USA, (in press) (1984).
27. G.M. Edelman, Science, 219 : 450–457 (1983).

28. J.E. Ellison and D.R. Garrod, J. Cell Sci., (in press)
     (1984).
29. R.M. Farb, R. Dykes and G.S. Lazarus, Proc. Natl. Acad. Sci.
     USA, 75 : 459-463 (1978).
30. M.G. Farquhar and G. Palade, J. Cell Biol., 19 : 375-412
     (1963).
31. W.W. Franke, H.P. Kaprell and H. Mueller, Eur. J. Cell Biol.,
     32 : 117-130 (1983).
32. W.W. Franke, R. Moll, D.L. Schiller, E. Schmid, J. Kartenbeck
     and H. Mueller, Differentiation, 23 : 115-127 (1982).
33. W.W. Franke, E. Schmid, C. Grund, H. Mueller, I. Engelbrecht,
     R. Moll, J. Stadler and E.D. Jarasch, Differentiation, 20 :
     217-241 (1981).
34. D.R. Garrod, Fortschritter der Zoologie, 26 : 184-195 (1981).
35. D.R. Garrod and P. Cowin, In: "Receptors in Tumour Biology",
     ed. C.M. Chadwick (in press) (1984).
36. D.R. Garrod and A. Nicol, Biol. Rev., 56 : 199-242 (1981).
37. B. Geiger, A.H. Dutton, K.T. Tokuyasu and S.J. Singer, J. Cell
     Biol., 91 : 614-628 (1981).
38. B. Geiger, E. Schmid and W.W. Franke, Differentiation, 23 :
     189-205 (1983).
39. B. Geiger, K.T. Tokuyasu, A.H. Dutton and S.J. Singer, Proc.
     Natl. Acad. Sci. USA, 77 : 4127-4131 (1980).
40. B.M. Geiger, K.T. Tokuyasu and S.J. Singer, Proc. Natl. Acad.
     Sci. USA, 76 : 2833-2837 (1979).
41. N.B. Gilula, In: "Intercellular Junctions and Synapses", ed.
     J. Feldman, N.B. Gilula and J.D. Pitts, (Chapman and Hall,
     London), pp. 1-22 (1978).
42. I.K. Gipson, M.S. Grill, S.J. Spurr and S.J. Brennan, J. Cell
     Biol., 97 : 849-857 (1983).
43. G. Gorbsky, S. Cohen and M.S. Steinberg, J. invest. Dermatol.,
     80 : 475-480 (1983).
44. G. Gorbsky and M.S. Steinberg, J. Cell Biol., 90 : 243-248
     (1981).
45. H. Green, O. Kehinde and J. Thomas, Proc. Natl. Acad. Sci.
     USA, 76 : 5665-5668 (1979).
46. J.P. Heath, In: "Cell Behaviour", ed. R. Bellairs, A.S.G.
     Curtis and G.A. Dunn, (Cambridge University Press), pp. 77-108
     (1982).
47. J.P. Heath and G.A. Dunn, J. Cell Sci., 29 : 197-212 (1978).
48. D. Henderson and K. Weber, Exp. Cell Res., 132 : 297-311
     (1981).
49. H. Hennings and K. Holbrook, Exp. Cell Res., 143 : 127-142
     (1983).
50. H. Hennings, K.A. Holbrook and S.H. Yuspa, J. Cell Physiol.,
     116 : 265-281 1983 a).
51. H. Hennings, K.A. Holbrook and S.H. Yuspa, J. invest.
     Dermatol., 81 : 50s-55s (1983 b).
52. H. Hennings, D. Michael, C. Cheng, P. Steinert, K. Holbrook
     and S.H. Yuspa, Cell, 19 : 245-254 (1980).

53. D.A. Herzlinger and G.K. Ojakian, J. Cell Biol., 98 :
     1777-1787 (1984).
54. C.S. Izzard and L.R. Lochner, J. Cell Sci., 21 : 129-159
     (1976).
55. M.A. Jakus, Invest. Ophtalmol., 1 : 202-225 (1962).
56. J.C.R. Jones, J. Arun, L.A. Staehelin and R.D. Goldman, Proc.
     Natl. Acad. Sci. USA, 81 : 2781-2785 (1984).
57. P. Karfunkel, Dev. Biol., 25 : 30 (1971).
58. J. Kartenbeck, W.W. Franke, J.G. Moser and U. Stoffels, EMBO
     J., 2 : 735-742 (1983).
59. J. Kartenbeck, E. Schmid, W.W. Franke and B. Geiger, EMBO J.,
     1 : 725-732 (1982).
60. J. Kartenbeck, K. Schwechleimer, R. Moll and W.W. Franke, J.
     Cell Biol., 98 : 1072-1081 (1984).
61. D.E. Kelly, J. Cell Biol., 28 : 51-73 (1966).
62. M.W. Klymkowsky, R.H. Miller and E.B. Lane, J. Cell Biol., 96
     : 494-509 (1983).
63. W.S. Krawczyk and G.F. Wilgram, J. Ultrastruct. Res., 45 :
     93-101 (1973).
64. E.B. Lane, J. Cell Biol., 92 : 665-673 (1982).
65. N.J. Lane, In: "Electron Microscopy" Vol. 3, State of the Art
     Symposia ed. J.M. Sturgess, (Imperial Press, Toronto), pp.
     673-691 (1978).
66. T.L. Lentz and J.P. Trinkaus, J. Cell Biol., 48 : 455-472
     (1971).
67. H.L. Malinoff and M.S. Wicha, J. Cell Biol., 96 : 1475-1479
     (1983).
68. D.L. Mattey and D.R. Garrod, J. Cell Sci., 67 : 171-188 (1984
     a).
69. D.L. Mattey and D.R. Garrod, J. Cell Sci., 67 : 189-202 (1984
     b).
70. D.L. Mattey and D.R. Garrod, in preparation, (1984 c).
71. N.S. McNutt and R.S. Weinstein, Prog. Biophys. Mol. Biol., 26
     : 45-101 (1976).
72. J. Mollenhauer and K. von der Mark, EMBO J., 2 : 45-50
     (1983).
73. J. Morgan and D.R. Garrod, J. Cell Sci., 66: 133-145 (1984).
74. H. Mueller and W.W. Franke, J. mol. Biol., 163 : 647-671
     (1983).
75. A. Nicol and D.R. Garrod, J. Cell Sci., 54 : 357-372 (1982).
76. C. Ocklind, U. Forsum and D. Obrink, J. Cell Biol., 96 :
     1168-1171 (1983).
77. B. Oesch and W. Birchmeier, Cell, 31 : 671-679 (1982).
78. S.I. Ogou, G. Yoshida-Noro and M. Takeichi, J. Cell Biol., 97
     : 944-948 (1983).
79. J. Overton, J. exp. Zool., 168 : 203-214 (1968).
80. J. Overton, Prog. Surf. Sci., 8 : 161-208 (1974).
81. J. Overton, Dev. Biol., 55 : 103-116 (1977).
82. G.E. Palade and M.G. Farquhar, J. Cell Biol., 27 : 215-224
     (1965).

83. N.C. Rao, S.H. Barsky, V.P. Terranova and L.A. Liotta,
     Biochem. Biophys. Res. Commun., 111 : 804-808 (1983).
84. D.G. Rayns, F.O. Simpson and J.M. Ledingham, J. Cell Biol., 42
     : 322-326 (1969).
85. D.A. Rees, R.A. Badley, C.E. Lloyd, D. Thomn and C.G. Smith,
     In: "Cell-cell Recognition" 32 nd. Symp. Soc. Exp. Biol. ed.
     A.S.G. Curtis, (Cambridge University Press, Cambridge), pp.
     241-260 (1978).
86. L.A. Represh and J.C. Oberpriller, Am. J. Anat., 159 : 187-208
     (1980).
87. J.G. Rheinwald and H. Green, Cell, 6 :331-344 (1975).
88. K. Rubin, M. Hook, B. Obrink and R. Timpl, Cell, 24 : 463-470
     (1981).
89. I.I. Singer, J. Cell Biol., 92 : 398-408 (1982).
90. C.J. Skerrow and G.A. Matoltsy, J. Cell Biol., 63 : 515-523
     (1974).
91. L.A. Staehelin, Int. Rev. Cytol., 39 : 191-283 (1974).
92. M.S. Steinberg, In: "Cellular Membranes in Development" ed. M.
     Locke, (Academic Press, New-York), pp. 321-366 (1964).
93. M.S. Steinberg, J. Exp. Zool., 173 : 395-434 (1970).
94. M.S. Steinberg, In: "Specificity of Embryological
     Interactions" ed. D.R. Garrod, (Chapman and Hall, London), pp.
     99-130 (1978 a).
95. M.S. Steinberg, In: "Cell-cell Recognition" 32 nd Symp. Soc.
     Exp. Biol. ed. A.S.G. Curtis, (Cambridge University Press,
     Cambridge), pp. 25-49 (1978 b).
96. M. Stoker, J. Cell Physiol., (in press) (1984).
97. S.P. Sugrue and E.D. Hay, Dev. Biol., 92 : 97-106 (1982).
98. A. Suhrbier and D.R. Garrod, in preparation, (1984).
99. L.G. Tilney and M.S. Mooseker, Proc. Natl. Acad. Sci. USA, 68
     : 2611-2615 (1971).
100. V. Trinkaus-Randall and I.K.Gipson, J. Cell Biol., 98 :
     1565-1571 (1984).
101. F.M. Watt and H. Green, Nature, 295 : 434-436 (1982).
102. F.M. Watt, D.M. Mattey and D.R. Garrod, J. Cell Biol., (in
     press) (1984).
103. R.S. Weinstein, F.B. Merk and J. Alroy, Adv. Cancer Res., 23 :
     23-89 (1976).
104. L.L. Wiseman and J. Strickler, J. Cell Sci., 49 : 217-224
     (1981).
105. A. Woods, M. Hook, L. Kjellen, C.G. Smith and D.A. Rees, J.
     Cell Biol., (in press) (1985).

ADHESIVE MOLECULES AND THEIR ROLE DURING THE ONTOGENY

OF THE PERIPHERAL NERVOUS SYSTEM

Jean-Loup Duband and Jean-Paul Thiery

Institut d'Embryologie du C.N.R.S.
et du Collège de France
49 bis, avenue de la Belle Gabrielle
94130 Nogent-Sur-Marne, France

INTRODUCTION

The body plan that arises at the end of early embryonic morphogenesis is mainly the result of temporally and locally coordinated events, that is proliferation, adhesion, migration, differentiation and death of cells. The behavior of each cell depends largely on its degree of interaction with its neighbours and with extracellular environment. These interactions involve the cytoskeleton, the plasma membrane, and the extracellular matrix (ECM). Interestingly, all of these components are intimately linked through specific molecules and constitute a continuous framework within which changes in any one component rapidly lead to consistent changes in the other components and, as a consequence, a consistent cell behavior (see Garrod, Pitts, Gilula, Reggio and Franke, this volume ; see also Fig. 1).

Although each cell type exhibits a specificity in its shape and function, it appears that only a limited repertoire of cellular structures and of molecules are responsible for the adhesive properties of the cells (see Garrod, Pitts and Gilula, this volume). Beside the subcellular structures which are involved in adhesion (cytoskeleton, tight and gap junctions, desmosomes, and focal plaques), a number of adhesive molecules have been identified either at the cell surface or in the ECM. The purpose of this review is to focus on the properties of some of these molecules known to control cell-to-substratum and cell-to-cell adhesion. Their possible influence during morphogenesis will be examined in a model system, the neural crest.

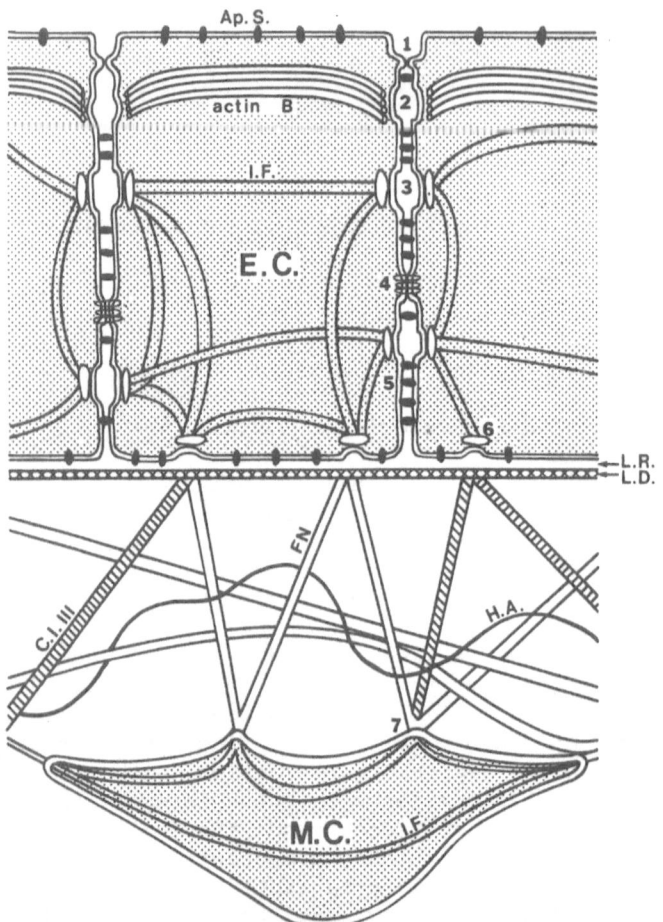

Fig. 1.   Diagrammatic representation of the different adhesive
          mechanisms found among epithelial and mesenchymal
          cells.
          Epithelia are polarized with an apical surface
          (Ap.S.), a basal surface limited by a basal lamina,
          and lateral surfaces in contact with neighbouring
          cells. At the ultrastructural level, the basal lamina
          is divided into a lamina rara (L.R.) and a lamina
          densa (L.D.) and contains mostly type IV collagen,
          laminin and fibronectin. Conversely, mesenchymal
          cells (M.C.) do not show any particular polarity;
          they have a few transitory contacts with their
          neighbours and adhere to the ECM to form transient
          plaques (7). The cytoskeleton and the extracellular
          components appear to be in continuity. A variety of
          membrane structures are involved : In epithelial

1. CELL-TO-SUBSTRATUM ADHESION MOLECULES

1.1. General features of the ECM

The ECM appears to be composed of a relatively small number of types of molecules. These include collagens, fibronectin (FN), hyaluronic acid (HA), glycosaminoglycans and proteoglycans, and laminin (LN), the latter being restricted to the basal surface of the epithelia (Fig. 2 and 3).

Both mesenchymal and epithelial cells establish direct contacts with a three-dimensional meshwork of ECM fibers. The two cell types differ principally in the organization and chemical composition of the ECM at their contact. However, with the exception of some highly specialized tissues (e.g. cartilages and bones ; Hewit et al., 1981 ; Termine et al., 1981 ; Poole et al., 1984), the overall composition of the ECM among many tissues is constant. Thus, it appears that the control of cell behavior and tissue organization does not depend so much on specific extracellular components, but rather on the three-dimensional organization of the meshwork, the relative concentration of the ECM molecules and the interactions of ECM with the cells.

Each molecule of the ECM plays a specific role with respect to the cell behavior. The different types of collagens constitute a scaffold which maintains the architecture of both epithelia and mesenchymes (Hay, 1981 ; Viidik and Vuust, 1980) ; fibronectin and laminin are involved in cell adhesion (Yamada, 1983) ; glycosaminoglycans were also found to mediate cell adhesion (Toole, 1981), but other studies contradict this hypothesis (Fisher and Solursh, 1979 ; Newgreen et al., 1982). Contrary to what was formerly claimed (Hay, 1973), collagens are not directly involved in cell adhesion. Cells do not generally adhere to collagen through specific receptors, but rather through other ECM components (for a

cells, the actin bundles (actin B) are end on intermediate junctions (2). Intermediate filaments (I.F.) are linked by desmosomes in the lateral surfaces (3) or are in continuity with the ECM through hemi-desmosomes (6) or transient plaques. Other specialized adhesive structures are the tight junctions (1), the gap-junctions (4) and the cell-adhesion molecules (5) the latter not being restricted to any particular surface of the epithelial cells. Type I and III collagens (C.I, III), fibronectin (FN), and hyaluronic acid (H.A.) are the common components of the ECM.

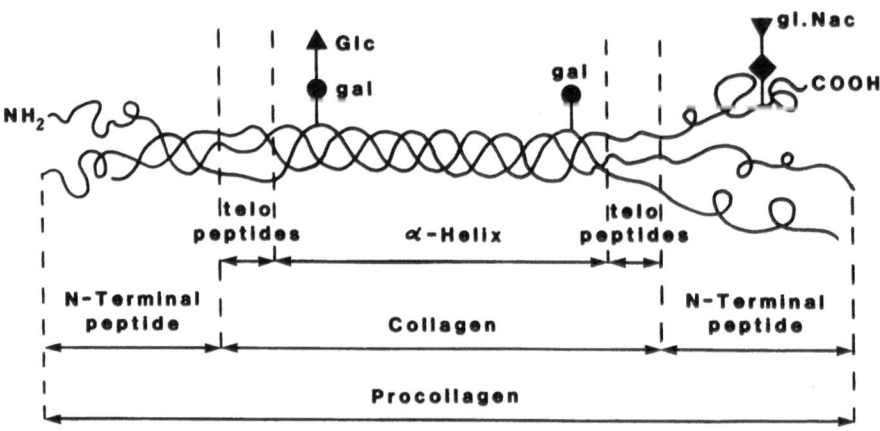

COLLAGENS

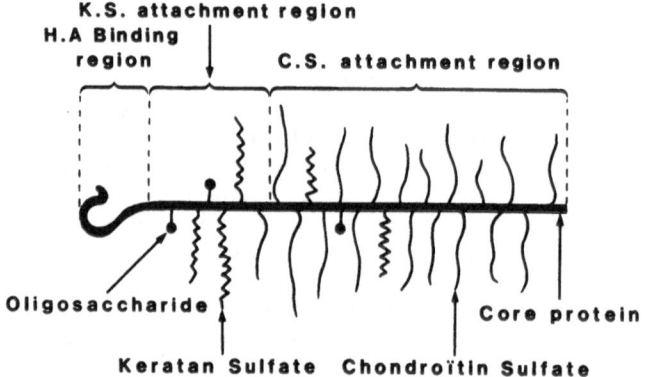

PROTEOGLYCAN

HYALURONIC ACID

review, see Yamada, 1983). An exception, that of the rat hepatocytes which can spread directly on collagen substrate, has been described (Rubin et al., 1981). Furthermore, the absence of type I collagen does not affect the early morphogenetic events which involve cell adhesion, since mouse embryos carrying an inactive type I collagen gene can reach day 12 of incubation (Lohler et al., 1984). In contrast, FN and LN have been clearly shown to mediate cell-to-substratum adhesion.

1.2. Fibronectin

The name of fibronectin comes from the most obvious characteristic of the molecule, that is to bind fibers. It consists of two almost identical polypeptide subunits of 220 kD linked by disulfide bonds (Fig. 3). The primary structure of FN has been nearly completely established (Petersen et al., 1983 ; Schwarzbauer

Fig. 2.   Structure of collagens, proteoglycans and hyaluronic acid.
          Collagens : This representation seems valid for type I, II, III and V collagens. A molecule of collagen is an helix composed of three α-chains. Each chain ends with globular domains at both the NH_2 and COOH terminal regions (telopeptides). The helical organization of the chains is due to the presence of a repetitive sequence (Glycine-X-Y, where X and Y are preferentially the amino-acid Proline and hydroxy-Proline). The chains of different type of collagen differ mostly in their content of hydroxy-Proline and in the glycosylation of Lysine (represented by Glc, gal, and gl.Nac). Collagens are synthesized from a precursor, the procollagen. Type IV collagen does not form fibers and is exclusively found under the procollagen form. The formation of the typical striated fibers of collagen concerns the telopeptides fragments.
          Glycosaminoglycans (e.g. hyaluronic acid, HA) and proteoglycans : they have structure based on a repeated sequence of disaccharidic units (the example of HA is given). The molecular weight of glycosaminoglycans varies between 1 and 20 kD, but can be 20 million daltons as in the case of hyaluronic acid. Except HA, glycosaminoglycans such as chondroitin sulfate (C.S.), and keratan sulfate (K.S.) are associated with proteins (core protein) to form high molecular weight complexes which are associated with HA to constitute megacomplexes.

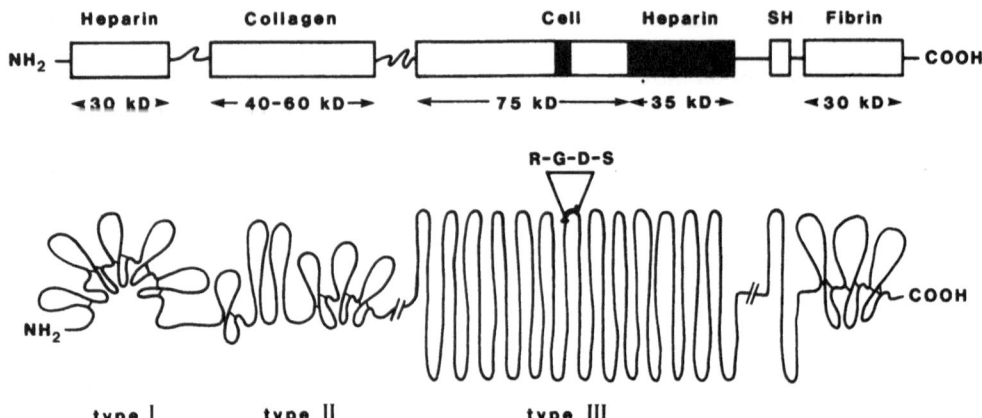

FIBRONECTIN

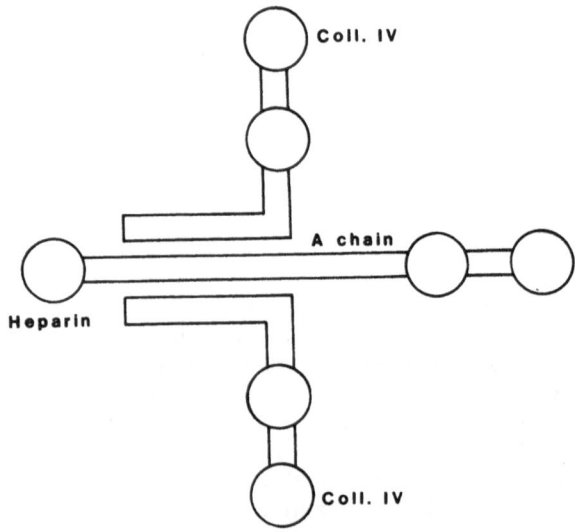

LAMININ

Fig. 3.   Structure of fibronectin (FN) and laminin (LN).
Fibronectin (see also text) : It is constituted of
two almost identical subunits (only one is
represented), linked by disulfide bonds. Five binding
domains have been precisely mapped along the
molecule. The structural domains are represented
(type I, II and III) along with the binding domains.

et al., 1983 ; Kornblihtt et al., 1984 a, b). FN is present in large amounts in the plasma (0.3 mg/ml) as a soluble dimer (plasma fibronectin). FN is also found as an insoluble form in 10 nm fibrils around mesenchymal cells and in the basal surface of epithelia. Plasma FN is synthesized in the liver (Tamkun and Hynes, 1982), while cellular FN is produced by a large variety of cell types (reviewed by Hynes and Yamada, 1982).

Several distinct structural and functional domains have been mapped along the molecule using limited proteolytic digestion (Furcht, 1983 ; Yamada, 1983 ; Petersen et al., 1983). A model of the secondary structure of FN has been deduced from these studies ; three major structural domains have been described (types I, II and III ; see Fig. 3). The functional domains mainly consist of binding sites for a variety of molecules such as collagen, heparin, fibrin, actin and possibly DNA. While the interest of the binding to collagen, heparin and fibrin is obvious, the significance of the binding to actin and DNA is not clear. Furthermore, a cell binding sequence has been precisely mapped. It consists of a peptide in which the very hydrophilic sequence Arg-Gly-Asp-Ser (RGDS) is absolutely required for the adhesion of cells to FN (Pierschbacher and Ruoslahti, 1984 ; Yamada and Kennedy, 1984). There is a partial correspondance between the structural and the functional domains. In particular, the cell binding sequence located at the top of a given loop of the type III domain is externally exposed, thus allowing interactions with the cell surface receptor (Fig. 3).

The nature of the FN receptor(s) has not yet been elucidated. A number of potential receptors has been proposed, including glycoproteins, gangliosides and heparan sulfate (Aplin et al., 1981 ; Perkins et al., 1982 ; Laterra et al., 1983 ; Akiyama and Yamada, 1984). Recently, the dissociation constant of FN_c to the cell surface of fibroblasts has been estimated to 0.8 x 10^{-6}M, which is not very high and the number of receptors estimated to be about 500,000 per cell (Akiyama and Yamada, 1984). FN could possibly interact with a variable affinity to a wide variety of cell surface molecules, according to the cell type or even to the physiological state of the cell. Interestingly, the RGDS sequence is found in several other molecules such as fibrinogen, the receptor for phage

---

Laminin (see also text) : The diagram has been constructed from information obtained in part through gene cloning (Hogan, personal communication). Two identical subunits carrying the type IV collagen binding site are associated with a long chain (A chain) which carries the heparin binding site. Together, they form a cross-shaped structure.

λ on E. coli and the Sendai virus coat protein. The wide
distribution of this short sequence suggests that it may act as a
"glue" sequence without any high specificity (Pierschbacher and
Ruoslahti, 1984).

As opposed to collagens, we do not know how cellular FN
self-associates to form polymers in some experimental conditions
(Vaheri and Mosher, 1978). A possible candidate responsible for the
polymerization of FN is a 90 amino-acid sequence which is only
present in cellular FN (Schwarzbauer et al., 1983 ; Kornblihtt et
al., 1984 a, b). Alternatively, the codistribution of FN and
collagens in fibers (Furcht et al., 1980) suggests that collagen
constitutes a framework for FN polymerization.

Due to its unique binding properties, FN is implied in the
adhesion, spreading and migration of cells, and in the organization
of the ECM (reviewed by Yamada, 1983). In the blood, plasma FN is
involved in the formation of the plasma clot through its binding to
fibrin (Ruoslahti and Vaheri, 1975) and in opsonization (in the
activation of phagocytosis by macrophages ; Van de Water et al.,
1981). The association of FN with the plama membrane and its
importance in the maintenance of cell architecture are still
controversial. Until recently, it was difficult to explain how FN
could promote two mechanisms as different as cell adhesion and cell
migration. It seems that the links between FN and the cell have a
crucial role. Indeed, the behavior of chick heart mesenchymal cells
depends on their synthesis of FN and of the relative concentration
of FN in the milieu. The anchorage of a non-migrating cell is
correlated with the synthesis by the cells themselves of a dense
fibrillar meshwork of FN. Conversely, exogenous FN, when present as
small plaques on the cell surface greatly enhances cell motility
(Couchman et al., 1982). Interestingly, migrating embryonic cells
lack the ability to produce FN but can bind to exogenous FN
(Newgreen and Thiery, 1980 ; Wylie and Heasman, 1982). Furthermore,
during migration, cells can degrade (Chen et al., 1984) or
reorganize (Rovasio et al., 1983) the FN-meshwork. So far, the
relationships between FN and migrating or stationnary cells appear
extremely complex.

1.3. Laminin

Laminin is a high molecular weight glycoprotein (900 to 1,000
kD) which is found in the basal laminae of epithelia (Timpl et al.,
1979 ; see Fig. 3). Like FN, LN is characterized by the presence of
binding sites for type IV collagen, heparin, and for cells. It does
not form fibrils either in vivo and in vitro. This is correlated
with its structure shaped as an asymetric cross (Ott et al., 1982 ;
Rao et al., 1982 ; Hogan, personal communication). The ends of the
arms possess a globular morphology and are the sites of the binding
domains for heparin and type IV collagen (Ott et al., 1982). The

cell binding site lies in the central part of the cross (see Fig. 3). In contrast to FN, the cellular receptor for LN has been characterized as a 70 kD protein (Rao et al., 1983 ; Terranova et al., 1983 ; Lesot et al., 1983).

The primary function of LN is to mediate adhesion of epithelial cells to type IV collagen (Terranova et al., 1980 ; Donaldson and Nahan, 1984). Additionally, several recent studies suggest a role of LN during the ontogeny of the nervous system. In vitro, normal astrocytes synthesize LN but not type IV collagen (Liesi et al., 1983) and in vivo, after injury, LN is found in the ECM surrounding these cells (Liesi et al., 1984). On the other hand, LN greatly enhances in vitro neuron attachment and growth cone motility as compared to FN (Rogers et al., 1983) through its heparin-binding domain (Edgar et al., 1984). This and other studies suggest that besides its structural role in epithelium architecture, LN can mediate cell adhesion in cases where FN plays this role.

2. CELL-TO-CELL ADHESION MOLECULES

Cell-to-cell adhesion involves numerous types of structures (e.g. desmosomes, gap, tight, and intermediate junctions, see Fig. 1). These structures can be detected at the ultrastructural level and play a major role in the shape and function of epithelial cells (see Garrod, Pitts, Gilula and Franke, this volume). Beside these organized structures, a number of molecules have also been found to be involved in cell adhesion (see also Burger, this volume). It appears that each of the adhesive mechanisms plays a specific role, but the biological relationships between them are not yet established. Particularly, we do not know whether the appearance of a specific adhesive mechanism is a prerequisite for the formation of an epithelium. However, it seems clear that they are not all required to maintain the epithelial architecture, at least in the embryo in which tissues rapidly reorganize. In the present section, we shall review the biochemical and biological properties of some cell-adhesion molecules (CAMs) which have been found in vertebrates. Examples of CAMs in invertebrates are described in another chapter of the volume (see Burger).

2.1. Isolation of CAMs. Different types of CAMs

Using an in vitro immunologically-based assay originally developped by Huesgen and Gerish (1975), it was possible to identify on chick embryo neurons a neural cell-adhesion molecule (N-CAM ; Brackenbury et al., 1977 ; Thiery et al., 1977). The same procedure allowed the identification of two other molecules in the chick : L-CAM originally isolated from embryonic liver cells (Gallin et al., 1983) and Ng-CAM isolated from neurons and known to mediate adhesion between neurons and glia (Grumet and Edelman, 1984). Using either the same assay or monoclonal antibodies, other adhesion molecules have been found in mammals : the proteins BSP-2 (Goridis et al., 1984) and D-2 (Jorgensen et al., 1980) are specific for neurons and share similarities with N-CAM (Jorgensen et al., 1978). Uvomurolin (Hyafil et al., 1981) and cadherin (Yoshida-Noro et al., 1984) perform similar functions to L-CAM ; and Ng-CAM has been found identical to L-1 (Lindner et al., 1983). A body of evidence indicates that other CAMs remain to be isolated (Edelman, 1983, 1984). However, it appears that only a limited number of CAMs are responsible for the adhesion of the various cell types (see also section 2.3.).

2.2. Structure and biochemistry of CAMs

The structure and biochemistry of N-CAM and L-CAM are being elucidated (Fig. 4 ; for a review, see Edelman, 1983, 1984). Both are glycoproteins. N-CAM is unusual in that it contains large amounts of sialic acid in an uncommon polymerized form. The protein chain of N-CAM of 160 kD consists of three domains linked by two stretches of chain that are very susceptible to enzyme proteolysis. N-CAM is an integral membrane protein : the aminoterminal domain extends away from the cell and includes a region that binds to the corresponding region of another N-CAM molecule (homophilic binding). A middle domain carries the polysialic acid and the carboxyl-terminal domain is associated with the cell membrane (Fig. 4 ; Hoffman et al., 1982 ; Cunningham et al., 1983).

L-CAM has a MW of 124 kD and lacks the unusual polysialic acid. Unlike N-CAM, L-CAM mediates cell adhesion only in presence of calcium and is rapidly degraded in its absence. However, the ligand of L-CAM remains to be determined. By enzymatic cleavage, L-CAM appears to have two domains, one projecting away from the membrane and the other (the COOH terminus) being associated to the membrane. It is not yet established whether L-CAM is an integral membrane protein (Fig. 4 ; Gallin et al., 1983).

Ng-CAM is also a glycoprotein but without unusual polysialic acid. Although N-CAM and Ng-CAM show a similar molecular weight, their peptide maps are different (Grumet et al., 1984). Until the counterpart ligand of Ng-CAM is isolated from glial cells, it will not be possible to define the mechanism of binding involving Ng-CAM, but it is heterophilic in contrast to N-CAM.

2.3. Distribution of CAMs during ontogeny

L-CAM and N-CAM appear early at the blastula stage of the chick embryo (Edelman et al., 1983). On the other hand, uvomurolin is present at the morula stage in mouse embryos. During gastrulation, L-CAM disappears from the mesoderm and neural plate

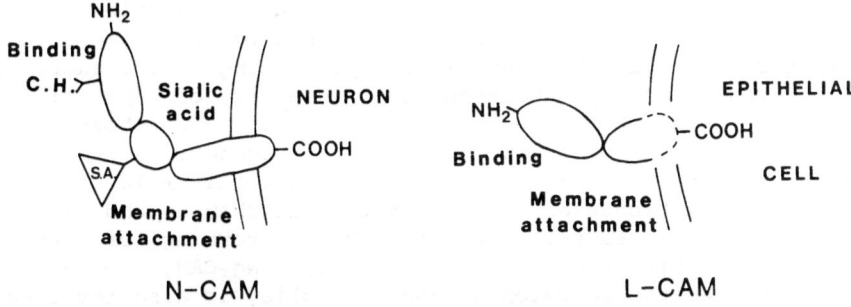

Fig. 4.   Structure of the neural and liver cell-adhesion mo-
          lecules (N-CAM and L-CAM).
          These representations are obtained   through   limited
          proteolytic digestions (see text). N-CAM is composed
          of three domains, one being integrated into the
          membrane, one carrying the polysialic chain (S.A.),
          and the last one, being the binding domain which also
          bears   carbohydrates (C.H.). L-CAM possesses two
          domains, one carrying the binding site and the other
          being associated with the membrane. Whether L-CAM is
          an integral membrane protein, is not yet determined
          (dashed lines).

and is confined to the ectoderm and endoderm. Conservely, N-CAM
increases in amount in the neural plate and remains expressed by
many mesodermal tissues (Thiery et al., 1982 b, 1984 a). N-CAM is
also expressed transitorily in non muscular and non nervous tissues
(e.g. kidney tubules) and its expression is correlated with
epithelium formation (Thiery et al., 1982 b). During embryogenesis,
the glycosylation level of CAM diminishes, a phenomenon which is
not observed for L-CAM (Rothbard et al., 1982 ; Hoffman and
Edelman, 1983). Since this decrease in glycosylation enhances the
binding strength of the molecule, it has been correlated with the
stabilization of the neuronal network (Chuong and Edelman, 1984).
As opposed to N-CAM, Ng-CAM is not expressed transitorily by
non-neuronal tissues ; it appears only after the start of axon
outgrowth. However, its appearance preceds and thus, does not seem
to be correlated with the differentiation of glial cells (Thiery et
al., 1984 b). In older embryos and in the adult, N-CAM is
maintained in the nervous tissues, in the heart and gonads, and is
restricted to the neuromuscular junction area in striated muscles ;
L-CAM is permanently associated with the ectoderm derivatives,
kidney tubules and all the endodermal derivatives.

2.4. Biological importance of CAMs

     The importance of CAMs in morphogenesis has been clearly

established by in vivo and in vitro perturbation experiments using
monovalent antibodies. Uvomurolin is required for the compaction of
the morula, a crucial step in the patterning of the extra- and
intra-embryonic tissues. The structural integrity of epithelia is
linked to the presence of L-CAM (Damsky et al., 1983 ; Yoshida-Noro
et al., 1984). N-CAM participates in the histogenesis of the retina
(Buskirk et al., 1980). It has a pivotal role in nerve
fasciculation (Rutishauser et al., 1978 ; Rutishauser and Edelman,
1980) and, as a consequence, in the patterning of neuronal
connections (Fraser et al., 1984). Finally, Ng-CAM, through its
role in the specific adhesion of nerve to glia, is also involved in
the formation of neuronal connections, as shown in the cerebellum
(Lindner et al., 1983).

3. PATTERNING OF THE TRUNK PERIPHERAL NERVOUS SYSTEM

The ganglia of the trunk peripheral nervous system (PNS)
derive totally from a single structure, the neural crest (for
reviews, see Le Douarin, 1982 ; Le Douarin et al., 1984 b). The
neural crest is an embryonic tissue lying along the entire dorsal
border of the neural tube. In the chick embryo, it appears after
the neural tube closure, first in the head and progressively in the
trunk. Thereafter, crest cells detach from the neural tube and
migrate to various sites of the embryo where they undergo their
differentiation (see Smith, this volume, for a review on the
differentiation of crest cells). Numerous questions arise when
considering the contrast between the location of the premigratory
crest cells and the distribution of their various derivatives. How
can a single embryonic structure generate so many different types
of cells in so many different places ? How do crest cells reach so
precisely their target sites ? What are the morphological cues that
direct crest cell migration and final localization ? A variety of
experiments using transplantation of the neural tube into host
embryos (Weston, 1963 ; Noden, 1975, 1978 ; Le Douarin et al.,
1984) have provided information regarding the routes of migration
and the fate of crest cells. It appears that they depend largely on
the level of the neural axis to which the cells are transplanted
(Le Douarin and Teillet, 1974).

In the following sections, we will mainly consider the
genesis of the ganglia of the PNS in the trunk (dorsal root,
sympathetic, and enteric ganglia). The formation of the trunk
peripheral ganglia can be considered as a three-step process : the
segregation of the neural crest from the neural tube, the migration
phase per se, and the aggregation phase of the cells into a
ganglion rudiment - this step being followed by the differentiation
of the cells into neurons and supportive cells. Thus, one can see
that the ontogeny of the PNS, at least in the trunk, can be
summarized as a series of successive epithelium-mesenchyme

interconversions which involve modulations of the adhesive
properties of crest cells.

3.1. Segregation of the neural crest from the neural tube

Prior to their migration, the presumptive crest cells are
integrated in the neuroectodermal epithelium. In order to migrate,
crest cells have to dissociate from the neural epithelium. This
highly complex mechanism is not yet elucidated. Its study is
greatly hampered by its overlap with other embryonic events (e.g.
closure of the neural tube and the separation of the ectoderm from
the neural tube itself). Furthermore, the conditions of the
segregation of crest cells vary from one species to another and
also at the different levels within an embryo. In the head of the
avian embryo, it occurs concomitantly with the closure of the
neural tube ; in the trunk, after the closure and in the mouse
embryo, prior to the closure, at the neural fold stage (Di Virgilio
et al., 1967 ; Tosney, 1978, 1982 ; Nichols, 1981 ; Duband and
Thiery, 1982 ; Thiery et al., 1982 a ; Erickson and Weston, 1983).

In all cases, the mechanism of release of crest cells from the
neuroepithelium involves at least a local disruption of the basal
lamina and the disappearance of the intercellular junctions. The
absence of electrical coupling in the dorsal side of the neural
tube suggests that the gap junctions are lost (Revel and Brown,
1975). By transmission electron microscopy, it has been impossible
to detect tight junctions between the premigratory crest cells
while, ventrally and more laterally in the neural tube, they are
commonly found (Tosney, 1978, 1982; Newgreen and Gibbins, 1982). In
the presumptive neural crest area, cells are irregular in shape,
unlike the highly elongate neural tube cells (Newgreen and Gibbins,
1982). Also numerous extracellular spaces separate the crest cells
(Tosney, 1978, 1982). The basal lamina bordering the neural tube is
frequently interrupted in the neural crest area (Tosney, 1978, 1982
; Newgreen and Gibbins, 1982). LN progressively disappears, while
FN and collagens remain between the ectoderm and the neural tube
(see Fig. 5 ; Duband and Thiery, 1982 ; Thiery et al., 1982 a ;
Duband et al., 1984 a, b). Finally, crest cells extend projections
into the ECM overlying the neural tube (Tosney, 1978, 1982) and are
progressively surrounded by FN (see Fig. 5 b ; Thiery et al., 1982
a). Their cytoskeleton becomes typical of migratory cells (Newgreen
and Gibbins, 1982). It thus appears that all the cellular and
extracellular components implied in cell adhesion are involved in
the emigration of crest cells from the neural tube.

The mechanism that triggers the disruption of the epithelial
architecture is not known. Previous studies on corneal epithelium
have shown that the disappearance of the basal lamina from under
epithelial cells induces them to send projections into the
underlying ECM and to separate from their neighbours (Sugrue and

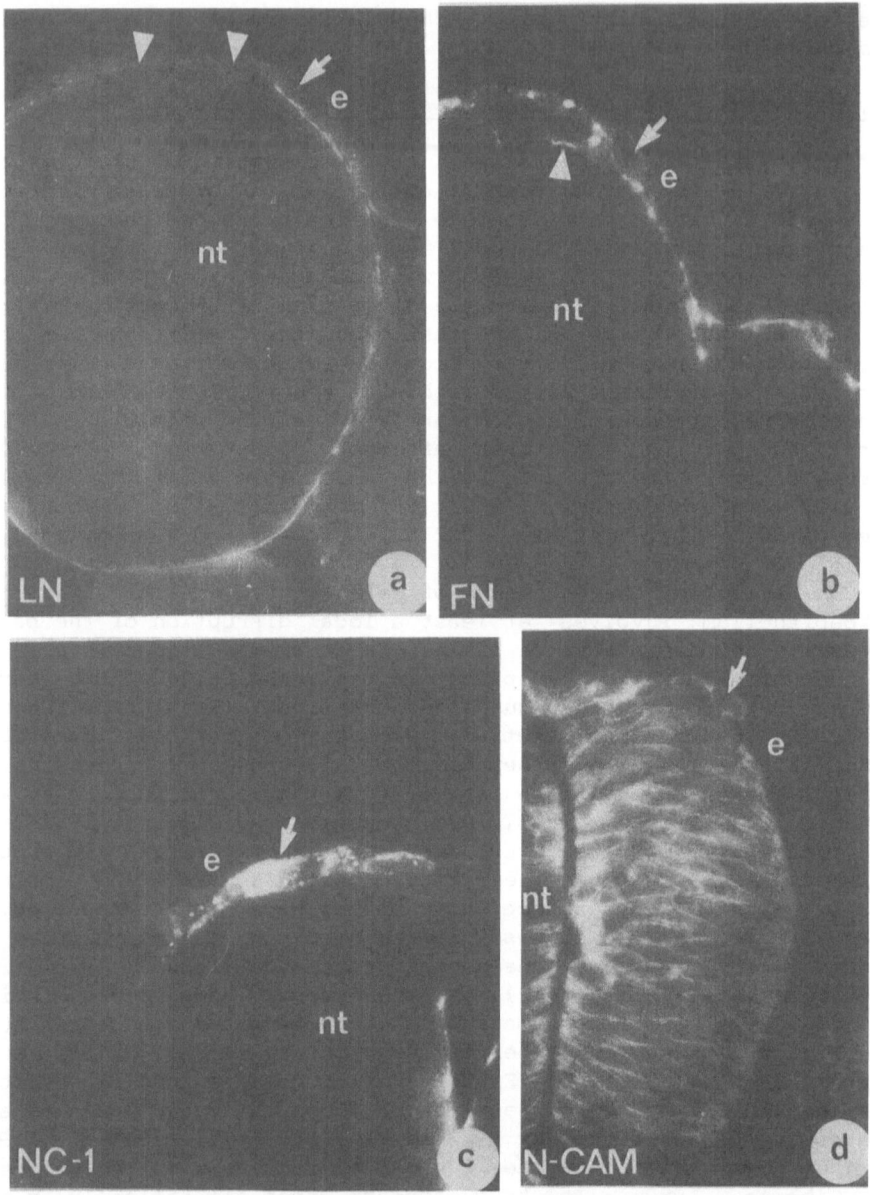

Fig. 5.   Segregation of the neural crest cells from the neural
           tube.
           a), b), c) and d) Transverse sections of 18 to 20
           somite stage chick embryo at the level of the 15th
           somite   stained   for   LN,   FN,   NC-1 and N-CAM
           respectively. Crest cells individualize at the dorsal
           border of the neural tube (nt). In this area, the
           basal lamina of the neural tube is interrupted as

Hay, 1981, 1982). The emigration of crest cells is probably due to the disappearance of the basal lamina. What causes the disruption of the basal lamina ? According to the species and to the level in the embryo, a wide variety of factors may participate to this mechanism (Fig. 6). Crest cells are located in a key area between the neural epithelium and the ectoderm. This region is the site of morphogenetic movements, particularly in the head, (folding and bending ; Di Virgilio et al., 1967 ; Karfunkel et al., 1974 ; Jacobson, 1981) which could create mechanical forces able to damage the basal lamina. Cephalic crest cells are actively proliferating at the onset of their migration (our unpublished results), this, too, could provoke a disorganization of the epithelial structure. Crest cells synthesize plasminogen activator (Valinsky and Le Douarin, 1984) ; plasmin and also collagenases could digest LN and type IV collagen of the basal lamina. However, in the trunk of the avian embryo and in mamals, no intense proliferation of crest cells has been observed and the emigration cannot be directly correlated with mechanical pressure. Furthermore, the departure of crest cells from the neural tube is time-dependant (Newgreen and Gibbins, 1982). The expression of new cell surface molecules is probably involved. Alternatively, the modification of cell surface molecules (e.g. glycosylation) or the modulation of their synthesis could simply diminish the binding between the cells. The role of CAMs in the segragation of crest cells from the neural tube is not clearly established. By immunofluorescence, N-CAM does not disappear at the onset of crest migration but rather its amount decreases progressively (Fig. 5 d). However, a small decrease of the number of N-CAM molecules, not detectable with immunofluorescence techniques, can induce an important decrease in the cell binding (Hoffman and Edelman, 1983). Recently, we have produced a monoclonal antibody, NC-1, which recognizes migrating crest cells and their derivatives (Vincent et al., 1983 ; Vincent and Thiery, 1984). The appearance of this marker at the onset of crest migration (Fig. 5 c) along with its presence during other embryonic processes suggest that it could be involved in separation of cells (Tucker et al., 1984 ; Tucker and Thiery, in preparation).

It is clear from these observations that the conversion of an

shown by staining for LN (arrow heads in a). Crest cells detaching from their neighbours are partly surrounded by FN (arrow head in b). Those in the course of migration bind NC-1 at their surface (c). Premigrating and migrating crest cells are labelled for N-CAM (d). Arrows point of migrating crest cells (c, d). Ectoderm, e.

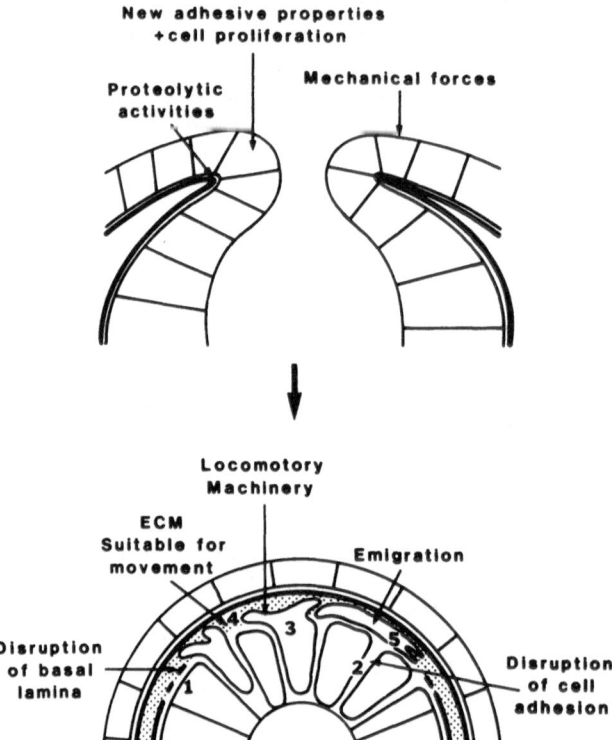

Fig. 6.   Model of the segregation of crest cells from the
          neural tube.
          The first part of the diagram shows the morphology
          of the neurectoderm prior to crest emigration and
          the possible factors which trigger the dissociation
          of crest cells from the neural epithelium. The
          second part of the diagram represents a possible
          reconstruction of the successive events occuring
          during crest emigration : 1. the disruption of the
          basal lamina; 2. the disruption of the intercellular
          adhesion; 3. the formation of cell projections
          allowing motility; 4. the presence of a substrate
          suitable for cell movement; 5. the beginning of
          crest migration.

epithelium into a mesenchyme is a highly complex process which
involves a number of factors. Unfortunately, an _in vitro_ model
system for studying crest cell emigration is not yet available.

3.2. Migration of crest cells

3.2.1. Substrate of migration

At the onset of migration, crest cells encounter a substrate suitable for migration. It is acellular and contains FN (see Fig. 7 a), type I and III collagens, hyalunoric acid and small amount of chondroitin sulfate (Derby, 1978 ; Tosney, 1978, 1982 ; Mayer et al., 1981 ; Duband and Thiery, 1982 ; Thiery et al., 1982 a ; Duband et al., 1984 a, b). As far as the migration phase is concerned, the composition of the ECM remains constant (Duband and Thiery, 1982 ; Thiery et al., 1982 a). It thus seems improbable that a gradient of adhesiveness (haptotactism ; Carter, 1967) is responsible for the orientation of migration of crest cells. The three-dimensional structure of the ECM does not appear to influence the direction of crest migration by a mechanism of contact guidance (Weiss, 1945). With the possible exception of the amphibian embryo (Löfberg et al., 1980), the meshwork does not seem to exhibit any particular orientation (Tosney, 1978).

The role of most of the ECM components in the crest cell migration has been established partly through in vivo observations, but mostly through in vitro experiments (Greenberg et al., 1981 ; Newgreen et al., 1982 ; Rovasio et al., 1983 ; Tucker and Erickson, 1984). FN alone, associated with collagen in two- and three-dimensional lattice, or deposited by fibroblasts greatly promotes the spreading and the movements of crest cells (Greenberg et al., 1981 ; Newgreen et al., 1982 ; Rovasio et al., 1983 ; Tucker and Erickson, 1984). In vitro crest cells do not attach to pure collagen deposited on two-dimensional substrate (Newgreen et al., 1982 ; Rovasio et al., 1983), but they are able to move in three-dimensional collagen gels, with the restriction that collagen must be native and at a low concentration. However, the speed of movement is low as compared to that in the presence of FN (Tucker and Erickson, 1984). In the embryo, the principal role of collagen during crest migration is probably to provide a framework which maintains the shape of the tissues and of the pathways of migration. Hyaluronate in two- and three-dimensional gels is a very poor substrate of migration (Newgreen et al., 1982 ; Tucker and Erickson, 1984), but due to its hydration properties, it expands spaces and indirectly enhances the speed of migration (Pratt et al., 1975 ; Pintar, 1978 ; Tucker and Erickson, 1984). From this point of view, it is interesting to note that crest cells synthesize hyaluronate at the onset of their migration (Greenberg and Pratt, 1977 ; Pintar, 1978). Likewise, chondroitin sulfate is not a good substrate for movement (Tucker and Erickson, 1984), and its presence has been correlated with the arrest of crest cells (Derby, 1978). Finally, LN is present in the basal laminae bordering crest cell pathways (Fig. 7 b ; Duband et al., 1984 a), but does not seem to play a role in the migration since crest cells

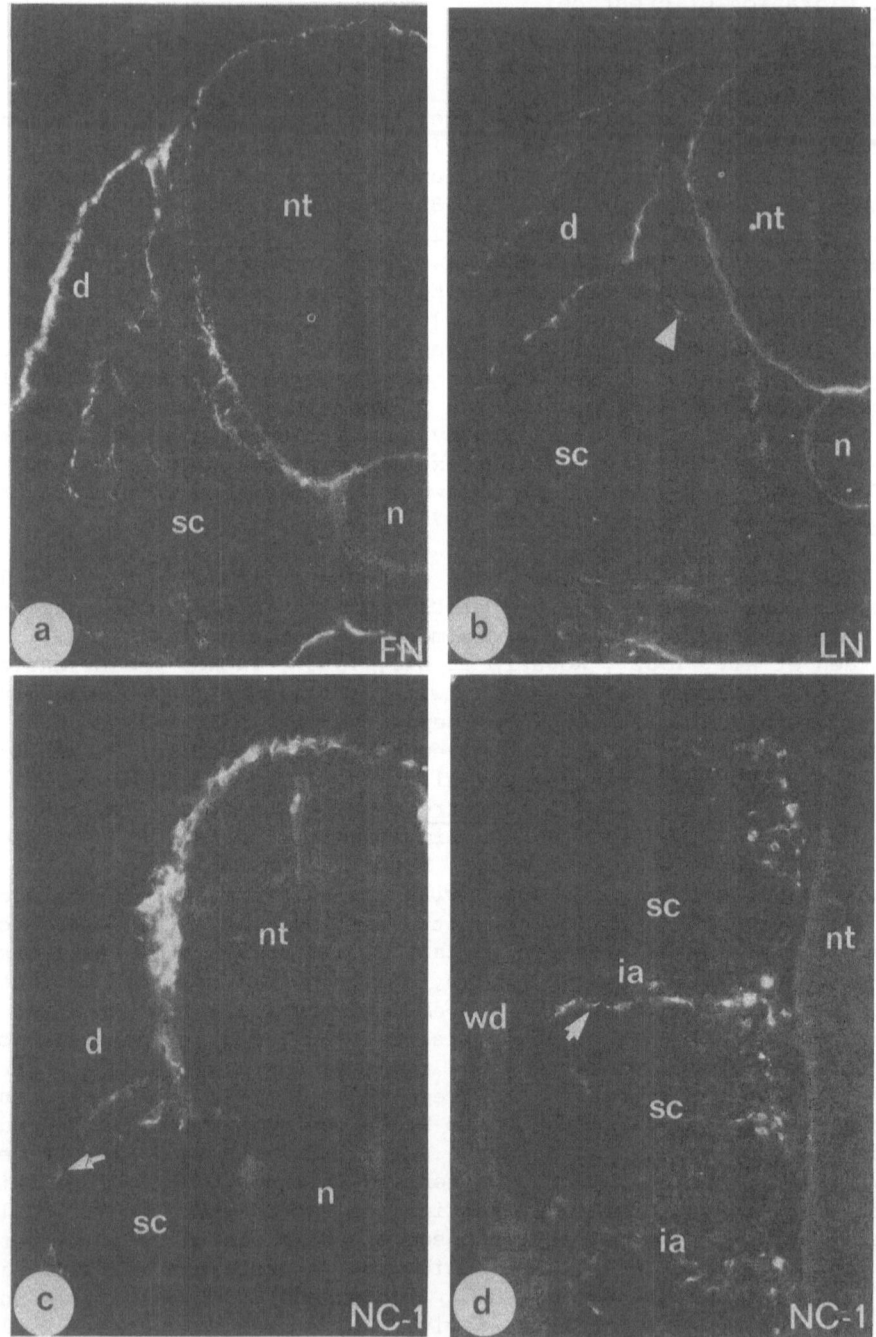

Fig. 7.   Crest cell migration at the   mid-trunk   level (15th
          somite).

are not frequently in direct contact with it. Furthermore, in vitro studies have revealed that the crest adhesion to LN is low during their migration (Rovasio et al., 1983).

The importance and the role of FN in crest cell displacement has been approached in an in vitro system and in perturbation experiments. When crest cells were confronted with alternative stripes of FN and coated serum proteins, they migrated exclusively on the FN ones. When LN stripes were used, few crest cells could move on LN and instead tended to aggregate on it (Rovasio et al., 1983). Monovalent antibodies directed against the cell binding site of FN reversibly block the migration of crest cells both in vitro and in vivo. When a decapeptide containing the RGDS sequence is used to compete with the cell binding site of FN, crest cells do not move in vitro ; in vivo, crest cells are not seen on the sides of the neural tube, but rather form a bulk in the neural tube lumen (Boucaut et al., 1984 ; Poole and Thiery, unpublished).

3.2.2. Pathways of migration

Various markers such as the quail nucleolar marker (Le Douarin, 1973), acetylcholinesterase (Cochard and Coltey, 1983) and the monoclonal antibody NC-1 (Vincent and Thiery, 1984) have provided complementary data on the routes of migration of crest cells (for a comparative study, see Le Douarin et al., 1984 a). The

---

a), b), c) Immunofluorescence staining for FN, LN and NC-1 respectively on transverse sections of a 30 somite stage embryo. FN is present in the ECM bordering the neural tube (nt), dermomyotome (d) and notochord (n) and around sclerotomal cells (sc), while LN is found in the basal laminae of the epithelia (neural tube and dermomyotome particularly). The crest population is located in a restricted area between the dermomyotome, neural tube and sclerotome and also along the basement membrane of the myotome (arrow in Fig. c). Note that crest cells are surrounded by FN and that the LN-rich basal lamina constitutes a channel (arrow head, Fig. b).
d) Immunofluorescence stained for NC-1 on a longitudinal section of a 30 somite stage embryo. Crest cells are found both along the neural tube (nt) and along the intersomitic arteries (ia) which run between consecutive sclerotomes (sc); the intersomitic pathway leads them to the ventral side of the embryo, near the Wolffian duct (wd). The arrow points to crest cells in the intersomitic area.

structure and topology of the pathways have been elucidated largely
through the use of markers of ECM (Derby, 1978 ; Duband and Thiery,
1982 ; Thiery et al., 1982 a). The routes of migration and the site
of arrest of crest cells are determined by the morphology of the
embryo and do not depend on crest cells themselves (Le Douarin and
Teillet, 1974). Even though they are greatly influenced by the
morphology of the embryo, the structure of the routes of migration
remains constant throughout the embryo. They are cell-free spaces
filled with an ECM and limited by one or two basal laminae of
epithelia. In this respect, it is interesting to note that, at the
onset of crest migration, epithelial structures are predominant in
the embryo. This is particularly obvious in the trunk where the
cessation of crest cell movement correlates with the conversion of
epithelia into mesenchymal cells (Thiery et al., 1982 a).

The presence of FN in acellular spaces is necessary but not
sufficient for crest migration. For example, crest cells do not
invade the notochordal area and the space between the ectoderm and
the dermatome, areas rich in FN and particularly in chondoitin
sulfate (Derby, 1978 ; Thiery et al., 1982 a).

On the other hand, various tissues may constitute obstacles or
guides to crest migration. In the head, the presence of local
thickening of the neural tube (optic vesicle) or of the ectoderm
(ectodermal placodes) prevent ventral migration of crest cells
which locally accumulate and then bypass these barriers caudally
and/or rostrally (Noden, 1975 ; Duband and Thiery, 1982 ; Cochard
and Coltey, 1983). In the trunk, the migration of crest cells is
mainly guided by metamerized structures, the somites (Thiery et
al., 1982 a). Depending on their location with respect to the
somite, crest cells move between the somite and neural tube or
between two consecutive somites (Fig. 7 c, d). Those using the
first pathway are rapidly arrested by the expanding sclerotome, a
mesenchymal structure which derives from the somite. Crest cells do
not appear to invade the sclerotome extensively although the latter
synthesizes FN. This absence of invasion is probably due to several
reasons : the cell density of the sclerotome, the three-dimensional
organization of the ECM around the sclerotomal cells, the migratory
properties of crest cells, and the presence of the remaining
LN-rich basal lamina along the medial side of the sclerotome (see
Fig. 7 b). As a consequence of this obstacle crest cells accumulate
and aggregate into the dorsal root ganglia (see section 3.4.) or
migrate more ventrally along the basement membrane of the myotome
(see Fig. 7 c). The intersomitic pathway leads crest cells to the
aorta where they will provide the sympathetic ganglia (see Fig. 7
d).

3.2.3. Behavior of crest cells

Does the final distribution of crest cells at the end of their

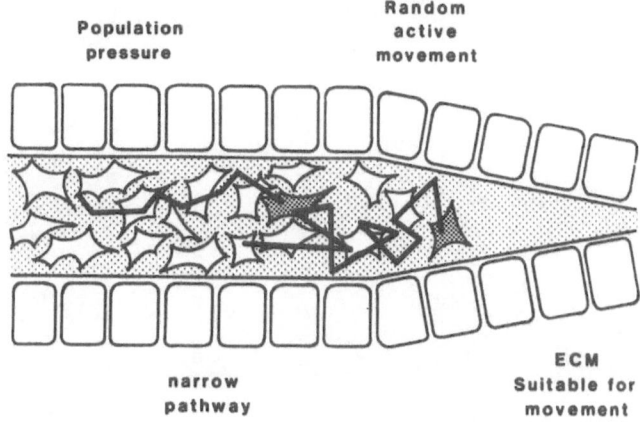

Fig. 8.   Model of migration of neural crest cells.
This diagram represents the various parameters that
have been found to govern crest cell migration and
ensure a proper directionality to the different
target sites. They concern crest cells themselves
(active cell migration, population pressure and
expansion of space by HA) or their immediate
environment (narrow pathways, ECM suitable for
migration and expansion of space).

migration result solely from the structure of the pathways or also
from specific behavior of crest cells themselves ? When crest cells
or their derivatives (pigment cells and ganglia) are grafted in
crest migratory pathways, they usualy distribute in the normal
sites of arrest. Conversely, embryonic fibroblastic cells from
somite, limb bud and lateral plate do not migrate extensively and
remain intact at the site of the graft. Interestingly, other motile
cells such as tumor cells exhibit the same behavior as crest cells
(Erickson et al., 1980). On the other hand, Wolffian duct cells
also known to migrate, when grafted in crest pathways, do not
locate in crest derivative sites but rather behave normally ; that
is, they form a duct (Poole, personal communication). Taken
together, these data strongly suggest that crest cells exhibit a
specific migratory behavior. However, recent experiments using
latex beads have provided evidence that crest displacement may be
passive (Bronner-Fraser, 1982, 1984).

In vitro studies using time-lapse microcinematography have
substantiated the results of the grafting experiments (Newgreen et
al., 1979 ; Rovasio et al., 1983). Migratory crest cells do not

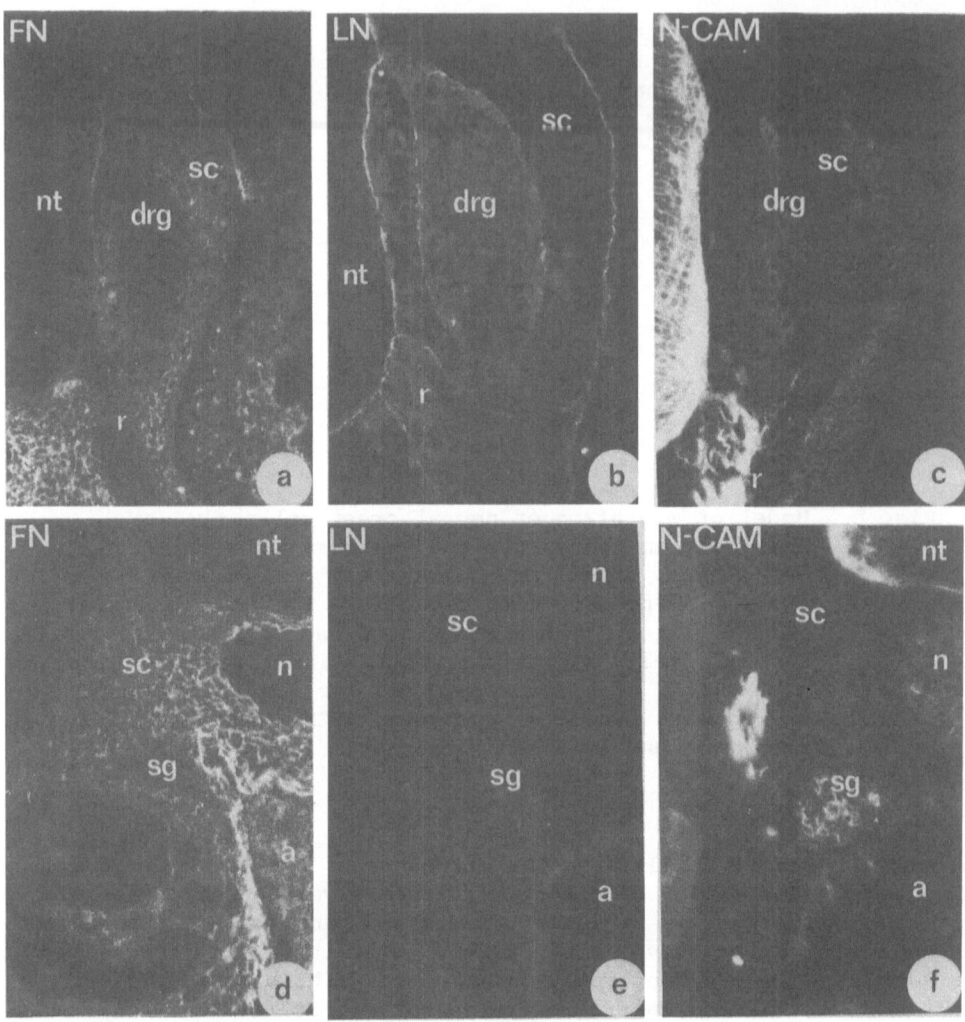

Fig. 9.   Aggregation of crest cells.
          a), b) and c) Formation of the dorsal root ganglia as
          seen by immunofluorescent staining for FN, LN and
          N-CAM respectively. Transverse section of chick
          embryos at 4 days of incubation. The primordium of
          the sensory ganglion (drg) is devoid of FN but is
          surrounded by a basal lamina stained by anti-LN
          antibodies. Conversely, the sclerotome is stained for
          FN and not for LN. Note that the part of the motor
          root (r) close to the neural tube (nt) is surrounded
          by LN. While the neural tube and motor root are
          heavily labelled for N-CAM, only a few differen-
          tiating cells in the sensory ganglion are stained.

show any polarized shape as fibroblasts do, and have many cellular expensions. On a suitable substrate, isolated crest cells move very actively but their effective distance is very small (Twitty and Niu, 1954 ; Newgreen et al., 1979 ; Rovasio et al., 1983). Indeed, they use successively each of their filopodia, which makes them moving randomly. Among a cell population, crest cells move in a precise direction. This may be due to contact inhibition (Abercrombie, 1970 ; Newgreen et al., 1982). Crest cells are only weakly contact inhibited ; the presence of numerous filopodia allow them to move backward very rapidly after a contact with another cell (Newgreen et al., 1983). Numerous studies on crest behavior together with the information given by the study of the pathways have led us to suggest that an efficient unidirectional translocation of crest cells is the result of several parameters. 1. Active individual motility ; 2. The presence of a suitable substrate ; 3. Active cell proliferation ; 4. Presence of narrow pathways of migration. A model of crest migration which accounts for the different parameters involved during the migration is given on Fig. 8.

Beside the specific motile behavior, one should mention other properties of crest cells which help them to reach their target sites. During their migration, crest cells synthesize important amounts of HA (Greenberg and Pratt, 1977 ; Pintar, 1978), which allow them to expand the space available for migration. Crest cells do not synthesize FN, an unusual behavior at a time when all embryonic tissues do it (Newgreen and Thiery, 1980). This behavior is consistent with good displacement in a FN-meshwork (see section 1.2.).

3.3. Aggregation of crest cells

As far as the PNS is concerned, crest cells mostly aggregate into ganglion rudiments after completing their migration. According to the conditions of crest migration (see Fig. 8), their arrest must be the result of a complete transformation either of the environment or of the behavior of the crest cells themselves. As

---

d), e) and f) Formation of the sympathetic ganglia; staining is for FN, LN and N-CAM, respectively, of a chick embryo at 3.5 days of incubation. The sympathetic ganglion (sg) forms in an area between the sclerotome (sc), the aorta (a) and the notochord (n). This area is very rich in FN and completely devoid of LN. As soon as crest cells aggregate, the sympathetic ganglion rudiment is thoroughly stained for N-CAM.

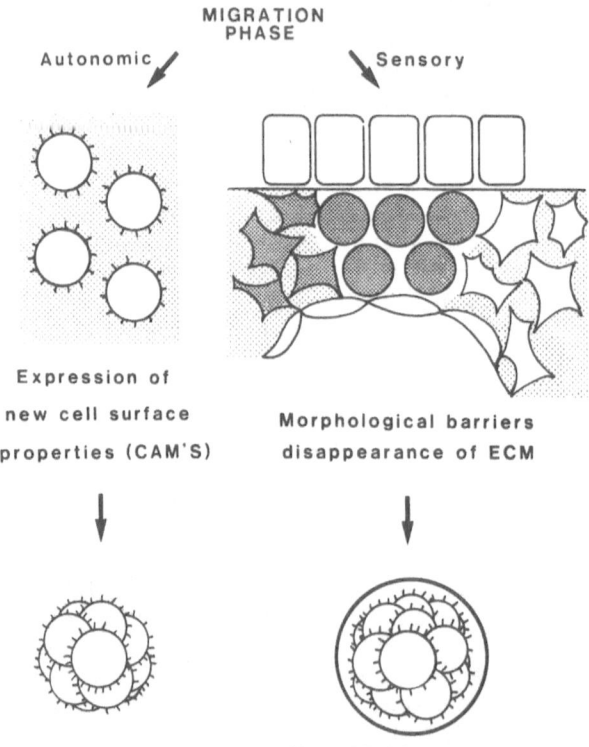

Fig. 10. Model of aggregation of crest cells into ganglia of
        the peripheral nervous system.
        One can distinguish two mechanisms of aggregation
        according to the fate of crest cells. The sensory
        ganglia appear as a consequence of the lack of space
        and substratum necessary for migration. Crest cells
        aggregate and progressively change their cell surface
        properties (adhesion for LN, synthesis of N-CAM). The
        autonomic   ganglia   form   as   a   consequence of a
        modification of their cell-to-cell adhesive proper-
        ties (precocious   synthesis   of   N-CAM);   the   ECM
        surrounding the autonomic ganglia is not modified.

crest cells undergo their migration, their local environment is
modified : numerous epithelia dissociate into dense mesenchymes
(see for example the sclerotome), cells proliferate, tissues expand
and, as a consequence, presumptive pathways become obstructed and,
the ECM itself is sometimes modified. This is particularly true in
the area where the sensory ganglia will to appear. The somite
dissociates into the sclerotome which rapidly expands ventrally,

thus obstructing the ventral pathway along the neural tube towards the notochord. Chondroitin sulfate increases in amount among crest cells (Derby, 1978), while FN, collagens and HA disappear (Fig. 9 a; Duband and Thiery, 1982 ; Duband et al., 1984 a, b). Soon after this, the still proliferating crest cells accumulate and become surrounded by LN (Fig. 9 b). In vitro, LN tend to induce crest aggregation (see section 3.3.1.) and, when crest cells are cultured for a long period, they progressively develop a higher affinity for LN than for FN (Rovasio et al., 1983). Finally, N-CAM is expressed, although with a delay, as neurons differentiate (Fig. 9 c). It thus seems probable that the presence of physical barriers and the lack of a substrate suitable for movement in a restricted area are responsible for the cessation of movement of crest cells which, in turn, induces a complete change in their surface properties (Fig. 10). Similarly, in the head, blood vessels may constitute physical barriers to crest migration ; numerous peripheral ganglia appear in the head along the cardinal vein (Noden, 1975 ; Duband and Thiery, 1982 ; Duband et al., 1984 c ; D'Amico-Martel and Noden, 1983).

As opposed to the sensory ganglia, the sympathetic ones form near the aorta in an area whose ECM does not seem to be modified (Fig. 9 d, e ; Thiery et al., 1982 a ; Duband et al., 1984 a, b). Furthermore, the mesentery located ventral to the aorta would provide a pathway of migration. We have observed that near the aorta, crest cells rapidly form small clusters with high amounts of N-CAM (Fig. 9 f). Such cells still divide while expressing neuronal markers (Rothman et al., 1978). This series of events has also been observed in the case of the autonomic ganglia in the gut. Thus autonomic ganglia could form as a consequence of a change of the cell surface properties of crest cells (Fig. 10). If they exist, environmental factors inducing this modification remain to be defined.

4. CONCLUDING REMARKS

Throughout development, a given cell population may express various adhesive properties that lead it to form either an epithelium or a mesenchyme. The neural crest is a striking example of such behavior, but many other embryonic tissues are similarly affected by modulations of their adhesive properties, although at a less spectacular level. Nevertheless, only a limited number of cells are capable of autonomous displacement at a precise time of embryogenesis. It seems that, in order to acquire motile properties, a cell population like the neural crest must benefit from a conjunction of favorable conditions. This does not exclude, however, the possibility that other cell types use different ways to migrate, involving other types of interactions (reviewed in Thiery, 1984).

Until now, a detailed model for the genesis of movement, integrating the different types of interactions between the extracellular environment, the cell surface and the cytoskeleton, is not yet available. The transduction of the signals and the regulatory mechanisms operating within and between these domains remain largely unknown.

ACKNOWLEDGEMENTS

We thank Drs Leslie Blair and Philippe Cochard for revising the manuscript and, Stephane Ouzounoff and Sophie Tissot for the illustrations. This work was supported by grants from the Centre National de la Recherche Scientifique, the Institut de la Santé et de la Recherche Médicale, the Ministère de l'Industrie et de la Recherche, the Ligue Nationale Française contre le Cancer and the Fondation pour la Recherche Médicale.

REFERENCES

Abercrombie M., 1970, Contact inhibition in tissue culture, In vitro, 6 : 128-142.
Akiyama S.K. and Yamada K. M., 1984, The interaction of plasma fibronectin with fibroblastic cells in suspension, J. Biol. Chem., in press.
Aplin J.D., Hughes R.C., Jaffe C.L. and Sharon N., 1981, Reversible cross-linking of cellular components of adherent fibroblasts to fibronectin and lectin-coated substrata, Exp. Cell Res., 134 : 488-494.
Brackenbury R., Thiery J.P., Rutishauser U. and Edelman G.M., 1977, Adhesion among neural cells of the chick embryo. 1. An immunological assay for molecules involved in cell-cell binding, J. Biol. Chem., 252 : 6835-6840.
Boucaut J.C., Darribere T., Poole R.J., Aoyama H., Yamada K.M. and Thiery J.P., 1984 b, Biologically active synthetic peptides as probes of embryonic development : a competitive inhibitor of fibronectin function inhibits gastrulation in amphibian embryos and neural crest cell migration in avian embryos, J. Cell Biol., 99 : 1822-1830.
Bronner-Fraser M., 1982, Distribution of latex beads and retinal pigment epithelial cells along the ventral neural crest pathway, Dev. Biol., 91 : 50-63.
Bronner-Fraser M., 1984, Latex beads as probe of a neural crest pathway. Effects of laminin, collagen and cell surface charge on bead translocation, J. Cell Biol., 98 : 1947-1960.
Buskirk D.R., Thiery J.P., Rutishauser U. and Edelman G.M., 1980, Antibodies to a neural cell adhesion molecule disrupt

histogenesis in cultured chick neural retinae, <u>Nature</u>, 285 :
    488-489.
Carter S.B., 1967, Haptotaxis and the mechanics of cell motility,
    <u>Nature</u>, 213 : 256-260.
Chen W.T., Olden K., Bernard B.A. and Chu F.F., 1984, Expression of
    transformation-associated protease(s) that degrade
    fibronectin at cell contact sites, <u>J. Cell Biol.</u>, 98 :
    1546-1555.
Chuong C.M. and Edelman G.M., 1984, Alterations in neural cell
    adhesion molecules during development of different regions
    of the nervous system, <u>J. Neurosci.</u>, 4 : 2354-2368.
Cochard P. and Coltey P., 1983, Cholinergic traits in the neural
    crest : acetylcholinesterase in crest cells of the chick
    embryo, <u>Dev. Biol.</u>, 98 : 221-238.
Couchman J.R., Rees D.A., Green M.R. and Smith C.G., 1982,
    Fibronectin has a dual role in locomotion and anchorage of
    primary chick fibroblasts and can promote entry into the
    division cycle, <u>J. Cell Biol.</u>, 93 : 402-410.
Cunningham B.A., Hoffman S., Rutishauser U., Hemperly J.J. and
    Edelman G.M., 1983, Molecular topography of the neural cell
    adhesion molecule N-CAM : Surface orientation and location
    of sialic acid-rich and binding regions, <u>Proc. Natl. Acad.
    Sci. USA</u>, 80 : 3116-3120.
D'Amico-Martel A. and Noden D.M., 1983, Contribution of placodal
    and neural crest cells to avian cranial peripheral ganglia,
    <u>Am. J. Anat.</u>, 166 : 445-468.
Damsky C.H., Richa J., Solter D., Knudsen K. and Buck C.A., 1983,
    Identification and purification of a cell surface
    glycoprotein mediating intercellular adhesion in embryonic
    and adult tissue, <u>Cell</u>, 34 : 455-466.
Derby M.A., 1978, Analysis of glycosaminoglycans within the
    extracellular enviroments encountered by migrating neural
    crest cells, <u>Dev. Biol.</u>, 66 : 321-336.
Di Virgilio G., Lavenda N. and Worden J.L., 1982, Sequence of
    events in neural tube closure and the formation of the
    neural crest in the chick embryo, <u>Acta Anat.</u>, 68 : 127-146.
Donaldson D.J. and Nahan J.T., 1984, Epidermal cell migration on
    laminin-coated substrates. Comparison with other
    extracellular matrix and non-matrix proteins, <u>Cell Tiss.
    Res.</u>, 235 : 221-224.
Duband J.L. and Thiery J.P., 1982, Distribution of fibronectin in
    the early phase of avian cephalic neural crest cell
    migration. <u>Develop. Biol.</u>, 93 :308-323.
Duband J.L., Timpl R. and Thiery J.P., 1984 a, Laminin in the early
    chick embryo : Correlation with epithelium-mesenchyme
    interconversion, <u>Dev. Biol.</u>, submitted.
Duband J.L., Grimaud J.A. and Thiery J.P., 1984 b, Collagens at the
    time of crest cell migration, <u>Cell Diff.</u>, submitted.
Duband J.L., Tucker G.C., Poole T.J., Vincent M., Aoyama H. and
    Thiery J.P., 1984 c, How do the migratory and adhesive
    properties of the neural crest govern ganglia formation in

the avian peripheral nervous system ?, J. Cellular Biochem.,
     in press.
Edelman G.M., 1983, Cell adhesion molecules, Science, 219 :
     450-457.
Edelman G.M., 1904, Cell adhesion and the molecular processes of
     morphogenesis, Ann. Rev. Biochem., in press.
Edelman G.M., Gallin W., Delouvée A., Cunningham B.A. and Thiery
     J.P., 1983, Early epochal maps of two different cell
     adhesion molecules, Proc. Natl. Acad. Sci. USA, 80 :
     4384-4388.
Edgar D., Timpl R. and Thoenen H., 1984, The heparin-binding domain
     of laminin is responsible for its effects on neurite
     outgrowth and neuronal survival, EMBO J., 3 : 1463-1468.
Erickson C.A., Tosney K.W. and Weston J.A., 1980, Analysis of
     migratory behavior of neural crest and fibroblastic cells in
     embryonic tissues, Develop. Biol., 77 : 142-156.
Erickson C.A. and Weston J.A., 1983, An SEM analysis of neural
     crest migration in the mouse, J. Embryol. exp. Morph., 74 :
     97-118.
Fisher M. and Solursh M., 1979, The influence of the substratum on
     the mesenchyme spreading in vitro, Exp. Cell Res., 123 :
     1-14.
Fraser S.E., Murray B.A., Chuong C.M. and Edelman G.M., 1984,
     Alteration of the retinotectal map in Xenopus by antibodies
     to neural cell adhesion molecules, Proc. Natl. Acad. Sci.
     USA, 81 : 4222-4226.
Furcht L.T., 1983, Structure and function of the adhesive
     glycoprotein fibronectin, In: "Modern Cell Biology", vol. 1,
     pp 53-117.
Furcht L.T., Smith D., Wendelschafer-Crabb G., Woodbridge P.A. and
     Foidart J.M., 1980, Fibronectin presence in native collagen
     fibrils of human fibroblasts : immunoperoxidase and the
     immunoferritin localization, J. Histochem. Cytochem., 28 :
     1319-1333.
Gallin W.J., Edelman G.M. and Cunningham B.A., 1983,
     Characterization of L-CAM, a major cell adhesion molecule
     from embryonic liver cells, Proc. Natl. Acad. Sci. USA, 80 :
     1038-1042.
Goridis C., De Agostini-Bazin H., Hirn M., Hirsch M.R., Rougon G.,
     Sadoul R., Langley O.K., Gombos G. and Finne J., 1984,
     Neural surface antigens during nervous system development,
     In:" Cold Spring Harbor Symposia on Quantitative Biology",
     vol. 48, pp 527-537.
Greenberg J.H. and Pratt R.M., 1977, Glycosaminoglycan and
     glycoprotein synthesis by cranial neural crest cells in
     vitro, Cell Diff., 6 : 119-132.
Greenberg J.H., Seppä H., Seppä S. and Tyl Hewitt A., 1981, Role of
     collagen and fibronectin in neural crest cell adhesion and
     migration, Develop. Biol., 87 : 259-266.
Grumet M. and Edelman G.M., 1984, Heterotypic binding between

neuronal membrane vesicles and glial cells is mediated by a specific cell adhesion molecule, J. Cell Biol., 98 : 1746-1756.

Grumet M., Hoffman S. and Edelman G.M., 1984, Two antigenically related neuronal cell adhesion molecules of different specificities mediate neuron-neuron and neuron-glia adhesion, Proc. Natl. Acad. Sci. USA, 81 :267-271.

Hay E.D., 1973, Origin and role of collagen in the embryo, Amer. Zool., 13 : 1085-1107.

Hay E.D., 1981, Cell biology of extracellular matrix, Plenum Press, New York.

Hewit A.T., Kleinman H.K., Pennyparker J.P. and Martin G.R., 1980, Identification of an adhesion factor for chondrocytes, Proc. Natl. Acad. Sci. USA, 77 : 385-388.

Hoffman S., Sorkin B.C., White P.C., Brackenbury R., Mailhammer R., Rutishauser U., Cunningham B.A. and Edelman G.M., 1982, Chemical characterization of a neural cell adhesion molecule purified from embryonic brain membranes, J. Biol. Chem., 257 : 7720-7729.

Hoffman S. and Edelman G.M., 1983, Kinetics of homophilic binding by embryonic and adult forms of the neural forms of the neural cell adhesion molecule, Proc. Natl. Acad. Sci. USA, 80 : 5762-5766.

Huesgen A. and Gerish G., 1975, FEBS lett., 56 : 46-49.

Hyafil F., Babinet C. and Jacob F., 1981, Cell-cell interactions in early embryogenesis : a molecular approach to the role of calcium, Cell, 26 : 447-454.

Hynes R.O. and Yamada K.M., 1982, Fibronectins : multifunctional modular glycoproteins, J. Cell Biol., 95 : 369-377.

Jacobson A.G., 1981, Morphogenesis of the neural plate and tube. In: "Morphogenesis and pattern formation", T.G. Connelly, ed., Raven Press, New York, pp 233-263.

Karfunkel P., 1974, The mechanism of neural tube formation, Intern. Rev. Cytol., 38 : 245-271.

Jorgensen O., Delouvée A., Thiery J.P. and Edelman G.M., 1980, The nervous system specific protein D2 is involved in adhesion among neurites from cultured rat ganglia, FEBS Lett., 111 : 39-42.

Kornblihtt A.R., Vibe-Petersen K. and Baralle F.E., 1984, Human fibronectin : molecular cloning evidence for two mRNA species differing by an internal segment coding for a structural domain, EMBO J., 3 : 221-226.

Kornblihtt A.R., Vibe-Petersen K. and Baralle F.E., 1984, Human fibronectin : cell specific alternative mRNA splicing generates polypeptide chains differing in the number of internal repeats, Nucleic Acid Res., in press.

Lattera J., Ansbacher R. and Culp L.A., 1980, Glycosaminoglycans that bind cold-insoluble globulin in cell-substratum adhesion sites of murine fibroblasts, Proc. Natl. Acad. Sci. USA, 77 : 6662-6666.

Le Douarin N.M., 1973, A biological cell labelling technique and
    its use in experimental embryology, Dev. Biol., 30 :
    217-232.
Le Douarin N.M., 1982, The Neural Crest, Cambridge University
    Press, Cambridge.
Le Douarin N.M. and Teillet M.A., 1974, Experimental analysis of
    the migration and differentiation of neuroblast of the
    autonomic nervous system and of neurectodermal mesenchymal
    derivatives using a biological cell marking technique, Dev.
    Biol., 41 : 162-184.
Le Douarin N.M., Cochard P., Vincent M., Duband J.L., Tucker G.C.
    Teillet M.A. and Thiery J.P., 1984 a, Nuclear cytoplasmic
    and membrane markers to follow neural crest cell migration.
    A comparative study. In: "The role of extracellular matrix
    in development", R.L. Treslstad, ed., Alan R. Liss, New
    York, pp 373-398.
Le Douarin N.M., Teillet M.A. and Fontaine-Perus J., 1984 b,
    Chimaeras in the study of the peripheral nervous system.
    In: "Chimaeras in Developmental biology", N. Le Douarin and
    A. McLaren eds., Academic Press Inc., London, pp. 313-352.
Lesot H., Kühl U. and Von der Mark K., 1983, Isolation of a laminin
    binding protein from muscle cell membranes, EMBO J., 2 :
    861-865.
Liesi P., Dahl D. and Vaheri A., 1983, Laminin is produced by early
    rat astrocytes in primary rat culture., J. Cell Biol., 96 :
    920-924.
Liesi P., Kaakkola S., Dahl D. and Vaheri A., 1984, Laminin is
    induced in astrocytes of adult brain by injury, EMBO J., 3 :
    683-686.
Lindner L., Rathjen F.G. and Schachner M., 1983, L1 mono and
    polyclonal antibodies modified cell migration in early post
    natal mouse cerebellum, Nature, 305 : 427-430.
Löfberg J., Ahlfors K., Fällstrom C., 1980, Neural crest cell
    migration in relation to extracellular matrix organization
    in the embryonic axolotl trunk, Dev. Biol., 75 : 148-167.
Löhler J., Timpl R. and Jaenish R., 1984, Lethal mutation of mouse
    collagen. I gene causes rupture of blood vessels and is
    associated with erythropoietic and mesenchymal cell death at
    day 12 of gestation, Cell, 38 : 597-607.
Mayer B.W., Hay E.D. and Hynes R.O., 1981, Immunocytochemical
    localization of fibronectin in embryonic chick trunk and
    area vasculosa, Dev. Biol., 82 : 267-286.
Newgreen D.F., Ritterman M. and Peters E.A., 1979, Morphology and
    behaviour of neural crest cells of chick embryo in vitro,
    Cell Tissue Res., 203 : 115-140.
Newgreen D.F. and Gibbins I.L., 1982, Factors controlling the time
    of onset of the migration of neural crest cells in the fowl
    embryo, Cell Tissue Res., 224 : 145-160.
Newgreen D.F. and Thiery J.P., 1980, Fibronectin in early avian
    embryos : synthesis and distribution along the migration

pathways of neural crest cells, Cell Tissue Res., 211 :
269-291.

Newgreen D.F., Gibbins I.L., Sauter J., Wallenfels B. and Wültz,
1982, Ultrastructural and tissue culture studies on the role
of fibronectin, collagen and glycosaminoglycans in the
migration of neural crest cells in the fowl embryo, Cell
Tissue Res., 221 : 521-549.

Nichols D.M., 1981, Neural crest formation in the head of the mouse
embryo as observed using a new histological technique, J.
Embryol. exp. Morph., 64 : 105-120.

Noden D.M., 1975, An analysis of the migratory behaviour of avian
cephalic neural crest cells, Dev. Biol., 42 : 106-130.

Noden D.M., 1978, The control of avian cephalic crest cell
cytodifferentiation, Dev. Biol., 67 : 296-329.

Ott U., Odermatt E., Engel J., Furthmayer H. and Timpl R., 1982,
Protease resistance and conformation of laminin, Eur. J.
Biochem., 123 : 63-72.

Petersen T.E., Thogersen H.C., Skortengaard K., Vibe-Pedersen K.,
Sahl P., Sottrup-Jensen L. and Magnusson S., 1983, Partial
primary structure of bovine plasma fibronectin : three types
of internal homology, Proc. Natl. Acad. Sci. USA, 80 :
137-141.

Perkins M.E., Ji T.H. and Hynes R.O., 1979, Cross-linking of
fibronectin to sulfated proteoglycans at the cell surface,
Cell, 16 : 941-952.

Pierschbacher M.D. and Ruoslahti E., 1984, Cell attachment activity
of fibronectin can be duplicated by small synthetic
fragments of the molecule, Nature, 309 : 30-33.

Pintar J.E., 1978, Distribution and synthesis of glycosaminoglycans
during quail neural crest morphogenesis, Dev. Biol., 67 :
444-464.

Poole A.R., Pidoux I., Reiner A., Choi H. and Rosenberg L.C., 1984,
Association of an extracellular protein (chondrocalcin) with
the calcification of cartilage in endochondral bone
formation, J. Cell Biol., 98 : 54-65.

Pratt R.M., Larsen M.A., Johnston M.C., 1975, Migration of cranial
neural crest cells in a cell-free hyaluronic rich matrix,
Dev. Biol., 44 : 298-305.

Rao N.C., Margulies I.M.L., Goldfarb R.H., Madri J.A., Woodley D.T.
and Liotta L.A., 1982, Differential proteolytic
susceptibility of laminin alpha and beta subunits, Arch.
Biochem. Biophys., 219 : 65-70.

Rao L.A., Barsky S.H., Terranova V.P. and Liotta L.A., 1983,
Isolation of a tumor cell laminin receptor, Biochem.
Biophys. Res. Commun., 111 : 804-808.

Revel J.P. and Brown S.S., 1975, Cell junctions in development with
particular reference to the neural tube. Cold Spring Harbor
Symp. Q. Biol., 40 : 443-455.

Rogers S.L., Letrourneau P.C., Palm S.L., McCarthy J. and Furcht
L.T., 1983,Neurite extension by peripheral and central

nervous system neurons in response to substratum-bound
    fibronectin and laminin, Dev. Biol., 98 : 212-220.
Rothbard J.B., Brackenbury R.B., Cunningham B.A. and Edelman G.M.,
    1982, Differences in the carbohydrate structures of neural
    cell-adhesion molecules from adult and embryonic chicken
    brains, J. Biol. Chem., 257 : 11064-11069.
Rothman T., Gershon M.D. and Holtzer M., 1978, The relationship of
    cell division to the acquisition of adrenergic
    characteristics by developing sympathetic ganglion cell
    precursors, Dev. Biol., 65 : 322-341.
Rovasio R.A, Delouvée A., Yamada K.M., Timpl R. and Thiery J.P.,
    1983, Neural crest cell migration : Requirement for
    exogenous fibronectin and high cell density, J. Cell Biol.,
    96 : 462-473.
Rubin K., Höök M., Obrink B. and Timpl R., 1981, Substrate adhesion
    of rat hepatocytes : mechanisms of attachment to collagen
    substrates, Cell, 24 : 463-470.
Ruoslahti E. and Vaheri A., 1975, Interaction of soluble fibroblast
    antigen with fibrinogen and fibrin. Identify with cold
    insoluble globulin of human plasma, J. exp. Med., 141 :
    497-501.
Rutishauser U. and Edelman G.M., 1980, Effects of fasciculation on
    the outgrowth of neurites from spinal ganglia in culture, J.
    cell Biol., 87 : 370-378.
Rutishauser U., Gall W.E. and Edelman G.M., 1978, Adhesion among
    neural cells of the chick embryo IV. Role of the cell
    surface molecule CAM in the formation of neurite bundles in
    cultures of spinal ganglia, J. Cell Biol., 79 : 382-393.
Schwarzbauer J.E., Tamkun J.W., Lemischka I.R. and Hynes R.O.,
    1983, Three different fibronectin mRNAs arise by alternative
    splicing within the coding region, Cell, 35 : 421-431.
Sugrue S.P. and Hay E.D., 1981, Response of basal epithelial cell
    surface and cytoskeleton to solubilized extracellular matrix
    molecules, J. Cell Biol., 91 : 45-54.
Sugrue S.P. and Hay E.D., 1982, Interaction of embryonic corneal
    epithelium with exogenous collagen, laminin and fibronectin
    : role of endogenous protein synthesis, Dev. Biol., 92 :
    97-106.
Tamkun J.W. and Hynes R.O., 1982, Plasma fibronectin is synthesized
    and secreted by hepatocytes, J. Biol. Chem., 258 :
    4641-4647.
Termine J.D., Kleinman H.K., Whitson S.W., Conn K.M., McGarvey M.L.
    and Martin G.R., 1981, Osteonectin, a bone specific protein
    linking mineral to collagen, Cell, 26 : 99-105.
Terranova V.P., Rohrbach D.H. and Martin G.R., 1980, Role of
    laminin in the attachment of PAM 212 (epithelial) cells to
    basement membrane collagen, Cell, 22 : 719-726.
Terranova V.P., Rao C.N., Kalebic T., Margulies I.M. and Liotta
    L.A., 1983, Laminin receptor on human breast carcinoma
    cells, Proc. Natl. Acad. Sci. USA, 80 : 444-448.

Thiery J.P, Brackenbury R., Rutishauser U. and Edelman G.M., 1977,
    Adhesion among neural cells of the chick embryo II
    Purification and characterization of a cell adhesion
    molecule from neural retina, J. Biol. Chem., 252 :
    6841-6845.
Thiery J.P., 1984, Mechanisms of cell migration in the vertebrate
    embryo, Cell Diff., 15 : 1-15.
Thiery J.P., Duband J.L. and Delouvée A., 1982 a, Pathways and
    mechanism of avian trunk neural crest cell migration and
    localization, Dev. Biol., 93 : 324-343.
Thiery J.P., Duband J.L., Rutishauser U. and Edelman G.M., 1982 b,
    Cell adhesion molecules in early chicken embryogenesis,
    Proc. Natl. Acad. Sci. USA, 79 : 6737-6741.
Thiery J.P:, Delouvée A., Gallin W., Cunningham B.A. and Edelman
    G.M., 1984 a, Ontogenetic expression of cell adhesion
    molecules : L-CAM is found in epithelia derived from the
    three primary germ layers, Dev. Biol., 102 : 61-78.
Thiery J.P., Delouvée A., Grumet M. and Edelman G.M., 1984 b,
    Initial appearance and regional distribution of the
    neuron-glia cell adhesion molecule (Ng-CAM) in the chick
    embryo, J. Cell Biol., in press.
Timpl R., Rohde H., Robey P.G., Rennard S.I., Foidart J.M. and
    Martin G.R., 1979, Laminin, a glycoprotein from basement
    membranes, J. Biol. Chem., 259 : 9933-9937.
Toole B.P., 1981, Glycosaminoglycans in morphogenesis. In: "Cell
    Biology of extracellular matrix", E.D. Hay ed., Plenum
    Press, PP 259-294.
Tosney K.W., 1978, The early migration of neural crest cells in the
    trunk region of the avian embryo. An electron microscopic
    study., Dev. Biol., 62 : 317-333.
Tosney K.W., 1982, The segregation and early migration of cranial
    neural crest cells in the chick embryo, Dev. Biol., 16 :
    78-106.
Tucker G.C., Aoyama H., Lipinski M., Tursz T. and Thiery J.P.,
    1984, Identical reactivity of monoclonal antibodies HNK-1
    and NC-1 : Conservation in vertebrates on cells derived from
    the neural primordium and some leukocytes, Cell Differ., 14
    : 223-230.
Tucker R.P. and Erickson C.A., 1984, Morphoplogy and behaviour of
    quail neural crest cells in artificial three-dimensional
    extracellular matrices, Dev. Biol., 104 : 390-405.
Twitty V.C. and Niu N.C., 1954, The motivation of cell migration
    studied by isolation of embryonic pigment cells singly and
    in small groups in vitro, J. Exp. Zool., 125 : 541-574.
Vaheri A. and Mosher D.F., 1978, High molecular weight cell surface
    associated glycoprotein (fibronectin) lost in malignant
    transformation, Biochem. Biophys. Acta, 516 : 1-25.
Valinsky J. and Le Douarin N.M., 1984, Production of plasminogen
    activator by migrating cephalic neural crest cells, in
    press.

Van der Water L., Schroeder S., Crenshaw B. and Hynes R.O., 1981,
     Phagocytosis of gelatin-latex particles by a murine
     macrophage line is dependant on fibronectin and heparin, J.
     Cell Biol., 90 : 32-39.
Viidik A. and Vuust J., 1980, Biology of collagen, Academic Press,
     London.
Vincent M., Duband J.L. and Thiery J.P., 1983, A cell surface
     determinant expressed early on migrating avian neural crest
     cells, Dev. Brain Res., 9 : 235-238.
Vincent M. and Thiery J.P., 1984, A cell surface marker for neural
     crest and placodal cells : Further evolution in peripheral
     and central nervous system, Dev. Biol., 103 : 468-481.
Weiss P., 1945, Experiments on cell and axon orientation in vitro :
     The role of colloidal exudates in tissue organization , J.
     Exp. Zool., 100 : 353-386.
Weston J.A., 1963, A radioautographic analysis of the migration and
     localization of trunk crest cells in the chick, Dev. Biol.,
     6 : 279-310.
Wylie C.C. and Heasman J., 1982, Effects of the substratum on the
     migration of primordial germ cells, Phil. Trans. R. Soc.
     Lond., B 299 : 177-183.
Yamada K.M., 1983, Cell surface interactions with extracellular
     materials, Ann. Rev. Biochem., 52 : 761-799.
Yamada K.M. and Kennedy D.W., 1984, Dualistic nature of adhesive
     protein function : Fibronectin and its biologically active
     peptide fragments can auto inhibit fibronectin function, J.
     Cell Biol., 99 : 29-36.
Yoshida-Noro C., Susuki N. and Takeichi M., 1984, Molecular nature
     of th calcium dependent cell-cell adhesion system in mouse
     teratocarcinoma and embryonic cells studied with a
     monoclonal antibody, Dev. Biol., 101 : 19-27.

DEVELOPMENT OF THE PERIPHERAL NERVOUS SYSTEM FROM THE NEURAL CREST:

ASPECTS OF CELL LINE SEGREGATION AND DIFFERENTIATION

Julian Smith and Nicole M. Le Douarin

Institut d'Embryologie du CNRS
et du Collège de France
49 bis, Avenue de la Belle Gabrielle
94130 Nogent-sur-Marne, France

The neural crest, which originates from the lateral ridges of the neural primordium in the Vertebrate embryo, constitutes a useful model for the study of a number of important developmental processes, including cell migration and differentiation. All the peripheral nervous system, apart from certain cranial sensory ganglia, is derived from the neural crest, and much interest currently centres on the mechanisms whereby a wide variety of differentiated cell types emerges from a structure that is, by classical cytological criteria, apparently homogenous.

One of the pivotal issues concerns the extent to which the developmental potentialities of crest cells are restricted when the latter leave the neural axis, for this problem encompasses the notions of predetermination, pluripotentiality and plasticity and bears upon the mechanisms of the differentiation processes themselves. Some of the experimental investigations of these questions are relevant to the theme of this volume in that they have provided strong evidence for the interaction of neural crest cells with their embryonic environment, even if the molecular basis for this phenomenon remains obscure at the present time.

The bird embryo possesses obvious practical advantages for examining ontogenetic events, and much of what is known of the initial steps of peripheral nervous system development is the result of in vivo and in vitro experiments with avian species. By means of a cell marking technique that enables quail and chick cells to be identified with ease in interspecific chimaeras, the normal and potential fates of neural crest cells at various levels of the embryonic neural axis have been elucidated. The combined results of homo- and heterotopic transplantation of neural primordium have revealed that the developmental potentiality of neural crest at all axial levels is significantly greater than that actually expressed during normal ontogeny, thus indicating an important role for the embryonic microenvironment during the early phases of nervous system development.

MULTIPOTENTIALITY   OF   THE   NEURAL   CREST   REVEALED   BY   IN   VIVO
EXPERIMENTS

     The introduction of the quail/chick chimaera system (Le Douarin, 1969, 1976, 1980) paved the way to the establishment of a fate map of the neural crest with respect to its ability to develop into the different cell types composing the peripheral nervous system. The construction of "neural tube chimaeras", in which sequential regions of the neural primordium are replaced by the equivalent portion from a quail embryo at the same stage of development, revealed a marked regionalisation of the neuraxis with respect to the origin of peripheral ganglion cells. For example, neural crest cells from the vagal and lumbosacral levels were shown to invade the splanchnopleure to form the myenteric plexuses. In contrast, ganglion cell precursors from the trunk level were found not to penetrate the dorsal mesentery ; instead, they remain restricted to the dorsal mesenchyme, where they give rise to the sympathetic trunks, aortic and adrenal plexuses and adrenomedullary cords.

     A second series of experiments showed that axial displacement of the neural primordium did not lead to phenotypic anomalies in the developing peripheral nervous system. Thus, heterotopic neural tube grafts resulted in apparently normal embryogenesis in which the neural crest derivatives differentiated into cholinergic, adrenergic or peptidergic cells according to their final location after migration rather than to their initial level of origin (Le Douarin and Teillet, 1974 ; Le Douarin et al., 1975 ; Fontaine-Pérus et al., 1982). It was concluded that the neural crest can, at all the levels considered, provide the entire spectrum of peripheral ganglia and adrenomedullary cells ; the phenotypic expression of crest cells from any given region of the neural primordium is thus not immutably fixed before they leave the summit of the neural tube, but is largely dependent on external signals received during or just after their migration phase. This conclusion, of course, concerns the neural crest as a population ; its applicability to individual crest cells is becoming increasingly questionable.

EVIDENCE FROM IN VIVO STUDIES   FOR THE SEGREGATION OF SENSORY   AND
AUTONOMIC CELL LINES IN THE NEURAL CREST

That the premigratory neural crest contains predetermined cell
lineages is suggested by the results of the transplantation, not of
neural primordium, but of embryonic peripheral ganglia. These
experiments were prompted by the wish to determine whether the
multipotentiality already demonstrated for neural crest was
retained in any of its derivatives and, if so, for how long.
Various developing quail peripheral ganglia of different ages were
thus transplanted into the trunk region of 2-day chick embryo
hosts, and the behaviour and fate of the grafted ganglion cells
were examined at various times thereafter (Le Douarin et al., 1978,
1979 ; Le Lièvre et al., 1980). The cells derived from the
different types of donor ganglia displayed similar patterns of
localisation in the host with one striking difference : whereas
cells from grafted dorsal root ganglia were able to colonise the
developing sensory ganglia of the host embryo, transplanted
sympathetic or parasympathetic ganglion cells never contributed to
these structures, either as neurons or satellite cells. In
contrast, cells derived from donor autonomic and sensory ganglia
alike were to be found in the host sympathetic ganglia and plexuses
and adrenal medulla, where they displayed adrenergic properties.

The results of these experiments demonstrate that all types of
embryonic peripheral ganglia contain cells that are capable of
expressing an autonomic phenotype. In contrast, autonomic ganglion
cells - unlike the neural crest from which they originate - do not
possess sensory potentialities, thus indicating an early,
irreversible segregation of the autonomic and sensory lineages.

Several lines of evidence lead to the conclusion that the
phenotypic changes described above concern exclusively cells
derived from the non-neuronal population of "back-transplanted"
ganglia. Firstly, a strict positive correlation was observed
between the ability of grafted dorsal root ganglia to provide cells
able to differentiate into sensory neurons and the presence in the
donor ganglion of mitotically active neuronal precursors (Schweizer
et al., 1983). Secondly, by grafting cranial sensory ganglia in
which either the satellite or neuronal cell population was
selectively labelled by the quail nuclear marker, it was possible
to show that only the former contributed cells to the peripheral
ganglia of the host (Ayer-Le Lièvre and Le Douarin, 1982). Finally,
ciliary ganglion neurons, labelled by ^{3}H-thymidine injection during
their last mitotic cycle, were found to die very shortly after
back-transplantation of the ganglion into a younger host (Dupin,
1984).

A model for the development of sensory and autonomic neurons and glia has been constructed on the basis of the results described above (Le Douarin et al., 1979 ; Le Lièvre et al., 1980 ; Le Douarin, 1984). According to this scheme, separate pools of precursors for sensory and autonomic cell lines exist in the neural crest but sensory neuron precursors can only express the "sensory" phenotype in close proximity to the central nervous system. If they migrate to more distal positions they die (or lose their sensory potentialities). Autonomic neuron precursors, on the other hand, do not have the same survival requirements and can be found in numerous peripheral ganglia, including those in which the autonomic phenotype is not normally expressed. However, if these cells find themselves in an appropriate environment, such as that provided by certain regions of a young embryo host (after back-transplantation), then their latent properties can be disclosed. The factors that repress the development of these cells during normal embryogenesis remain to be discovered. On the other hand, rather more is known of the parts played by extrinsic agents in the differentiation of autonomic neurons.

ENVIRONMENTAL INFLUENCES ON THE DEVELOPMENT OF AUTONOMIC NEURON PRECURSORS

Although a certain category of neural crest progeny apparently possesses an "autonomic" imprint at a very early stage, a wide range of developmental possibilities is still available to cells initially segregated in this way, and terminally differentiated autonomic ganglion and paraganglion cells display a considerable degree of biochemical and morphological diversity. Much evidence exists implicating environmental agents in the establishment of chemical heterogeneity in such cells (see recent reviews by Le Douarin and Smith, 1983 ; Smith and Le Douarin, 1984), and this section will consider some of the factors that have been shown to affect specifically the development of neurotransmitter-associated phenotypes. Particular emphasis will be placed on recent results concerning the differentiation of adrenergic cells in cultured avian neural crest.

a. Differentiation of Sympathoblast Precursors in vitro

In the avian embryo in vivo, sympathetic ganglion cell precursors leave the cervico-dorsal region of the neural primordium and the majority follow a fibronectin-rich pathway corresponding to the space between consecutive somites (Thiery et al., 1982). Migrating rapidly down this channel, the cells become distributed along the aorta, where they condense to form the sympathetic trunks and aortic plexuses. Catecholamine (CA)-specific fluorescence, absent from crest cells before or during their migration (Allan and Newgreen, 1977), can first be detected at 3.5 days of incubation, just after the primary sympathetic chains have formed (Enemar et al., 1965 ; Allan and Newgreen, 1977 ; our own unpublished observations).

In culture, the adrenergic differentiation process can be reproduced by culturing migrating neural crest cells originating from the trunk level of a 3-day quail embryo in a medium supplemented with 15% foetal calf serum and 2% 9-day chick embryo extract (Fauquet et al., 1981). Crest cells migrating in this region of the embryo can only be isolated in intimate association with the sclerotomal component of the somite, which is, in fact, the normal microenvironment for sympathetic ganglion development in vivo. When this combination of tissues is placed in culture, CA synthesis from ^{3}H-tyrosine can be detected within 24 hours, i.e., at the time it would normally have occurred in the developing embryo. After a few days, ganglion-like structures and plexuses can be seen. Many of the cells composing them appear to be adrenergic neurons, as indicated by their morphological features, their immunoreactivity for tyrosine hydroxylase (the rate-limiting enzyme in CA biosynthesis), their cytofluorescence after glyoxylic acid treatment and their content of dense-core vesicles visible by electron microscopy after permanganate fixation.

Neural cells developing in culture under these conditions also display another characteristic feature of sympathetic ganglia, i.e., somatostatin-like immunoreactivity (SLI). Here, again, the appearance of the neuropeptide in neural crest/sclerotome cultures has been found to respect the in vivo chronology (Garcia-Arraras et al., 1984). Immunoreactivity was demonstrated in cells, isolated and in groups of various sizes, very similar to those previously shown to be adrenergic. In fact, subsequent experiments showed that the great majority of CA- or tyrosine hydroxylase-positive cells in these cultures also possess SLI and vice versa (Garcia-Arraras et al., manuscript in preparation). Similar coexpression of adrenergic and peptidergic properties - already well documented in mammals in vivo (Lundberg et al., 1982) - has recently been described in another type of neural crest culture (Maxwell et al., 1984).

b. Effects of Glucocorticoid Hormones on the Differentiation of Neural Crest in vitro.

Although the appearance of the adrenergic (and peptidergic) cells in neural crest/sclerotome cultures is "spontaneous" in vitro, the subsequent expression of the phenotype can be modulated significantly by the addition of extrinsic factors to the culture medium. We have recently conducted a study designed to examine the effects of glucocorticoid hormones on this process (Smith and Fauquet, 1984). Our investigation was prompted by previous observations that these hormones can increase the content of CA and the activity of CA-synthesising enzymes in neural crest derivatives in vivo and in vitro (e.g., Eränkö and Eränkö, 1972 ; Costa et al.,

1974 ; Otten and Thoenen, 1976). Perhaps even more relevant to the
initial stages of autonomic neuron development, they had also been
shown to stimulate selectively the expression of catecholaminergic
properties in cell populations that display both adrenergic and
cholinergic features (Fukada, 1980 ; McLennan et al., 1980).

We found the system comprising migrating neural crest plus
sclerotomal mesenchyme to be responsive to glucocorticoids : when
hydrocortisone or corticosterone (10^{-6}M) was added at the beginning
of the culture period, a 2-to 3-fold stimulation of CA synthesis
and accumulation could be measured radiochemically 7 days later.
Detailed analysis of the various CAs produced revealed a
preferential stimulation of dopamine production (suggesting an
initial effect of hormone on the activity of tyrosine hydroxylase),
although noradrenaline remained the major labelled product. Only a
very small quantity of adrenaline was produced.

The minimal concentration of a variety of natural and
synthetic glucocorticoids that had a measurable effect on CA
production was 10^{-7}M, and the maximal stimulation was attained at
10^{-6}M. Of the other steroid hormones tested, only progesterone
produced an effect, but did so at a concentration 10-fold higher
than the glucocorticoids.

It should be emphasised that none of the hormones studied
produced any detectable effect either on acetylcholine synthesis
and accumulation or on choline acetyltransferase activity -
cholinergic properties that are always associated with these neural
crest cultures (Fauquet et al., 1981). Consequently, the action of
glucocorticoids is selective for the adrenergic phenotype in this
system.

When the cultures were examined at the ultrastructural level,
the effects of hormone treatment were immediately apparent. After
exposure to hydrocortisone, the number and, especially, the size of
the CA storage vesicles in adrenergic cells were strikingly
increased. This phenomenon seemed to affect all of the
granule-containing cells : after hormone treatment, in contrast to
control conditions, no such cell was ever found to contain
exclusively small-sized (up to 60 nm diameter) vesicles. Indeed,
some cells possessed virtually no small-diameter granules after
exposure to glucocorticoid. These results, confirming the
radiochemical data, clearly indicate that corticosteroids increase
the intracellular pool of CA in neural crest derivatives
differentiating in the presence of sclerotomal cells.

It therefore appeared pertinent to determine whether the
adrenergic differentiation occurring in these cultures in the
absence of exogenous glucocorticoids was, in fact, caused by the
presence of such hormones in the complex components of the medium.

Accordingly, we cultured these embryonic rudiments in medium containing serum and embryo extract from which the endogenous steroid had been removed by treatment with charcoal : neither adrenergic nor cholinergic differentiation was significantly modified under such conditions. Consequently, it seems that the triggering of adrenergic differentiation is independent of glucocorticoids. In order to confirm this supposition, we cultured premigratory neural crest, excised microsurgically at an earlier stage of embryonic development, i.e., at 2 days of incubation. Cultured in the absence of heterologous tissues, these neural crest cells underwent negligible neuronal differentiation and virtually no CA synthesis could be detected even after 7 days in culture. The addition of hydrocortisone or corticosterone at the start of the culture did not notably improve this synthetic ability. Similarly, no cells of neuronal appearance were seen and no CA-specific fluorescence was ever observed after glyoxylic acid fixation.

Taken as a whole, these results showed that adrenergic differentiation of neural crest cells that had migrated to the sclerotomal moiety of the somite could be stimulated in vitro by exogenous glucocorticoid, whereas premigratory crest was unresponsive. This could reflect the maturation of crest cells during the course of their migration and/or the presence of a favorable environment (i.e., the sclerotomal mesenchyme). It was thus interesting to determine whether premigratory crest would also react to glucocorticoid if cultured in a propitious milieu. We had previously found (Fauquet et al., 1981) that, when the tips of neural folds from the trunk level of 2-day embryos were grown with somitic mesenchyme and notochord from embryos of the same age, they gave rise to cells that produced small but measurable quantities of CA (although no fluorescent cells could be detected by appropriate cytochemical tests).

When glucocorticoids were added to such cocultures, a 10- to 12-fold stimulation of CA production could be measured 7 days later. This selective enhancement of adrenergic differentiation (as before, acetylcholine synthesis was not modified by hormone treatment) was found to be accompanied by the development of CA stores that could be revealed after reaction with glyoxylic acid : numerous fluorescent cell bodies and processes were apparent in cultures treated in this way.

Glucocorticoids can thus act on premigratory neural crest cells, but only in the presence of heterologous tissue. This suggests that the latter may, in fact, be the initial target, although an alternative explanation is that crest cell progeny become receptive to hormone only in the presence of non-neural tissues. Irrespective of the mechanism of action of glucocorticoids in this system, our results indicate that neural crest derivatives are potentially receptive to corticosteroids at the moment they undergo gangliogenesis : i.e., when a number of important developmental processes are set in motion. The intervention of glucocorticoids in vivo at very early embryonic stages remains to be established, however. Furthermore, it is clear that these hormones cannot be considered as initiators of adrenergic differentiation ; their putative role more probably consists in modulating adrenergic expression in cells that have already been subjected to a triggering signal of another kind. The intra-embryonic origin and the nature of this signal has been the object of studies in several laboratories, as is mentioned below.

Finally, it is noteworthy that the conclusions reached concerning the effects of glucocorticoids on adrenergic differentiation also apply to SLI-containing cells : hormone treatment was shown to result in an approximately 2-fold increase in the neuropeptide content of the neuron-like cells developing in neural crest/sclerotome cultures (Garcia-Arraras et al., manuscript in preparation). SLI- and CA-related phenotypes can thus be regulated concomitantly. This constitutes one of the first demonstrations of the comodulation of a "classical" neurotransmitter and a neuropeptide present in the same cell. It will be of interest to determine to what degree coregulation occurs in cells producing other combinations of neuroactive substances.

c. The Role of Soluble Factors in the Initiation of Adrenergic Differentiation

The stimulatory effect of various young embryonic tissues (neural tube, somites and notochord) on the induction of the adrenergic phenotype in vivo is well documented (Cohen, 1972 ; Norr, 1973 ; Teillet et al., 1978 ; Teillet and Le Douarin, 1983). These studies have been reviewed recently (Le Douarin and Smith, 1983 ; Smith and Le Douarin, 1984) and will not be discussed in detail here. In fact, recent work on cultured neural crest shows that the requirement for heterologous tissue in order to trigger sympathoblast differentiation can be replaced by one or more soluble factors present in the young avian embryo. If a sufficiently high (10%-15%) concentration of chick or quail embryo extract is used to supplement the medium, extensive adrenergic differentiation can occur in cultures of neural crest grown in the virtual absence of other cell types (Cohen, 1977 ; Kahn et al., 1980 ; Maxwell et al., 1982 ; Howard et al., 1982 ; Fauquet et al., manuscript in preparation).

The cells that differentiate under these conditions are similar to those appearing in neural crest/sclerotome cultures and can easily be evidenced by radiochemical and cytochemical techniques. It is worth emphasising, however, that adrenergic cells appear in cultures of neural crest alone only after a lag period of 4-6 days. Furthermore, the development of adrenergic cells in

cultured   neural   crest/sclerotome   does   not   require   high
concentrations of embryo extract (Fauquet et al., 1981), suggesting
that the relevant triggering events have already occurred when the
explants are put into culture.

The young embryo, then, contains one or more active factors
which can elicit adrenergic differentiation in neural   crest
cultured _in vitro_ in the absence of heterologous tissue. The
biochemical nature of the agent(s) responsible and the mechanisms
involved remain to be elucidated.

CONCLUDING OBSERVATIONS

Some examples of the action of the environment on the
differentiation of neural crest-derived precursors of the
peripheral nervous system have been briefly described. External
factors could act on pluripotential neural crest progeny, or on
cells that are already predetermined on leaving the neural tube.
The _in vivo_ approaches presented in this chapter have led to the
conclusion that sensory and autonomic cell lines are already
segregated within the neural crest population. This hypothesis does
not deny the importance of extrinsic signals in the ultimate
manifestation of the respective phenotypes : their role would
consist in repressing or permitting their expression according to
the particular location of the peripheral ganglion cell precursors
in the developing embryo.

Within the broad pattern of predetermination existing in a
class of neural crest cells, other options, such as
neurotransmitter choice, may remain open longer. Here, again,
positional effects undoubtedly play an important role in the
emergence of the definitive neurotransmitter-related phenotype, as
is attested by the results of neural crest and peripheral ganglion
grafting studies.

Concerning the appearance of the catecholaminergic phenotype
in sympatho-adrenal precursors, evidence has been obtained that
this phenomenon may be mediated by as yet uncharacterised soluble
factors present in a total extract of young avian embryos. In
contrast, glucocorticoids most probably do not play a part in
neurotransmitter choice in the system described here ; on the other
hand, they may, by modulating the activity of CA-synthesising
enzymes and the rate of accumulation of intracellular CA, aid in
the   establishment   and   maintenance   of heterogeneity in the
adrenergic cell population itself (Landis and Patterson, 1981).

It is not yet clear whether the factors involved in initiating
adrenergic expression act in a permissive or an instructive manner.
However, results obtained in cultures of post-mitotic superior

cervical   neurons   (Patterson,   1978)   and   examples   of   apparent
chemical plasticity of sympathetic neurons in vivo (Landis and
Keefe, 1983) suggest that the definitive choice of neurotransmitter
may be a relatively late event in autonomic neuron development.   By
extrapolation,   it is not unreasonable to suggest that the initial
acquisition of neurotransmitter-related properties may be   dictated
to   a   cell,   already endowed with an "autonomic" label, by
appropriate external cues encountered before or during ganglion
formation.

ACKNOWLEDGMENTS

     Financial   support has been provided by the Centre National de
la Recherche Scientifique, Research Grant R01 DE0 42 57 04 CBY from
the National Institutes of Health and by Basic Research Grant 1-866
from March of Dimes Birth Defects Foundation.

REFERENCES

Allan I.J. and Newgreen D.F., 1977, Catecholamine accumulation in
          neural crest cells and the primary sympathetic chain, Am. J.
          Anat., 149 : 413-421.
Ayer-Le Lièvre C.S. and Le Douarin N.M., 1982, The early
          development of cranial sensory ganglia and the
          potentialities of their component cells studies in
          quail-chick chimaeras, Dev. Biol., 94 : 291-310.
Cohen A.M., 1972, Factors directing the expression of sympathetic
          nerve traits in cells of neural crest origin, J. Exp. Zool.,
          179 : 167-182.
Cohen A.M., 1977, Independent expression of the adrenergic
          phenotype by neural crest cells in vitro, Pro. Natl. Acad.
          Sci. USA, 74 : 2899-2903.
Costa M., Eränkö O. and Eränkö L., 1974, Hydrocortisone-induced
          increase in the histochemically demonstrable catecholamine
          content of sympathetic neurons of the newborn rat, Brain
          res., 67 : 457-466.
Dupin E., 1984, Cell division in the ciliary ganglion of quail
          embryos in situ and after back-transplantation into the
          neural crest migration pathways of chick embryos, Dev.
          Biol., 105 : 288-299.
Enemar A., Falck B. and Hakanson R., 1965, Observation on the
          appearance of norepinephrine in the sympathetic nervous
          system of the chick embryo, Dev. Biol., 11 : 268-283.
Eränkö L. and Eränkö O., 1972, Effect of hydrocortisone on
          histochemically demonstrable catecholamines in the
          sympathetic ganglia and extra-adrenal chromaffin tissue of
          the rat, Acta Physiol. Scand., 84 : 125-133.
Fauquet M., Smith J., Ziller C. and Le Douarin N.M., 1981,

Differentiation of autonomic neuron precursors in vitro :
cholinergic and adrenergic traits in cultured neural crest
cells, J. Neurosci., 1 : 478-492.

Fontaine-Pérus J., Chanconie M. and Le Douarin N.M., 1982,
Differentiation of peptidergic neurons in quail-chick
chimaeric embryos, Cell Different, 11 : 183-193.

Fukuda K., 1980, Hormonal control of neurotransmitter choice in
sympathetic neurone cultures, Nature, 287 : 553-555.

Garcia-Arraras J., Chanconie M. and Fontaine-Pétrus J., 1984, In
vivo and in vitro development of somatostatin-like
immunoreactivity in the peripheral nervous system of quail
embryos, J. Neurosci., 4 : 1549-1558.

Howard M.J., Bronner-Fraser M. and Tomosky-Sykes T., 1982,
Adrenergic differentiation of neural crest cells in vitro
without the neural tube, Soc. Neurosci. Abstr., 8 : 257.

Kahn C.R., Coyle J.T. and Cohen A.M., 1980, Head and trunk neural
crest in vitro : autonomic neuron differentiation, Dev.
Biol., 77 : 340-348.

Landis S.C. and Keefe D., 1983, Evidence for neurotransmitter
plasticity in vivo : developmental changes in properties of
cholinergic sympathetic neurons, Dev. Biol., 98 : 349-372.

Landis S.C. and Patterson P.H., 1981, Neural crest cell lineages,
Trends in Neurosci., 4 : 172-175.

Le Douarin N., 1969, Particularités du noyau interphasique chez la
caille japonaise (Coturnix coturnix japonica). Utilisation
de ces particularités comme "marquage biologique" dans les
recherches sur les interactions tissulaires et les
migrations cellulaires au cours de l'ontogenèse, Bull. Biol.
Fr. Belg., 103 : 435-452.

Le Douarin N., 1976, Cell migration in early vertebrate development
studied in interspecific chimaeras. In: "Embryogenesis in
Mammals", Ciba Foundation Symposium, Elsevier-Excerpta
Medica-North-Holland, Amsterdam, pp. 71-101.

Le Douarin N.M., 1980, The ontogeny of the neural crest in avian
embryo chimaeras, Nature, 286 : 663-669.

Le Douarin N.M., 1984, A model for cell line divergence in the
ontogeny of the peripheral nervous system, In: "Cellular and
molecular biology of neuronal development", I. Black ed.,
New York, Plenum, pp. 3-28.

Le Douarin N.M. and Smith J., 1983, Differentiation of avian
autonomic ganglia, In: "Autonomic ganglia", ed. Elfvin L.G.,
Wiley, 427-452.

Le Douarin N.M. and Teillet M.A., 1974, Experimental analysis of
the migration and differentiation of neuroblasts of the
autonomic nervous system and of neurectodermal mesenchymal
derivatives, using a biological cell marking technique, Dev.
Biol., 41 : 162-184.

Le Douarin N.M., Le Lièvre C.S., Schweizer G. and Ziller C.M.,
1979, An analysis of cell line segregation in the neural
crest, In: "Cell lineage, stem cells and cell

determination", INSERM Symp., n° 10, Ed. N. Le Douarin, Elsevier/North-Holland Biomedical Press.

Le Douarin N.M., Renaud D., Teillet M.A. and Le Douarin G.H., 1975, Cholinergic differentiation of presumptive adrenergic neuroblasts in interspecific chimaeras after heterotopic transplantations, Proc. Natl. Acad. Sci. USA, 72 : 728-732.

Le Douarin N.M., Teillet M.A., Ziller C. and Smith J., 1978, Adrenergic differentiation of cells of the cholinergic ciliary and Remak ganglia in avian embryo after in vivo transplantation, Proc. Natl. Acad. Sci. USA, 75 : 2030-2034.

Le Lièvre C.S., Schweizer G.G., Ziller C. and Le Douarin N.M., 1980, Restrictions of developmental capabilities in neural crest cell derivatives as tested by in vivo transplantation experiments, Dev. Biol., 77 : 362-378.

Lundberg J.M., Hökfelt T., Anggard A., Terenius L., Elde R., Markey K., Goldstein M. and Kimmel J., 1982, Organizational principles in the peripheral sympathetic nervous system : subdivision by coexisting peptides (somatostatin-, avian pancreatic polypeptide-, and vasoactive intestinal polypeptide-like immunoreactive materials), Proc. Natl. Acad. Sci. USA, 79 : 1303-1307.

Maxwell G.D., Sietz P.D. and Jean S., 1984, Somatostatin-like immunoreactivity is expressed in neural crest cultures, Dev. Biol., 101 : 357-366.

Maxwell G.D., Sietz P.D. and Rafford C.E., 1982, Synthesis and accumulation of putative neurotransmitters by cultured neural crest cells, J. Neurosci., 2 : 879-888.

McLennan I.S., Hill C.E. and Hendry I.A., 1980, Glucocorticoids modulate transmitter choice in developing superior cervical ganglion, Nature, 283 : 202-207.

Norr S.C., 1973, In vitro analysis of sympathetic neurons differentiation from chick neural crest, Dev. Biol., 34 : 16-38.

Otten U. and Thoenen H., 1976, Selective induction of tyrosine hydroxylase and dopamine β-hydroxylase in sympathetic ganglia in organ culture : role of glucocorticoids as modulators, Mol. Pharmacol., 12 : 353-361.

Patterson P.H., 1978, Enviromental determination of autonomic neurotransmitter functions, Ann. Rev. Neurosci., 1 : 1-17.

Schweizer G., Ayer-Le Lièvre C. and Le Douarin N.M., 1983, Restrictions of developmental capacities in the dorsal root ganglia during the course of development, Cell Differentiation, 13 : 191-200.

Smith J. and Fauquet M., 1984, Glucocorticoids stimulate adrenergic differentiation in cultures of migrating and premigratory neural crest, J. Neurosci., 4 : 2160-2172.

Smith J. and Le Douarin N.M., 1984, In vivo and in vitro studies on the development of the peripheral nervous system, In: "Organizing Principles of Neural Development", Ed. Sharma S.C., Plenum, 1-19.

Teillet M.A. and Le Douarin N.M., 1983, Consequences of neural tube
     and notochord excision on the development of the peripheral
     nervous system in the chick embryo, Dev. Biol., 98 :
     191-211.
Teillet M.A., Cochard P. and Le Douarin N.M., 1978, Relative roles
     of the mesenchymal tissues and of the complex neural
     tube-notochord on the expression of adrenergic metabolism in
     neural crest cells, Zoon., 6 : 115-122.
Thiery J.P., Duband J.L. and Delouvée A., 1982, Pathways and
     mechanisms of avian trunk neural crest cell migration and
     localization, Dev. Biol., 93 : 324-343.

NEURONAL MIGRATION IN THE HIPPOCAMPAL LAMINATION

DEFECT (Hld) MUTANT MOUSE

Richard S. Nowakowski

Department of Anatomy
University of Mississipi Medical Center
Jackson, Mississipi 39216, U.S.A.

INTRODUCTION

Movement of cells is an important feature of the development
of many tissues and organs in a variety of species (for reviews see
Trinkaus, 1976 ; Abercrombie, 1980). In some cases, cell movement
is the result of a passive displacement of the entire tissue
including the environment of a cell. In other cases, however, a
cell can be considered to migrate in that it actively participates
in its displacement and its movement can be considerable relative
to the constituents of its environment. This distinction is
important because it emphasizes two things. First, a migrating cell
generates the forces necessary for locomotion, and it interacts
with its environment in order to move through it. Second, the
displacement of cells by morphogenetic movements or the growth of a
cell without a concomitant displacement of the cell body, e.g., the
extension of an axon by a neuron, should not be considered to be
cell migration (Weiss, 1961 ; Carter, 1967 ; Trinkaus, 1976).
During the development of the vertebrate embryo there is a large
amount of cell migration both of single cells and populations of
cells from one part of the embryo to another. The four best
understood examples are : 1) the colonization of the developing
gonads by the primordial germ cells, 2) the development of the
hematopoietic organs, 3) the extensive migrations of derivatives of
the neural crest, and 4) the migration of young neurons in the
developing central nervous system.

The major feature shared by these four migrations is that a
population of cells moves a great distance from one part of the
embryo to another. How do cells manage to move such long distances
and what guides them ? Three basic guidance mechanisms have been

133

proposed : 1) chemotaxis, 2) contact guidance and 3) haptotaxis.
Chemotaxis is locomotion guided by a soluble, hence diffusible,
substance. The substance is produced in a particular location or
locations and a concentration gradient is produced by diffusion. A
locomoting cell is presumed to be sensitive enough to discern the
direction of the gradient and migrates towards (or away from) the
source of the substance. Contact guidance is locomotion along a
substrate (Weiss, 1961). In a nonhomogeneous environment the cells
are presumed to be selectively adhesive to particular constituents
of their milieu. Also, it needs to be emphasized that contact
guidance provides alignment but does not provide polarity for cell
movement (e.g., if a cell is oriented along a fiber, it could move
in either direction) but that other influences such as chemotaxis
or population pressures must do this. Haptotaxis is locomotion
along a gradient of adhesion (Carter, 1967). This hypothetical
mechanism is similar to chemotaxis except that the gradient is not
diffusible but bound to some fixed constituent of the environment
of the cell. In contrast to contact guidance, if the substrate is
oriented (e.g., a fiber) the adhesion gradient provides a signal
for polarizing the migration cell.

     The guidance mechanisms utilized and the nature of the
pathways followed by various migrating populations appear to be
different. The primordial germ cells migrate through the mesentery
(Heasman and Wylie, 1981). The migrating hematopoietic stem cells
enter the blood stream and are then passively carried to their
destinations (Le Douarin, 1978). Although both the primordial germ
cells and the hematopoietic stem cells are believed to use
chemotaxis for their proper targeting (Le Douarin, 1978, 1984 ; Ben
Slimane, 1983), they differ in that the germ cells colonize only
one target, whereas the hematopoietic stem cells colonize several
targets. The migration of the neural crest derivatives and of young
neurons in the CNS at first appears to be quite similar in that
both migrating populations begin their migration from a defined
region of cell proliferation, move along specific pathways, and
ultimately take up residence in particular locations. These two
migrations differ, however, in at least two ways. First, the neural
crest derivatives continue to proliferate after they begin their
migration, whereas in the CNS the migrating young neurons are, with
few exceptions, permanently postmitotic. Second, the neural crest
derivatives are guided in their migration by the presence of
fibronectin in the extracellular matrix (Rovasio et al., 1983 ;
Duband and Thiery, 1985), whereas in the CNS most migrating young
neurons are guided by an apposition to a radial glial fiber (Rakic,
1978, 1982). A further comparison of the migration of neural crest
derivatives and CNS neurons will be given later.

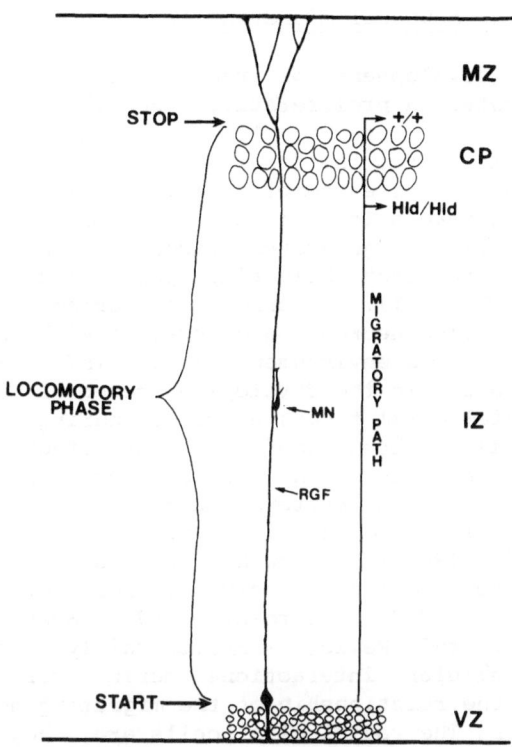

Fig. 1.   Schematic   diagram   of the migration   path   of young
           neurons in area CA3c of the developing   hippocampus.
           A migrating neuron (MN) goes through three "phases":
           1) it must leave the ventricular zone (VZ) and start
           its   migration,   2) it must propel itself across the
           intermediate   zone   (IZ)   and   through   the   cortical
           plate   (CP)   using   a   radial glial fiber (RGF) as a
           guide, and 3) it must recognize that it has   reached
           its   final position at the top of the cortical plate
           (CP) at the border with the marginal zone   (MZ)   and
           detach   itself from the radial glial fiber (RGF) and
           stop migrating. The   arrows   and   bracket   on   the
           left-hand   side of the diagram indicate the portions
           of the hemispheric   wall   in   which   each   of these
           phases   occur.   The   normal   stopping   point   of
           late-generated migrating neurons   in   the   +/+   is
           indicated   by   the   arrow on the top of the cortical
           plate. The   stopping   point · of   the   late-generated
           pyramidal cells in the Hld/Hld mouse is indicated by
           the arrow on   the   bottom   of   the   cortical   plate.
           Abbreviations : CP, cortical plate; IZ, intermediate
           zone; MN, migrating neuron; MZ, marginal zone;   RGF,
           radial glial fiber; VZ, ventricular zone.

General features of neuronal migration in the cerebral cortex

During the development of the mammalian cerebral cortex neurons are generated in proliferative zones that line the walls of the lateral ventricles (see Fig. 1). These proliferative zones are located some distance from the final positions of the neurons they produce. Consequently, after completing their last mitotic division young neurons must take a long migration across the intermediate zone and past previously generated neurons in the deep portions of the cortical plate to reach their final position at the top of the cortical plate (Fig. 1). In principle, during the course of its migration the migrating neuron could interact with any of a variety of cell types in its environment. Rakic (1971, 1972) discovered that migrating neurons in the developing cerebral and cerebellar cortices use radial glial fibers as guides during their migration. This finding has since been confirmed for other parts of the cerebral cortex and it is now widely accepted that during the development of the cerebral cortex young neurons are intimately apposed to radially aligned glial fibers that provide guidance (Rakic, 1971, 1978, 1982 ; Nowakowski and Rakic, 1979) and may provide the scaffolding for the columnar organization of the adult cortex (Rakic, 1978, 1982 ; Mountcastle,1978 ; Smart and McSherry, 1982 ; Eckenhoff and Rakic, 1984).Certainly, one of the most important intercellular interactions during cerebral cortical development is the relationship of the migrating neurons with the radial glial fiber. The radial glial cells are, however, not the only type of cell present in the environment of the migrating neurons. For example, migrating neurons could also interact with each other, with cells that continue to proliferate, with previously generated cells, or with axons arising from cell bodies that reside in distant parts of the brain. The process of neuronal migration in the cerebral cortex can be considered to have three separate phases (Fig. 1). First, a young neuron starts its migration. During this phase a cell in the proliferating population becomes a post-mitotic cell (i.e., a neuroblast becomes a young neuron), becomes apposed to a radial glial fiber and establishes an axis of polarity away from the ventricular surface.In the second, or locomotory phase, a young neuron propels itself along the surface of the radial glial cell, while maintaining its apposition to the radial fiber and its axis of polarity. Finally, in the third phase of its migration, a young neuron somehow "recognizes" that it has reached its final destination and stops its migration. During this phase the specific attachment of the migrating neuron to the radial glial fiber is lost and the young neuron can continue to differentiate. Also, as a result of this detachment the surface of the radial glial fiber is available for the guidance of subsequently generated neurons that migrate past those that were previously generated (Pinto-Lord et al., 1982).

Mutations that effect neuronal migration

At present it is unresolved whether the successful completion of a migration requires that a migrating neuron interact with the various constituents of its environment or if its relationship to the guiding radial glial fiber provides sufficient information during all three phases of the life history of the migrating neuron. The approach being used in this laboratory to try to answer these questions by exploiting the availability of mutations that effect the migration of neurons during the development of the cerebral cortex of the mouse. Currently there are five such mutations available (Nowakowski, 1985). Potentially, mutations could affect any portion of the migratory process, so initial investigations in this laboratory have been directed at mutations that affect the position of neurons in the mature nervous system. The rationale is that these mutations are likely to affect the "stopping" of young neurons either by acting on the normal signal for stopping or by disrupting the locomotory phase of migration. One mutation that appears to be particularly promising is the Hippocampal lamination defect (Hld). This autosomal dominant mutation (Nowakowski, 1984 a) produces an inversion in the laminar organization of the pyramidal cell layer of area CA3c of the hippocampus. In mice that are homozygous for Hld (i.e., in Hld/Hld mice) the late-generated neurons do not bypass the early-generated ones but instead take up a position below the early-generated neurons (Vaughn et al., 1977 ; Nowakowski, 1984 b). This is in contrast to the usual pattern in which the late-generated hippocampal pyramidal neurons are located superficial to the early-generated pyramidal cells. During normal development the young neurons are generated in proliferative zones that line the lateral ventricles (Nowakowski and Rakic, 1981) ; the newly produced neurons then migrate across a broad intermediate zone and past previously generated neurons to reach their final position at the top of the cortical plate (Angevine, 1965 ; Nowakowski and Rakic, 1979, 1981).

Anatomy of normal (+/+) and mutant (Hld/Hld) hippocampus

In the +/+ hippocampus the pyramidal cells of area CA3c are sandwiched between two bundles of mossy fibers, a suprapyramidal mossy fiber layer (sup in Fig. 2 A) and an infrapyramidal mossy fiber layer (inf in Fig. 2 A). Within the pyramidal cell layer of area CA3c of the +/+ mouse (Fig. 2 A) the early-generated neurons (i.e., those generated before embryonic day 14, E14) occupy the deep portions and the late-generated neurons (i.e., those generated on E15 or E16) occupy the superficial portions. The small arrow in the pyramidal cell layer in Fig. 2 A indicates this "inside-to-outside spatiotemporal gradient" in neuron position. (For a complete discussion of the concept of spatiotemporal gradients see Angevine, 1965.) In the pyramidal cell layer of area

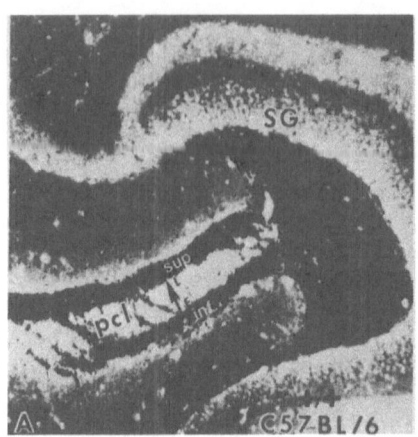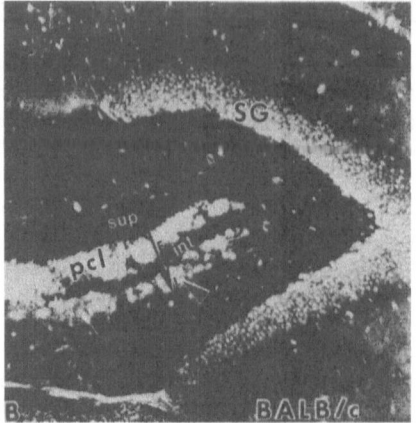

Fig. 2.   Photomicrographs of Timm's stain preparations of area
          CA3c and the adjacent dentate gyrus in the normal
          (+/+) mouse (A) and in the Hld mutant mouse (B).   In
          these preparations the axons of the granule cells
          (usually called mossy fibers) of the stratum
          granulosum of the dentate gyrus (SG) are stained
          black due to their high content of heavy metals.
          A. In area CA3c of the +/+ mouse, the mossy fibers
          divide   into two bundles such that the pyramidal cell
          layer (pcl) is sandwiched between a suprapyramidal
          mossy   fiber   bundle   (sup) that runs along its
          superficial face and an infrapyramidal mossy fiber
          bundle (inf) that runs along its inferior face.
          Within the pyramidal cell layer the early-generated
          neurons   occupy   the   deeper   layers   and   the
          late-generated neurons occupy the more   superficial
          layers (indicated by the arrow and the letters "E"
          for early-generated and "L" for late-generated.
          B. In area CA3c of the Hld/Hld mouse, there is a
          population of neurons (arrows in B) below the
          infrapyramidal mossy fiber bundle (inf).   These
          neurons are the late-generated neurons that in the
          +/+ mouse would occupy the superficial-most portion
          of   the   pyramidal   cell   layer just below the
          suprapyramidal mossy fiber bundle (Vaughn et al.,
          1977). Note that the positions of the early-generated
          neurons and late-generated neurons are essentially
          the reverse of that found in the +/+ mouse (indicated
          by the arrow and the letters "E" for early-generated
          and "L3 for late-generated). Abbreviations : inf,
          infrapyramidal mossy fiber layer; int, intrapyramidal
          mossy fiber layer; pcl, pyramidal cell layer; SG,
          stratum granulosum of the dentate   gyrus;   sup,
          suprapyramidal mossy fiber layer.

CA3c of the Hld/Hld mouse (Fig. 2 B) the early-generated cells are found above the <u>int</u>rapyramidal mossy fiber layer (int in Fig. 2 B), whereas the late-generated cells are found below the <u>intrapyramidal</u> mossy fiber layer. The small arrow in the pyramidal cell layer in Fig. 2 B indicates that this is an "outside-to-inside spatiotemporal gradient" in neuron position, which is the reverse of that found in the +/+ mouse. It is important to note that the reversal in relative position found in area CA3c of the Hld/Hld mouse is not present in area CA1 (Vaughn et al., 1977).

METHODS

The goal of the experiments presented here is to determine : 1) <u>when</u> during development and 2) <u>where</u> along the migratory pathway the migration of the late-generated pyramidal neurons in the Hld/Hld mouse is disrupted. This has been done by comparing the migration paths of late-generated neurons in Hld/Hld and +/+ hippocampus. In addition, the migration of the late-generated neurons destined for area CA3c of the Hld/Hld hippocampus has been compared to the migration of early-generated neurons destined for area CA3c and also to the migration of late-generated neurons destined for area CA1. To follow the migrating neurons from their site of origin in the ventricular zone to their final position in the cortical plate tritiated thymidine (^{3}H-TdR) was injected into pregnant mice either at embryonic day 13 (E13) or E14 to label the early-generated pyramidal cells or at E15 or E16 to label the late-generated pyramidal cells (see Nowakowski and Rakic, 1981). The labelled offspring were then sacrificed after various survival times ranging from 1 hour after ^{3}H-TdR injection to postnatal day 7 (P7). All animals were sacrificed under deep anesthesia by intracardiac perfusion with a mixed aldehyde fixative and their brains were processed for autoradiography (Rakic and Nowakowski, 1981). The Hld/Hld mice were all of the BALB/cByJ inbred strain, and the +/+ mice were all of the C57BL/6J inbred strain (see Nowakowski, 1984 b).

RESULTS

Migration of early-generated pyramidal cells

As might be expected from the observation that the final position of the early-generated cells is not affected in Hld/Hld mice (Vaughn et al., 1977), it was found that the early-generated neurons reach their final position at a normal time in both +/+ and Hld/Hld mice. For this reason, only the situation in the Hld/Hld hippocampus is illustrated. In Fig. 3 A an autoradiogram of Hld/Hld hippocampus made from a mouse embryo which received ^{3}H-thymidine at embryonic day 14 (E14) and which was sacrificed at E17 is shown. It

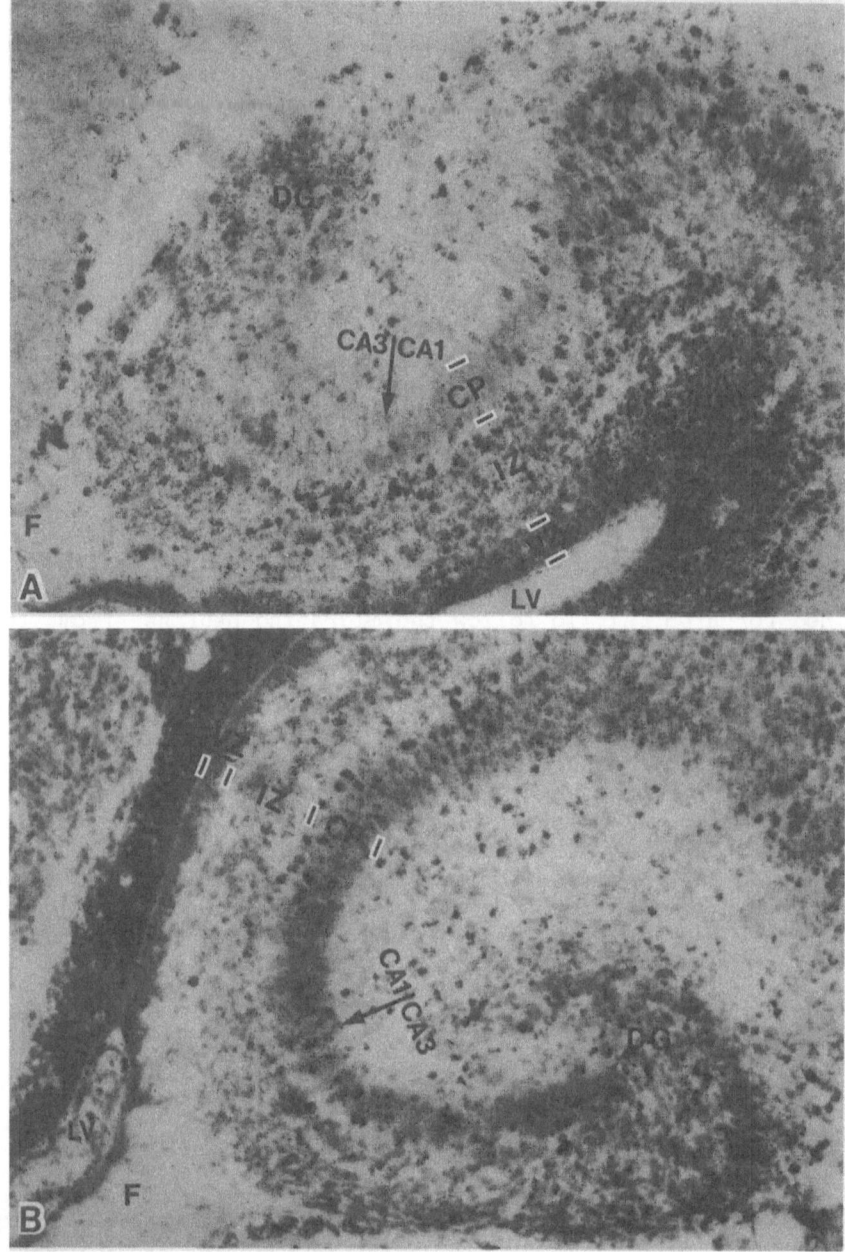

Fig. 3. Two autoradiograms illustrating the migration of
early-generated neurons in Hld/Hld embryos. In A the
hippocampus of an Hld/Hld embryo that was exposed to
^{3}H-thymidine on embryonic day 14 (E14) and
sacrificed on embryonic day 17 (E17) is shown. It

can be seen that in the three days since their generation, the early-generated neurons have left the ventricular zone (VZ) and moved approximately two-thirds of the way across the intermediate zone (IZ). In Fig. 3 B, the progress of migration by early-generated young pyramidal cells at the day of birth (PO), i.e., six days after being generated, is shown. By this time the early-generated neurons are in their final position in the cortical plate (CP) in both CA3 and CA1. These data indicate that it takes approximately 5 or 6 days for early-generated pyramidal cells to complete their migration.

Migration of late-generated pyramidal cells.

In Fig. 4 A, the position of the late-generated neurons in the +/+ mouse at PO is shown, whereas in Fig. 4 B the position of the late-generated neurons in the Hld/Hld mouse at PO is shown. In both +/+ and Hld/Hld mice the labelled late-generated migrating neurons are still confined to the intermediate zone and appear to have completed only about one-half of their migration. Thus, in contrast to the early-generated pyramidal cells, the late-generated pyramidal cells have NOT reached their final position in either CA3 or CA1 by the day of birth (PO). Note, however, that since the late-generated pyramidal cells were generated on embryonic day 16, only about 3 days have elapsed since these cells began their migration and that on the basis of the rate of migration of the early-generated pyramidal cells the expected arrival time of the late-generated pyramidal cells is not until P3.

By P3 the late-generated pyramidal cells have NOT reached their final position at the top of the cortical plate in CA3 but they have reached the top of the cortical plate in CA1 (Fig. 5). Comparison of Fig. 5 A and 5 C shows that the migratory progress of the late-generated neurons to area CA3c is similar in +/+ and Hld/Hld hippocampus and that in both the normal and mutant mouse the labelled migrating neurons are still in the intermediate zone

can be seen that the migrating neurons have moved about two-thirds of the way through the intermediate zone (IZ). In B the hippocampus of an Hld/Hld pup that was exposed to ^{3}H-thymidine on E13 and sacrificed on the day of birth (PO) is shown. By PO the early-generated pyramidal cells are already in their final position in the cortical plate (CP). The arrow indicates the approximate boundary between CA3 and CA1. Abbreviations : CP, cortical plate; DG, dentate gyrus; F, fimbria; IZ, intermediate zone; LV, lateral ventricle; VZ, ventricular zone.

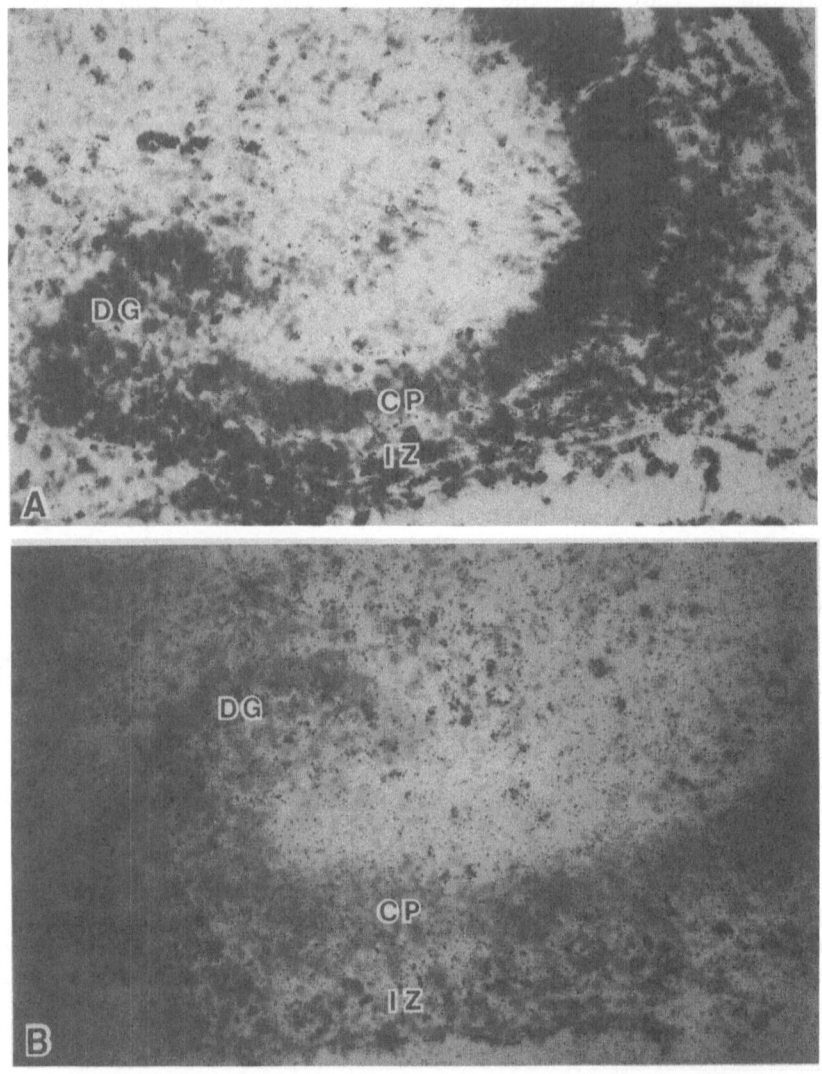

Fig. 4.   Two autoradiograms   illustrating the progress of the
          migration of late-generated neurons in +/+   (A)   and
          Hld/Hld (B) mice. In A hippocampus of a +/+ pup
          which was exposed to ³H-thymidine on E16   and
          sacrificed on P0 is shown. In B the hippocampus of
          an Hld/Hld pup which was exposed to ³H-thymidine on
          E16 and sacrificed on P0 is shown. In both the +/+
          and the Hld/Hld mice the migrating neurons are still
          found   below   the   cortical plate (CP) in the
          intermediate zone (IZ). Abbreviations : CP, cortical
          plate; DG, dentate gyrus; IZ, intermediate zone; VZ,
          ventricular zone.

below the developing cortical plate. In contrast, Fig. 5 B and 5 D
show that the late-generated neurons destined for area CA1 are at
the top of the cortical plate in both normal (B) and mutant (D).
Thus, the CA1 neurons have completed their migration in about six
days and also before the CA3 neurons have. However, in the Hld/Hld
mouse these labelled CA3c neurons will not pass through the
cortical plate, whereas in the +/+ mice they will (see Fig. 6).

By P5 the late-generated pyramidal cells in the +/+ mouse have
reached their final position at the top of the cortical plate in
CA3c (Fig. 6). At this stage the migrating late-generated neurons
in the Hld/Hld mouse are still in the same position as they were at
P3 (Fig. 5 B). Thus, in the +/+ mouse the late-generated pyramidal
cells migrating to area CA3c need about 8 days to complete their
migration, whereas in the Hld/Hld mouse the late-generated
pyramidal cells migrating to area CA3c complete their abbreviated
journey in about 6 days.

DISCUSSION

The major difference observed between the migration of young
neurons to area CA3c in the Hld/Hld mouse as compared to the +/+
mouse is that the migrating neurons in the mutant traverse most of
the intermediate zone but stop migrating at the inferior border of
the cortical plate (see Fig. 1) without traversing through the
cortical plate to their normal position. Therefore, they complete
most of their migration successfully ; moreover, they arrive at the
inferior border of the cortical plate at the same time as coevally
generated neurons in the +/+ mouse indicating that they migrate at
approximately the normal speed. This means that the influence of
the Hld locus on the migration of the late-generated pyramidal
cells occurs at a particular time and place during development.
Thus, it seems likely that Hld influences the stopping point of
migration and does not, for example, act by slowing down the rate
of migration of young neurons moving to area CA3c.

The fact that only the late-generated neurons destined for
area CA3c of the hippocampus are abnormally positioned whereas
late-generated pyramidal cells destined for other subdivisions of
the hippocampus, e.g., CA1, reach their final positions
successfully, is, perhaps, the most remarkable aspect of the
laminar reorganization produced by the Hld mutation. Thus, whatever
stops the migration to area CA3c must not affect the neurons
migrating to area CA1. The experiments described here have shown
that one difference between these two populations of neurons is the
time of arrival at their final position. Thus, the late-generated
neurons in the normal (+/+) mouse require approximately 8 days to
reach their final position in area CA3, but only about 6 days to
reach their final position in area CA1. Obviously, an event

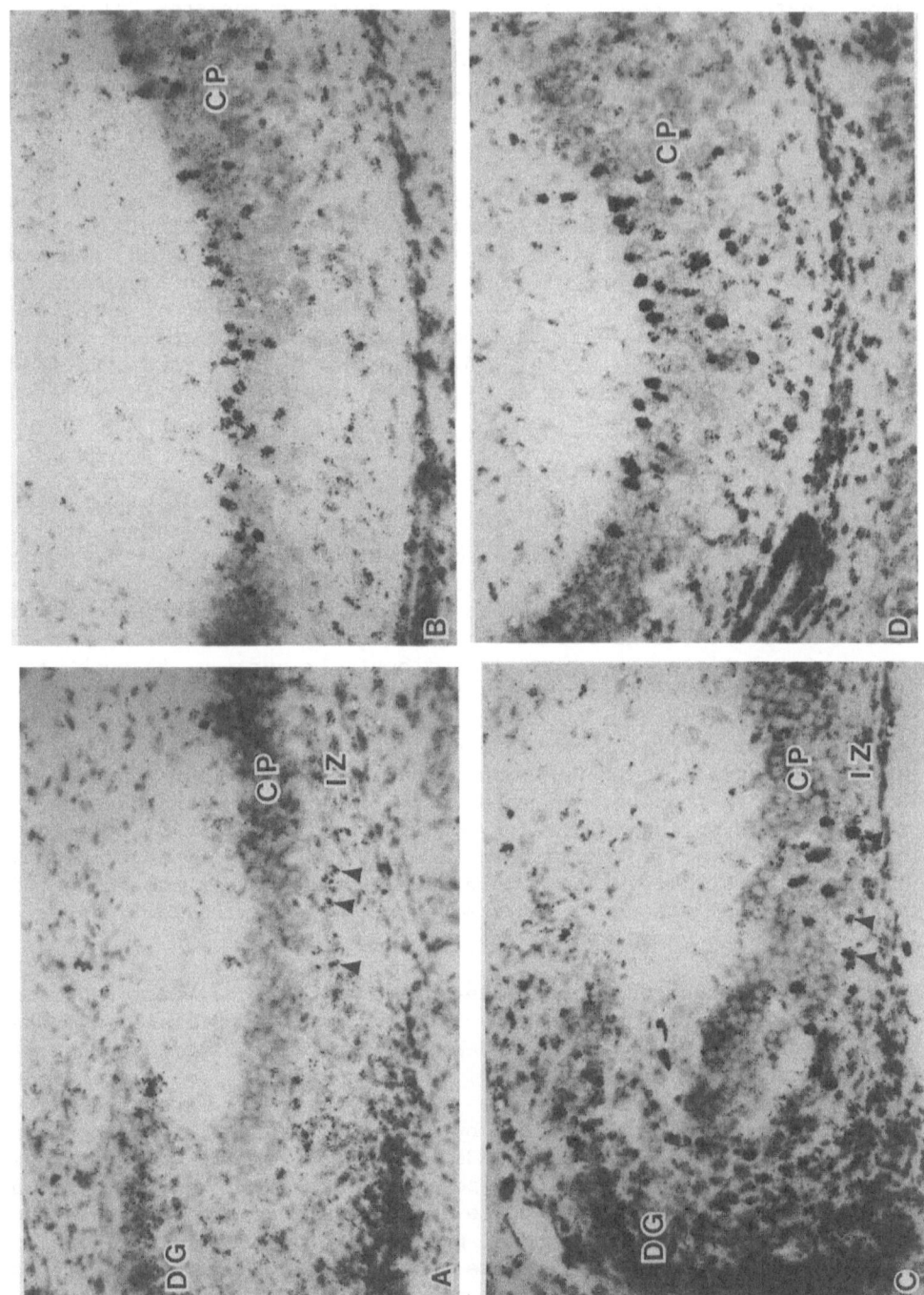

Fig. 5.   Four autoradiograms illustrating the progress of the migration of late-generated neurons to area CA3c of +/+ (A) and Hld/Hld (C) mice and to area CA1 of +/+ (B) and Hld/Hld (D) mice. In A and B areas CA3c and CA1 of the hippocampus of a +/+ pup which was exposed to ^{3}H-thymidine on E16 and sacrificed on P3 is shown. In C and D areas CA3c and CA1 of the hippocampus of an Hld/Hld pup which was exposed to ^{3}H-thymidine on E16 and sacrificed on P3 is shown. In area CA3c of both the +/+ and the Hld/Hld mice the migrating neurons (arrowheads) are still found below the cortical plate (CP) in the intermediate zone (IZ) or in the lower-most portions of the cortical plate. In contrast, however, in area CA1 of the both the +/+ and the Hld/Hld mice the late-generated neurons are already at the top of the cortical plate (CP). Abbreviations : CP, cortical plate; DG, dentate gyrus; IZ, intermediate zone.

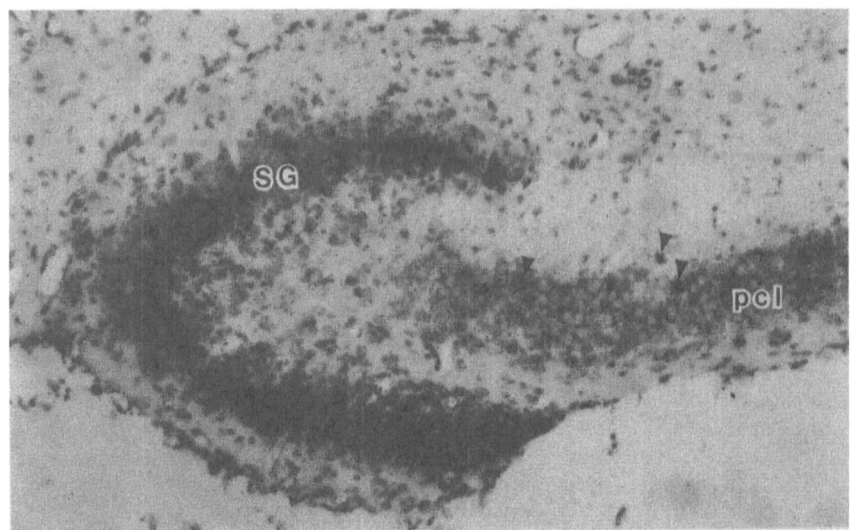

Fig. 6.   An autoradiogram from area CA3c of the hippocampus of
          a +/+ mouse which was exposed to ^{3}H-thymidine on   E15
          and   sacrificed   on P5 is shown. The arrowheads point
          to   the   position   of   late-generated   neurons   which
          already   completed their migration and are at the top
          of the pyramidal   cell   layer   (pcl).   Abbreviations:
          pcl,   pyramidal   cell   layer;   SG,   stratum granulosum of
          the dentate gyrus.

occuring during the two day interval between   the   arrival   of   the
late-generated   neurons in area CA1 and the arrival of the coevally
generated neurons destined for area CA3c   could   affect   area   CA3c
without   affecting area CA1. It should also be noted, however, that
the migration path of the late-generated neurons destined for   area
CA3   lengthens   considerably during the period that the neurons are
migrating, whereas the migration path of the neurons   destined   for
area   CA1   does   not change significantly (Fig. 7). Presumably, the
change in the distance from the ventricular zone   to   the   cortical
plate   of   area   CA3 lengthens due to the growth of the fimbria. It
seems likely that this increase in   the   length   of   the   migratory
pathway   contributes to the difference between the arrival times of
CA3 and CA1 neurons to their final   position   at   the   top   of   the
cortical   plate,   and,   therefore,   that   the difference in arrival
times at their final position may simply reflect the differences in
migratory distances.

          Another   obvious   difference   between   area   CA3c   and   other hippocampal subdivisions is the presence of the infrapyramidal mossy fiber layer (see Fig. 2). It is most attractive to consider that it is the mossy fibers themselves which influence the migrating neurons to stop. For example, in the +/+ mouse the migrating neurons continue along the radial glial fibers until they meet the mossy fibers at the top of the cortical plate. At issue is whether or not the mossy fibers play a role in stopping the migrating neurons or if their presence is unrelated. If the mossy fibers do provide information about when to stop migrating then it is possible that the Hld locus influences the mossy fibers to grow into area CA3c too early and that the presence of the mossy fibers signals the late-generated neurons to stop their migration. This idea is being tested by directly determining the time of ingrowth of the mossy fibers into area CA3c of +/+ and Hld/Hld mice. Also, experiments are presently underway using the shaker-short tail (gene symbol : sst/sst) mutant mouse in which the granule cells of the dentate gyrus are never formed and hence their axons, the mossy fibers, never grow into area CA3c (Wahlsten et al., 1983 ; Nowakowski and Wahlsten, 1985). Mice which are homozygous for both sst and Hld are being produced to determine if the mossy fibers contribute to producing the laminar reorganization of pyramidal cells characteristic of the Hld phenotype. These experiments will allow the possibility of an influence of the mossy fibers to be separated from the influence of the distance migrated and the time of arrival.

Implicit in most of the above discussion is that the influence of the Hld locus on neuronal migration is on the same mechanisms which normally act to control the stopping point of the migrating neurons. This is, however, not necessarily the case. In theory, there are many ways that the final position of pyramidal cells in area CA3c might be influenced by the action of the Hld locus. For example, in the reeler mutant mouse in which there is also an inversion in relative cell position in the cerebral cortex it seems likely that the early-generated neurons fail to detach from the radial glial fibers so that the migration of the late-generated neurons is obstructed because the space on the surface of the radial glial fibers is occupied (Pinto-Lord et al., 1982 ; Goffinet, 1984). In other words, in reeler the late-generated migrating neurons detach from the radial glial fiber and stop migrating because their further progress is obstructed. Thus, the disruption of migration in the reeler mouse is effected during the locomotory phase of migration (but not on the locomotory processes per se). Similar mechanisms for producing the Hld phenotype have not yet been eliminated, and there are, in addition, a variety of other, equally plausible ways through which the Hld locus could disrupt neuronal migration without involving the normal physiological signal(s) to stop migrating.

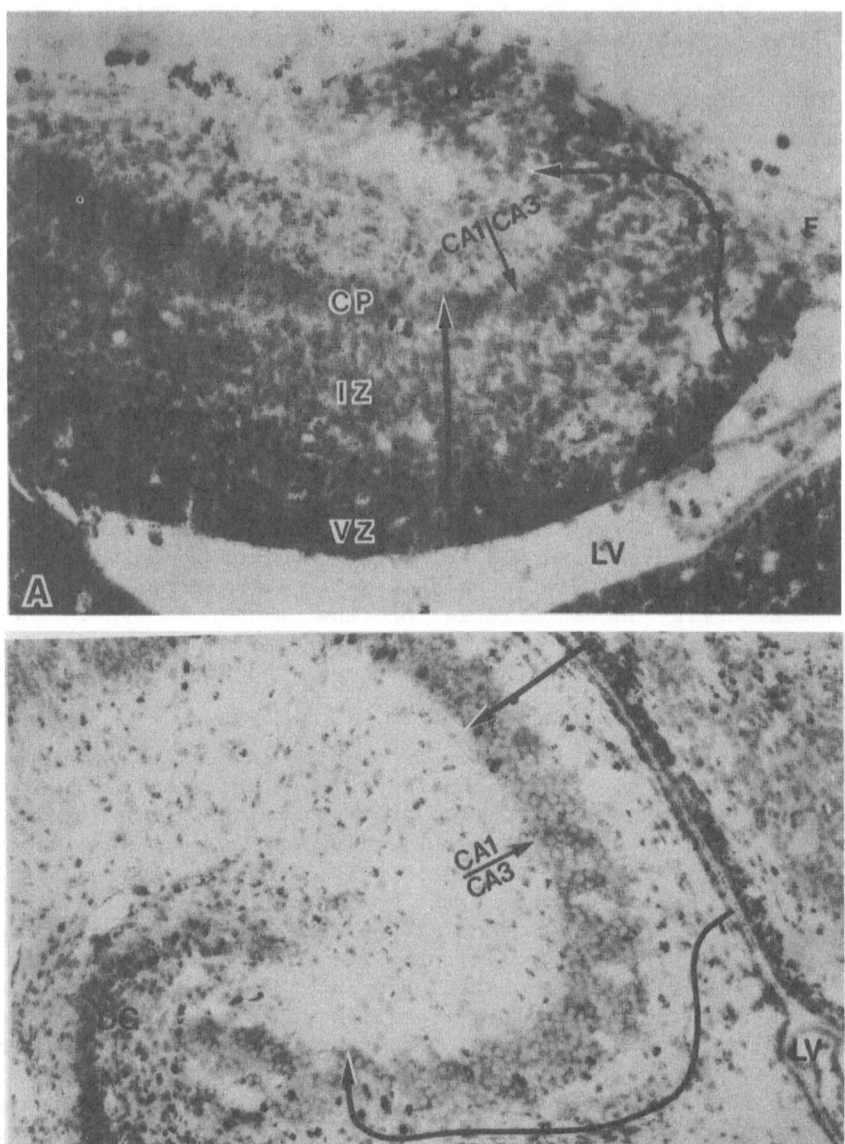

Fig. 7.   Photographs of the hippocampus of the mouse at E16
          (A) when   the   late-generated neurons   are   just
          beginning   their migration and   at P3 (B) when the
          late-generated neurons are close to finishing their
          migration. The arrows extending from the ventricular
          zone to the top of the cortical plate indicate   the

Table 1. A comparison of the phases of migration of neural crest derivatives and young neurons in the cerebral cortex. For details see the text.

	Neural crest derivatives	Young neurons in cortex
Signal to start migration	?	?
Provide and maintain polarity	Population Pressures	?
Guidance (i.e., direction)	Cell-extracellular matrix interactions (Fibronectin)	Apposition with radial glial fibers
Signal to stop migration	Morphological barrier or presence of N-CAM	Interaction with radial glial fibers, axons or dendrites, etc...

Comparison with migration of neural crest derivatives

In Table 1 what is known about the controlling influences on the phases of migration for migrating cortical neurons is compared to similar data for migrating cells derived from the neural crest. For the most part, the data for the neural crest cells are derived from the presentation by Duband and Thiery (1985) at this ASI.

---

approximate migration path of neurons migrating to area CA3c and to area CA1. (The approximate boundary between CA3 and CA1 is indicated.) Note that the lenght of the migration path to area CA3c increases considerably presumably because of the growth of the fimbria (F) whereas the length of the migration path to area CA1 remains approximately the same and may even get shorter. Abbreviations: CP, cortical plate; DG, dentate gyrus; F, fimbria; IZ, intermediate zone; LV, lateral ventricle; VZ, ventricular zone.

As they start to migrate both the cells derived from the neural crest and the young neurons in the CNS must first become separated from adjacent non-migratory cells. In the case of the neural crest cells the migratory cells become separate from the neuroepithelium of the dorsal part of the neural tube (Newgreen and Gibbins, 1982 ; Erickson and Weston, 1983). In the cerebral cortex the migrating neurons must separate from the proliferative zones (i.e., either the ventricular zone or the subventricular zone). Also, the sites of origin of both populations of cells line surfaces and the cells move away from the surface in only one direction and, thus, they are polarized as they begin to migrate. However, many of the migrating cells derived from the neural crest continue to proliferate during their migration (Le Douarin, 1982), whereas in the CNS most migrating neurons (with a few exceptions) are permanently postmitotic.

Two aspects of the locomotory phase must be considered. First, the guidance of the migrating cells is provided by an interaction with fibronectin in the extracellular matrix for the neural crest cells and with radial glial fibers for the cortical cells. Duband and Thiery (1985) point out that the migrating neural crest cells are not aligned with the orientation of the meshwork of the extracellular matrix. In contrast, in the cerebral cortex the migrating young neurons are clearly aligned with the radial glial fibers (Rakic, 1972, 1978, 1982 ; Nowakowski and Rakic, 1979). Moreover, Hatten et al. (1984) have shown that in tissue culture migrating granule cells in the developing cerebellum will migrate along fascicles of glial fibers in either direction. These experiments suggest that the relationship between migrating neurons and radial glial fibers is one of contact guidance rather than haptotaxis. Also, these data suggest that in vivo the guidance provided to the migrating young neurons by virtue of their alignment with radial fibers may need to be supplemented by other influences to maintain the polarity of the migrating cells. For the neural crest derivatives population pressures and contact inhibition from cells following along behind the migrating cells seem to be sufficient to maintain polarity (Rovasio et al., 1983 ; Duband and Thiery, 1985) ; in the cerebral cortex similar influences may be acting to maintain the polarity of the migrating neurons.

Signals to stop migration may arise from a variety of sources. For the neural crest cells, two different signals to stop have been suggested. First, a morphological barrier in the form of the sclerotome and a concomitant modification in the makeup of the extracellular matrix seems to correlate with the cessation of the migratory phase of the neural crest cells that will make up the sensory ganglia (Duband and Thiery, 1985). Second, the onset of expression of N-CAM by the migrating neural crest cells themselves (induced by unknown factors) seems to promote their aggregation into the autonomic ganglia and thereby inhibits their further migration (Duband and Thiery, 1985). In the cerebral cortex analogous mechanisms may well exist. For example, an analogy with the suggested mechanism of formation of the sensory ganglia would suggest that the surface of the radial glial fiber is a mosaic and that there is a change in its makeup at the site of termination of neuronal migration at the top of the cortical plate. It is conceivable that the distribution or concentration of a cell adhesion molecule (Edelman, 1983) is altered, and there is some evidence that such CAMs may be involved in neuronal migration in the CNS. Recently, it has been reported that antibodies to L1 antigen can block the migration of cerebellar granule cells along Bergmann glial fibers (Lindner et al., 1983) and also that migrating neurons may have Ng-CAM (which may be identical to L1 (Lindner et al., 1983)) on their cell surface (Grumet et al., 1983 ; Grumet and Edelman, 1984). Similarly, an analogy with the suggested mechanism of formation of the autonomic ganglia would suggest that the migrating neurons begin to produce N-CAM or a similar molecule either as a result of some intrinsic changes or as an aftereffect of an interaction with the axons and dendrites of previously generated cells. It is important to note, however, that regardless of whether the relationship of the migrating neurons with the radial glial fibers is one of contact guidance or haptotaxis, the signal to stop migrating could be resident either on the surface of the radial glial fiber in the form of a loss of adhesion for the migrating neuron or on the surface of some other constituents of the developing cortex either in the form of a greater adhesion for the migrating cell than the radial glial fiber has or by inducing a change in the migrating cell itself.

Concluding remarks

The migration of the young neurons in the developing central nervous system is an important feature of its normal development. There remain, however, a number of unanswered questions concerning the precise mechanisms of the interactions of the migrating neurons with the radial glial fibers as well as with the other constituents of their environment. In particular, the sources of the influences on the proliferating populations to start migration and on the migrating neurons themselves to stop migrating are unknown. Also, the capacity of neurons for establishing normal axonal connections and dendritic arborizations after they have migrated to "inappropriate" positions is still incompletely explored (e.g., Caviness and Rakic, 1978 ; Nowakowski and Davis, 1983, 1985). Further analysis of the disruption of neuronal migration and their sequelae in Hld/Hld mice and also in other mutants (Nowakowski, 1985) could shed light on these important issues.

ACKNOWLEDGEMENTS

    Supported by NSF Grant BNS-8120050 and NIH Biomedical Research
Support Grant 5S07RR05386.

REFERENCES

Abercrombie M., 1980, The crawling movement of metazoan cells,
        Proc. R. Soc. Lond., B 207 : 129-147.
Angevine J.B., 1965, Time of neuron origin in the hippocampal
        region : an autoradiographic study in the mouse, Exp.
        Neurol. Supp., 2 : 1-71.
Ben Slimane S., Houllier F, Tucker G. and Thiery J.P., 1983, In
        vitro migration of avian hemopoeitic cells to the thymus:
        preliminary characterization of a chemotactic mechanism,
        Cell Diff., 13 : 1-24.
Carter S.B., 1967, Haptotaxis and mechanism of cell motility,
        Nature, 213 : 261-264.
Caviness V.S., Rakic Jr. and P., 1978, Mechanisms of cortical
        development : a view from mutations in mice, Ann. Rev.
        Neurosci., 1 : 297-326.
Duband J.L. and Thiery J.P., 1985, Adhesive molecules and their
        role during the ontogeny of the peripheral nervous system,
        this volume.
Eckenhoff M.F. and Rakic P., 1984, Radial organization of the
        hippocampal dentate gyrus : a Golgi, ultrastructural, and
        immunocytochemical analysis in the developing rhesus monkey,
        J. Comp. Neurol., 223 : 1-21.
Edelman G.M., 1983, Cell adhesion molecules, Science, 219 :
        450-457.
Erickson C.A. and Weston J.A., 1983, An SEM analysis of neural
        crest migration in the mouse, J. Embryol. Exp. Morph., 74 :
        97-118.
Goffinet A.M., 1984, Events governing organization of postmigratory
        neurons : studies on brain development in normal and reeler
        mice, Brain Res. Rev., 7 : 261-296.
Grumet M. and Edelman G.M., 1984, Heterotypic binding between
        neuronal membrane vesicles and glial cells is mediated by a
        specific cell adhesion molecule, J. Cell Biol., 98 :
        1746-1756.
Grumet M., Rutishauser U. and Edelman G.M., 1983, Neuron-glia
        adhesion is inhibited by antisera to neural determinants,
        Science, 220 : 60-62.
Hatten M.E., Liem R.K.H. and Mason C.A., 1984, Two forms of
        cerebellar glial cells interact differently with neurons in
        vitro, J. Cell Biol., 98 : 193-204.
Heasman J. and Wylie C.C., 1981, Contact relations and guidance of
        primordial germ cells on their migratory route in embryos of
        Xenopus laevis, Proc. R. Soc. Lond., B 213 : 41-58.

Le Douarin N.M., 1978, Ontogeny of hematopoietic organs studied in
    avain embryo interspecific chimeras, In: "Differentiation of
    normal and neoplastic hematopoietic cells", Cold Spring
    Harbor Symposium, Cold Spring Harbor, New York, pp. 5-31.
Le Douarin N.M., 1982, The neural crest, Cambridge University
    Press, Cambridge.
Le Douarin N.M., 1984, Cell migrations in embryos, Cell, 38 :
    353-360.
Lindner J., Rathjen F.G. and Schachner M., 1983, L1 mono- and
    polyclonal antibodies modify cell migration in early
    postnatal mouse cerebellum, Nature, 305 : 427-430.
Mountcastle V.B., 1978, An organizing principle for cerebral
    function. The unit module and distributed system. In: "The
    Mindful Brain" by G. Edelman and V.B. Mountcastle, MIT
    Press, Cambridge, Mass., pp. 7-50.
Newgreen D.F. and I.L. Gibbins, 1982, Factors controlling the time
    of onset of the migration of neural crest cells in the fowl
    embryo, Cell Tissue Res., 224 : 145-160.
Nowakowski R.S., 1983, Single gene inheritance of an abnormality in
    lamination in area CA3c of the hippocampus of BALB/c mice,
    Soc. Neurosci. Abstr., 9 : 833.
Nowakowski R.S., 1984 a, The migration of pyramidal cells to area
    CA3c of the hippocampus of mice carrying the mutation
    "hippocampal lamination defect", Soc. Neurosci. Abstr., 10 :
    47.
Nowakowski R.S., 1984 b, The mode of inheritance of a defect in
    lamination in the hippocampus of the BALB/c mouse, J.
    Neurogenetics, 1 : 249-258.
Nowakowski R.S., 1985, Traffic regulation for the migrating neuron.
    In preparation.
Nowakowski R.S. and Davis T.L., 1983, A Golgi study of abnormally
    positioned neurons in area CA3c of BALB/c mice, Anat. Rec.,
    205 : 145 A.
Nowakowski R.S. and Davis T.L., 1985, Dendritic arbors and
    dendritic excrescences of abnormally positioned neurons in
    area CA3c of mice carrying the mutation "Hippocampal
    lamination defect", J. Comp. Neurol., accepted for
    publication.
Nowakowski R.S. and Rakic P., 1979, The mode of migration of
    neurons to the hippocampus : a Golgi and electron
    microscopic analysis in foestal rhesus monkey, J.
    Neurocytol., 8 : 697-718.
Nowakowski R.S. and Rakic P., 1981, The site of origin and route
    and rate of migration of neurons to the hippocampal region
    of the rhesus monkey, J. Comp. Neurol., 196 : 129-154.
Nowakowski R.S. and Wahlsten D., 1985, Anatomy and development of
    the hippocampus and dentate gyrus in the shaker-short tail
    (sst) mutant mouse, Anat. Rec., in press.
Pinto-Lord M.C., Evrard P. and Caviness V.S. Jr., 1982, Obstructed
    neuronal migration along radial glial fibers in the

neocortex of the reeler mouse : a Golgi-EM analysis, Dev.
    Brain Res., 4 : 379–393.
Rakic P., 1971, Neuron–glia relationship during granule cell
    migration in developing cerebellar cortex. A Golgi and
    electron microscopic study in macacus rhesus, J. Comp.
    Neurol., 141 : 283–312.
Rakic P., 1972, Mode of cell migration to the superficial layers of
    fetal monkey neocortex, J. Comp. Neurol., 145 : 61–83.
Rakic P., 1978, Neuronal migration and contact guidance in the
    primate telencephalon, Postgrad. Med. J., 54 : 25–40.
Rakic P., 1982, Early developmental events : cell lineages,
    acquisition of neuronal positions, and areal and laminar
    development, Neurosci. Res. Prog. Bull., 20 : 439–451.
Rakic P. and Nowakowski R.S., 1981, The time of origin of neurons
    in the hippocampal region of the rhesus monkey, J. Comp.
    Neurol., 196 : 99–128.
Rovasio R.A., Delouvée A., Yamada K.M., Timpl R. and Thiery J.P.,
    1983, Neural crest migration : requirements for exogenous
    fibronectin and high cell density, J. Cell Biol., 96 :
    462–473.
Smart I.H.M. and McSherry G.M., 1982, Growth patterns in the
    lateral wall of the mouse telencephalon : II. Histological
    changes during and subsequent to the period of isocortical
    neuron production, J. Anat., 134 : 415–442.
Trinkaus J.P., 1976, On the mechanisms of metazoan cell movements.
    In: "The cell surface in animal embryogenesis and
    development" G. Poste and G.L. Nicolson eds., North-Holland,
    Amsterdam, pp. 225–329.
Vaughn J.E., Matthews D.A., Barber R.P., Wimer C.C. and Wimer R.E.,
    1977, Genetically-associated variations in the development
    of hippocampal pyramidal neurons may produce differences in
    mossy fiber connectivity, J. Comp. Neurol., 173 : 41–52.
Wahlsten D., Lyons J.P. and Zagaja W., 1983, Shaker short-tail, a
    spontaneous neurological mutant in the mouse, J. Hered., 74
    : 421–425.
Weiss P., 1961, Guiding principles in cell locomotion and cell
    aggregation, Exp. Cell Res., Suppl., 8 : 260–281.

CELL INTERACTIONS IN THE ONTOGENY OF THE HEMOPOIETIC SYSTEM

Françoise Dieterlen-Lièvre

Institut d'Embryologie du CNRS
et du Collège de France
49 bis, Avenue de la Belle Gabrielle
94130 Nogent-sur-Marne, France

In vivo and in vitro functional tests demonstrate the existence in adult animals of progenitor cells which differentiate into various mature blood cells, provided specific factors are available to them. These progenitors themselves are thought to derive from pluripotent or totipotent stem cells. It is still a major questionmark whether the differentiating factors determine the commitment of progenitors towards one line of differentiation or whether these factors merely have a trophic action on cells committed through stochastic processes. Whichever the mechanism, the differentiating factors are provided to the progenitors within the microenvironments of hemopoietic organs. Thus functioning of the hemopoietic system is characterized by the interaction of two distinct cell lineages, on one hand the stromal cells which constitute the microenvironments, on the other hand the blood cell progenitors.

This dichotomy exists from the beginning of embryonic life. In the avian and mammalian embryo, it has been shown that stem cells form outside the hemopoietic organ rudiments and begin colonizing them at a precise stage of development. The thymus is even invaded by distinct waves of precursors ; these waves are equal in duration and separated by equal non-receptive periods. This periodicity, demonstrated a few years ago in the quail embryo, has now been revealed in the chick and in the mouse.

Cell movements between the various compartments of the hemopoietic system are detected by exchanging territories or anlage between embryos differing through a marker system. In birds, quail/chick tissue combinations are used a lot, since cells of

155

these two species may be distinguished through a nuclear
difference. Other cytological or biochemical characters make it
possible to identify cells in homospecific combinations. Examples
are sex chromosomes, immunoglobulin allotypes, major
histocompatibility antigens, or monoclonal antibodies ; the last
may be the both species and lineage specific.

Since stem cells form outside hemopoietic organs, their
origin is elusive. It was once thought that they emerged from the
yolk sac. However in chimeras associating a quail embryo to a chick
yolk sac, it appeared that intra embryonic stem cells relayed the
early blood cells formed in the yolk sac. The latter give rise to a
pool of erythrocytes but never contribute to the permanent stem
cell pool. Intraembryonic stem cells form in the ventral wall of
the embryo's aorta. The potentialities of these cells have been
established in inter-specific grafts using monoclonal antibodies
which recognize the quail hemangioblastic lineage (endothelial and
hemopoietic cells) and no cells of the chick. These cells are
capable of undergoing erythropoiesis or granulopoiesis and of
colonizing a thymus rudiment.

In the mouse, the potentialities of yolk sac stem cells have
been studied in vitro in organ culture or in a clonal test (BFU-E).
It was shown that the isolated yolk sac gives only embryonic red
cells. If erythropoietin is provided, then adult red cells are
produced. The yolk sac is also capable of producing adult red cells
in a BFU-E assay, after being dissociated into individual cells.
However in the latter system burst-promoting-activity must be
provided as well as EPO.

Many questions are still unanswered, for instance, in the
avian embryo, are the cells which form in the ventral wall of the
aorta those which give rise to the whole hemopoietic system or are
they in turn only an intermediate population ? In the mouse embryo,
is the yolk sac the only site where definitive stem cells form ?
One conclusion is obvious from the host of experiments which have
been carried out in the last ten years : many processes are very
similar in birds and mammals so that facts established through
experiments in the avian embryo should serve as guide lines in the
unravelling of mammalian development, where the feasible approaches
are very restricted.

REFERENCES

Dieterlen-Lièvre F. and Martin C., 1981, Diffuse intraembryonic
        hemopoiesis in normal and chimeric avian development,
        Develop. Biol., 88 : 180-191.
Jotereau F.V. and Le Douarin N.M., 1982, Demonstration of a cyclic
        renewal of the lymphocyte precursor cells in the quail

thymus during embryonic and perinatal life, J. Immunol.,
129: 1869-1877.

Labastie M.C., Thiery J.P. and Le Douarin N.M., 1984, Mouse yolk
sac and intraembryonic tissues produce factors able to
elicit differentiation of erythroid burst-forming units and
colony-forming units, respectively, Proc. Natl. Acad. Sci.
USA, 81 : 1453-1456.

Le Douarin N.M., Dieterlen-Lièvre F. and Oliver P., 1984, Ontogeny
of primary lymphoid organs and lymphoid stem cells, American
J. Anatomy, 170 : 261-299.

MORPHOLOGICAL BASES FOR CELL-TO-CELL AND CELL-TO-SUBSTRATE

INTERACTION STUDIES IN CEPHALOPOD EMBRYOS

H.-J. Marthy

Laboratoire Arago - U.A. 117 C.N.R.S.
Université Pierre et Marie Curie (Paris VI)
66650 Banyuls-sur-mer, France

INTRODUCTION

Cephalopod embryos not only are "remarkable" animals from anatomical, physiological, behavioural and, last but not least, aesthetical points of view ; at pre-organogenetic stages, they also are "a suitable material for cell and tissue interaction studies" (1). In the present study, performed on embryos of the Mediterranean squid Loligo vulgaris, further evidence is given, that the embryos at early organogenetic stages (Fig. 1) are valuable as well for studying cell-to-cell and cell-to-substrate interactions. By means of the Scanning Electron Microscopy (SEM) good insight can be obtained as to the tissue architecture of the germ layers, the spatial distribution of the cells, the cells' forms, surface structures and connections as well as to the surface structures of the cells' common substrate, the yolk syncytium. Many of the SEM-illustrations speak for themselves from the "cell biological" aspect alone ; however, regarding them, one should constantly keep in mind, that the various structures also directly express the actual determination state of the embryos at well defined steps of early organogenesis. The (more common) cell biological phenomena to be shown are, of course, also intimately linked to purely (squid-specific) "embryological" problems.

PREPARATION OF THE EMBRYOS

Embryos of the common squid Loligo vulgaris are readily available in our marine station from November until June. For studying the internal organisation of the embryos, the ectodermal cell layer is removed with fine needles from living specimens.

159

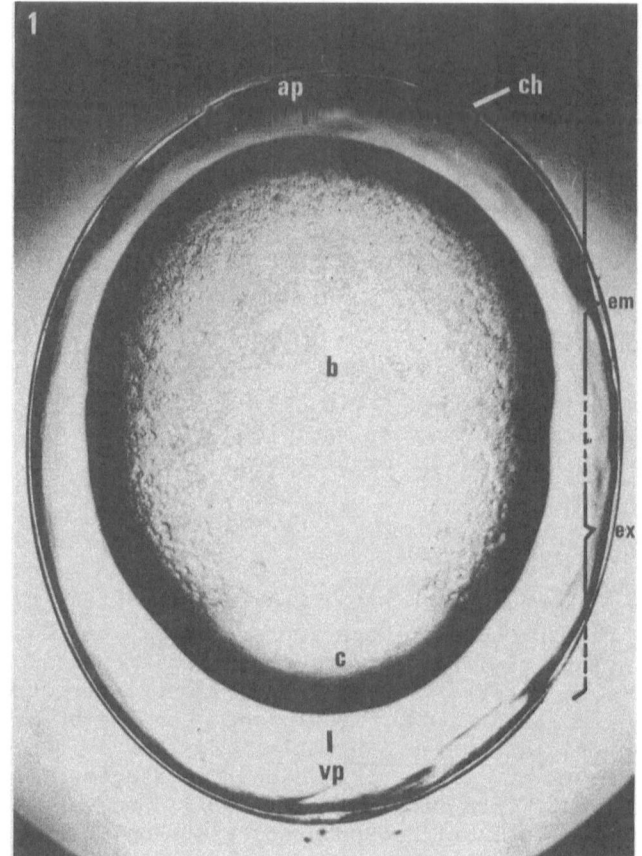

Fig. 1.   Loligo vulgaris embryo, at about stage VII. The cel-
          lular blastoderm (b) covers about four fifth of the
          egg surface.
          em : embryonic area of the blastoderm ; ex : extra-
          embryonic area of the blastoderm ; ap : animal pole ;
          vp : vegetal pole ; c : egg cortex, not yet covered
          by the blastoderm ; ch : chorion. Embryo : 2.2 mm x
          1.6 mm.

Immediately after the operation is performed, the embryo is fixed
with 1% OsO_4/sea water, added dropwise. After about 15 minutes, the
embryo, slightly fixed, is transferred into a accurate 1% OsO_4/sea
water solution for 30 minutes. For the observation by SEM
(Cambridge   250), the embryos are treated following standard
methods.

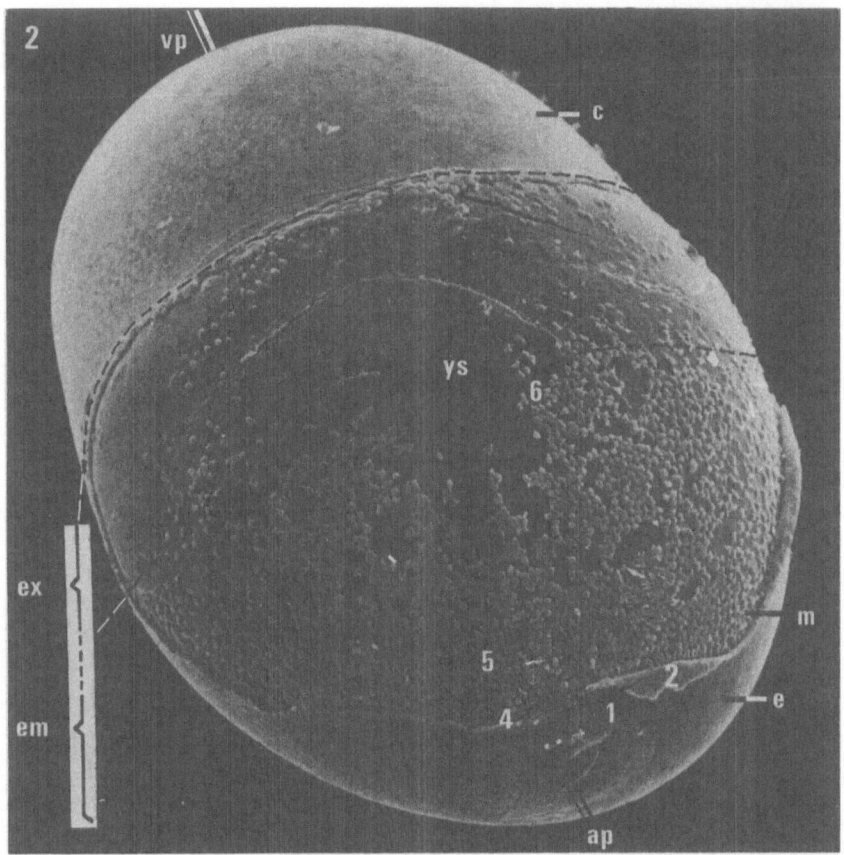

Fig. 2.   Ventrolateral view of an operated embryo at about
          stage VI. A large area of the ectodermal part of the
          blastoderm has been removed giving access to the
          mes-endoderm and the yolk syncytium. Dotted lines
          show approximate borders of the extra-embryonic
          area.
          ap : animal pole ; vp : vegetal pole ; c : egg cor-
          tex ; e : ectoderm ; m : mes-endoderm ; ys : yolk
          syncytium (syncytial substratum) ; em : embryonic
          part of the blastoderm ; ex : extra-embryonic part
          of the blastoderm.
          1: see Fig. 3 and 4 ; 2: see Fig. 5 ; 3: see Fig. 6
          and 7 ; 4: see Fig. 8 ; 5: see Fig. 9 ; 6: see Fig.
          10 and 11 ; 7: see Fig. 12 ; 8: see Fig. 13 and 14.
          Bar : 1 mm / 47 x.

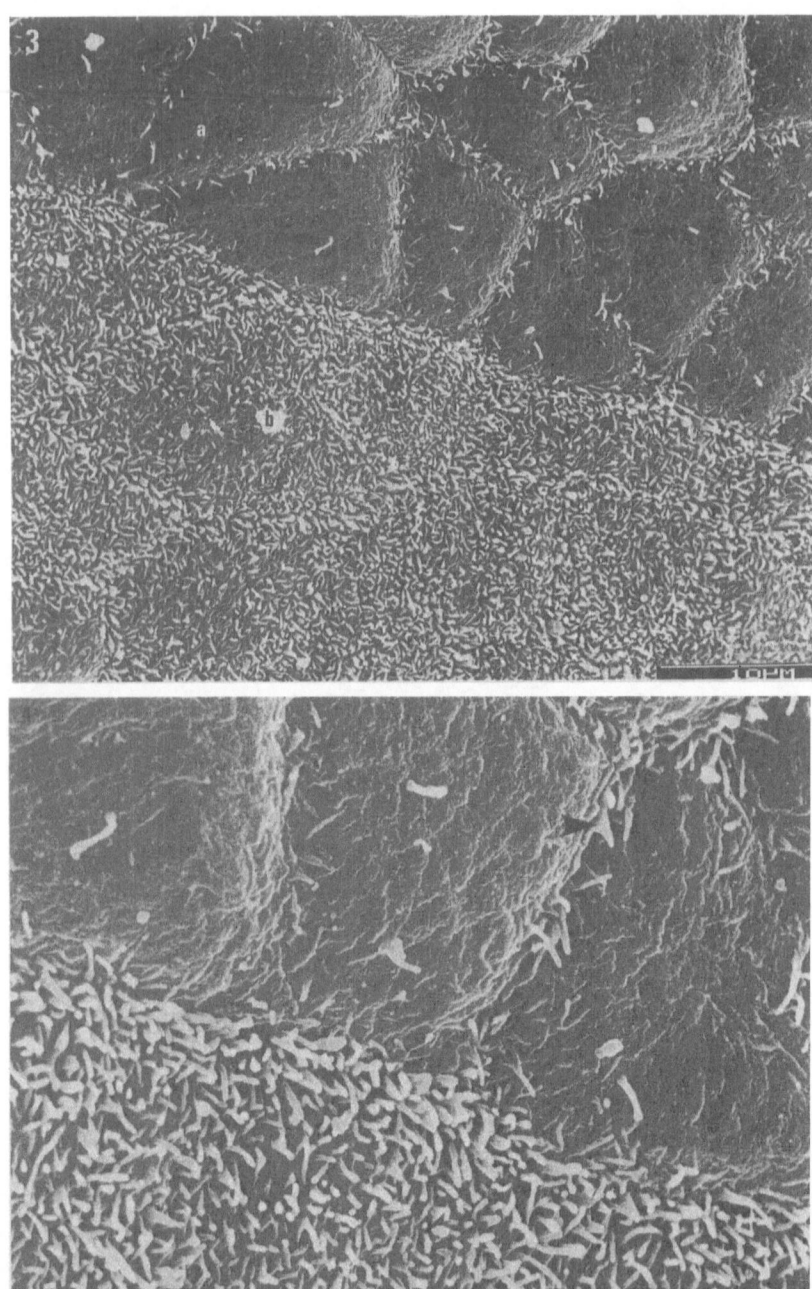

TISSUES, CELLS AND SUBSTRATE

    The   various   tissue,   cell   and   substrate   structures   are
described for three well defined developmental stages near and
during early organogenesis, that is for A) about stage VI (2), B)
stage VII and C) stage VIII-IX. The   presentation   of   the
SEM-pictures is made within the context of various additional
(published or unpublished) observations made on living embryos.

A) Tissues and cells of an embryo at about stage VI

    When the blastoderm extends over 50 to 60% of the egg surface,
an embryo normally has reached "stage VI" ; that is within the
blastoderm an embryogenetic and an extraembryonic part can be
vaguely distinguished in an approximate proportion of 2 : 1. The
anterio-dorsal side of the embryo is recognized by the position of
the polar bodies ; no well defined organ primordium is seen yet. On
the living embryo the extension of the blastoderm is a relatively
unsafe staging criterion and, it is not easy at all to define
precisely a stage VI. Since egg fragments containing an intact
cleavage territory also develop into "normal" dwarf embryos,
synchronously with normal sized eggs (3), we argue that the actual
developmental state is primarily recognized by the differentiation
state of the cells (4, 5). There exists one differentiation
phenomenon, which can be used as the best criterion for defining
stage VI in a living embryo : the germ layer affinity. As long as
the ectoderm can be removed from the underlying inner germ layer(s)
without detaching any of those cells, a "pre-organogenetic" state
is still present. When the ectoderm can no longer be removed
without detaching adhering mes-endodermal cells, an   "organogenetic

←─────────────────────────────

Fig. 3.   Detail n^0. 1 from Fig. 2. Zone of contact between the
          two cell populations representing the   embryonic
          ectoderm.
          a : cells showing on their rather "smooth" surface
          primary cilia (arrows) ; b : cells showing a dense
          villosity, which constitute the shell gland rudiment
          (and possibly also a part of the mantle rudiment).
          Numerous protrusions are present in the contacting
          zone between the cells (arrow head).
          Bar : 10 µm/1.900 x.

Fig. 4.   Detail from Fig. 3. Microvilli and protrusions (arrow
          head), interdigitating at the contact zones of cells,
          partially also extending over the cell surfaces.
          Bar : 4 µm/4.800 x.

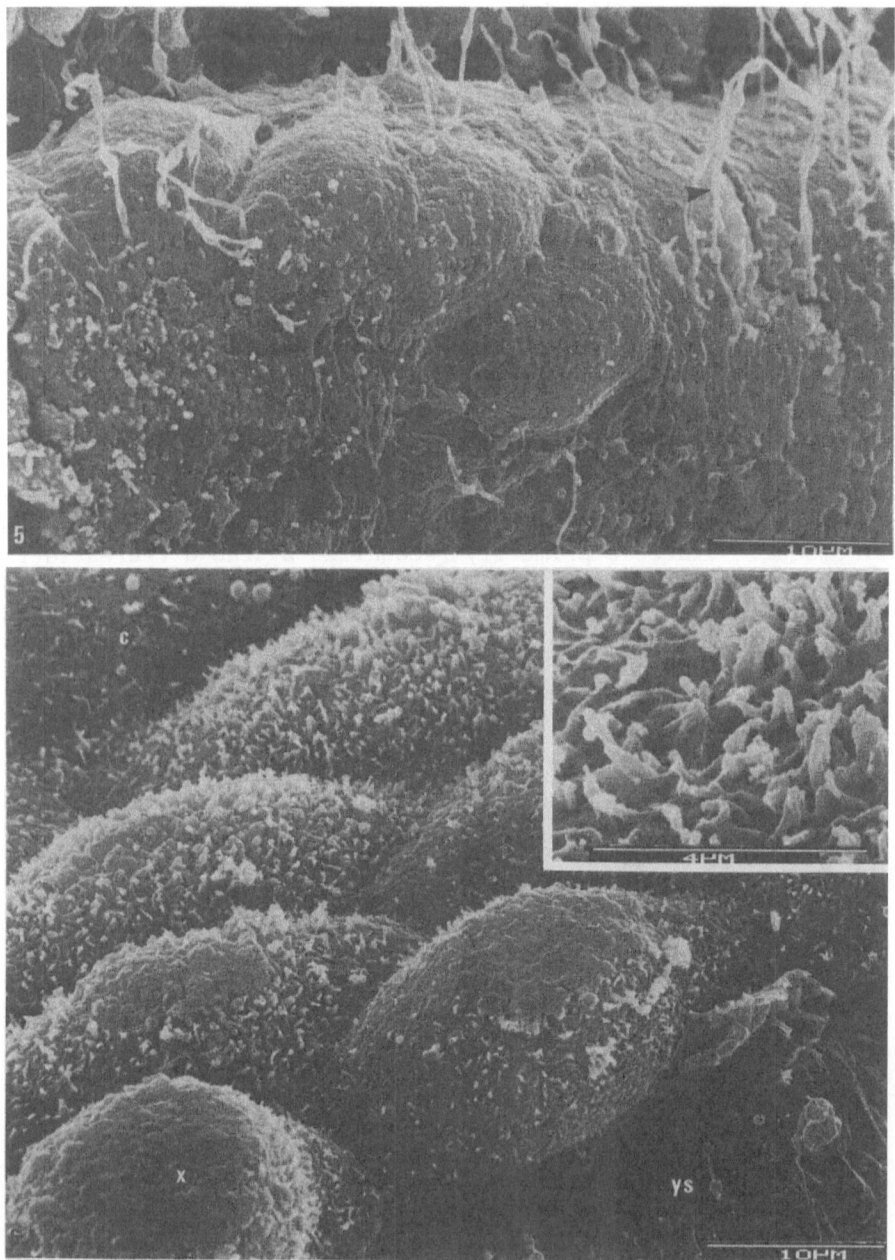

state" is reached. Therefore, the very latest moment, at which the germ layers can still be totally separated but a slight increase in germ layer affinity is observed, can be termed as "stage VI". An embryo of stage VI still may be considered as a late pre-organogenetic stage ; morphological and physiological characteristics show that both cell and organ-specific differentiation is engaged at this time.

Analysis of the SEM-pictures (Figs. 2-14) :

Figure 2 gives a general overview of an embryo at stage VI. A large area of ectoderm is removed giving access to the intact mes-endoderm and the yolk syncytium. Some remaining ectodermal parts still show the essential cell and tissue structures in the area of the future organogenetic region (Fig. 2 : 1, 2) and the extra-embryonic part (Fig. 2 : 3). Whereas the ectoderm is a solid monolayer of cells, the underlying mes-endoderm appears as a rather lose complex of large interconnected cells and of scattered single cells (Fig. 2 : 4, 5, 8). The common substrate of both germ layers is the yolk syncytium, delimiting the yolk mass. Parts of the original egg cortex are not yet covered by cells (Fig. 2 : 7). The series of pictures, taken at higher magnification (Fig. 3-14), provides more details.

The ectoderm of the embryonic region :

a) The outer surface. As seen from surface structures, the

Fig. 5.   Detail n⁰ . 2 from Fig. 2. Inner surface of the ecto-
           dermal cell layer. An "extracellular or interdermal
           material" forming "droplets and granules" (open
           arrows) and numerous elongated surface protrusions
           (arrows) and filamentlike "threads" (arrow heads) are
           observed. Bar : 10 µm/2.100 x.

Fig. 6.   Detail n⁰ . 3 from Fig. 2. Peripheral ectodermal cells
           of the blastoderm (extra-embryonic area). Numerous
           irregularly arranged microvilli form the surface of
           the cell.
           x : precipitate on the cell surface ; c : cortex ;
           ys : yolk syncytium. Bar : 10 µm/2.100 x.

Fig. 7.   Detail from Fig. 6. Microvilli on the surface of the
           large ectodermal cells forming the "outer yolk sac
           envelope". Bar : 4 µm/8.100 x.

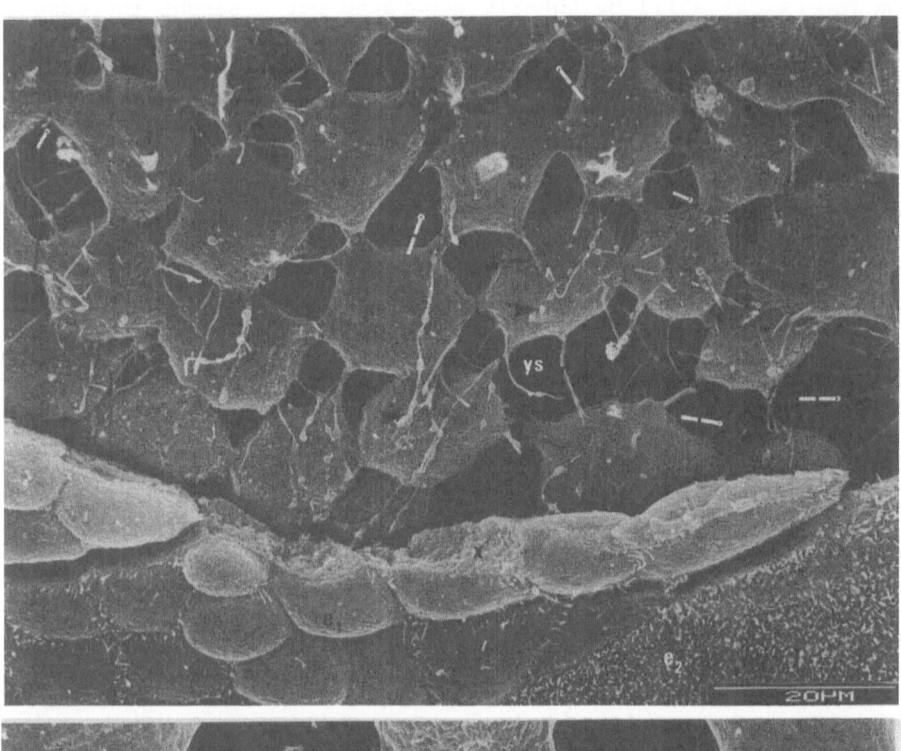

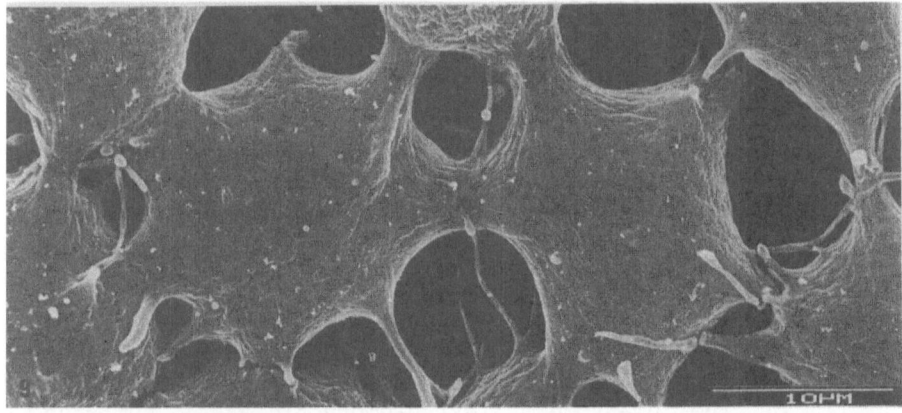

Fig. 8. Detail n⁰.4 from Fig. 2. Mes-endodermal cells inter-
         connected by cytoplasmic bridges (arrows). e_1 and
         e_2 : embryonic ectoderm (see also Fig. 3 and 4). x :
         fracture zone on an ectodermal cell ; arrow heads :
         "threads" and granules on and between the cells ;
         dotted arrows : filopodia anchored on the yolk
         syncytium (or on cell surfaces) ; m : mesodermal
         cell ; ys : yolk syncytium. Bar : 20 µm/1.090 x.

ectoderm of the future embryo is composed of two types of cells
(Fig. 2 : 1 ; Figs. 3, 4). One is observed in a well delimited zone
near the animal egg pole. These cells are polygonal, showing
densely and irregularly arranged microvilli. (This zone is the very
first manifestation of an organprimordium, the shell gland
primordium. The cells derive presumably from the two smallest
blastomeres at the fourth cleavage stage (2 : Table 1) ; however,
the cell lineage has not (yet) be shown experimentally.) The second
cell type constitutes all the rest of the ectoderm. These cells are
tri- to hexagonal in shape and present a "smooth" surface. On their
borders, there are interdigitating folds and villi. In particular
the folds extend occasionally towards the central part of the cell
surfaces. The most characteristic phenomenon of these cells however
is a short cilium (in some cases two cilia are found). It extrudes
from the cell surface in a slight depression. This cilium is
considered to by a "primary (rudimentary, solitary) cilium", which
is known from various in vivo and in vitro systems as a centriolar
specialisation of cells during interphase (i.e., for reference :
6). During preorganogenetic stages, no "primary cilia" have been
observed ; however, we may anticipate, that they are found
throughout early organogenesis.

   b) The inner surface. In a region, where the ectodermal layer
is turned back, its inner surface also can be studied (Fig. 2 : 2).
It is this surface, which is opposite to the underlying
mes-endoderm and, locally, to the yolk syncytium too. Figure 5
shows more details. No damaging of the cell surface, due to the
mechanical removal from the mes-endodermal contact is shown. Also
no injuries are found on the surfaces of the cells of the inner
germ layer (Fig. 8). No intimate membranous contact and no intimate
links (or cytoplasmic bridges) can therefore exist between the
cells of the outer and the inner germ layer. The low affinity of
adhesion of the germ layers observed in life is thus confirmed
structurally. However, one observes the presence of an
"interdermal" or "extracellular material". Whereas this material
appears to be an accidental deposit on the surface of the
mes-endodermal cells, it is abundant on the inner surface of the
ectoderm. There is good reason to locate the origin of this
material in the ectodermal and not the mes-endodermal cells.
Morphologically one distinguishes "droplets", granules, spherical
bodies, "rosary-like threads", thin filaments and blebs. Regional
differences as to the composition and the amount of accumulation of
the interdermal material seem to exist. Besides the extracellular

Fig. 9.   Detail n⁰.5 from Fig. 2. Large mesodermal cells in-
          terconnected by cytoplasmic bridges (x).
          Bar : 10 μm/2.100 x.

or interdermal material, the inner surface of the ectoderm is
characterized by filopodia-like outgrowths. They are either
oriented tangentially to the surface or towards the mes-endoderm
and the yolk syncytium. On the one hand, the cells of the ectoderm
are in direct contact with the cells of the mes-endoderm and
(trough the intercellular spaces within the mes-endoderm) with the
yolk syncytium, via these protrusions. On the other hand, one has
to assume that the interdermal material largely (selectively ?)
inhibits the direct membrane contact between the cells of the outer
and the inner germ layer. As to the role of the interdermal
material, various speculations could be made. (Could it be a
"glue", first necessary for bringing together ectodermal and
mes-endodermal cells before more definite junctions can be
formed?)

The ectoderm of the extra-embryonic region

        The border of the blastoderm is the most adhering part of the
ectoderm to the yolk syncytium and the cortex. It therefore is
generally not removed when removing the ectoderm. (The border of
the blastoderm has an important morphogenetic role : it guarantees,
prior to organogenesis, the maintenance of the well ordered spatial
cellular arrangement of the blastoderm as it derives following germ
layer formation (4, 5).) Some border cells are shown in Figs. 6 and
7 (see also Fig. 2 : 3). Cell shapes and surface structures
correspond to those observed everywhere in the extra-embryonic
ectoderm. The cells are large, somewhat fusiform with their longer
axis perpendicular to the egg axis. The surface is composed of
irregularly arranged microvilli of about the same size as those
described for cells in the embryonic region (1-1.5 μm). Locally
they appear also to produce folds.

The mes-endodermal complex

        The "typical" mes-endoderm of the pre-organogenetic develop-
mental phase is an annular complex of slightly shifted cells (1 :
p.230). At about stage V-VI it appears as a) a monolayer of
interconnected cells and b) single cells, spread out between the
yolk syncytium and the ectodermal cover (Fig. 2 : 4-6, 8). The
network-like cellular arrangement corresponds to the embryo proper;
the area of the individual cells is the envelope of the future
outer yolk sac.

        a) The area of interconnected cells. The cells of this area
are large (25-30 μm in diameter in a fixed state) and each cell is
interconnected with its neighbours by several cytoplasmic bridges
(Fig. 8, 9). This is particularly the case in the more posterior
regions, whereas in the peripheral or rostral zone (the equatorial
region of the egg) the cells are less connected (Fig. 10). In this

zone individual groups of a few interconnected cells are also observed. (Cytoplasmic bridges have first been described in a Loligo species by Arnold (9), which then led to further studies (10, 11).) Besides the plasmatic bridges, the cells are also reciprocally connected by various filopodia and surface protrusions. Numerous filopodia also extend within the intercellular spaces towards the yolk syncytium. The cell surfaces are "smooth". On their surfaces however and within the intercellular spaces an "interdermal material" (filaments, "threads", granules etc...) is found. Despite the fact, that the monolayer of interconnected cells is disrupted in some zones, absolutely no mes- or endodermal organ primordium can be defined yet. No distinction could even be made between the endodermal and the mesodermal component of the inner germ layer.

b) The area of individual cells. There seems to be no doubt, that the population of individual cells observed in embryos at stage VI derives from the most peripheral mes-endodermal cells observed in embryos at pre-organogenetic stages (1). As a final consequence, we believe, that they are descendants from those cells formed from the blastocones (7) as the process of germ layer formation proceds (stage II-III). Individual cells are first observed in embryos at about stage IV-V. At stage VI, they are irregularly scattered within the space formed by the syncytium and the ectoderm. In some cases, the most peripheral cells are close to the blastoderm border (Fig. 2 : 8 ; Fig. 13) ; in other cases they may be far behind. The cells are anchored to the substrate by numerous filopodia. Further details are given within the description of the yolk syncytium below.

The yolk syncytium

In Loligo vulgaris about 20% of the yolk syncytium is formed during the cleavage of the blastodisc : it is the innermost cytoplasmic region "delimiting" the yolk droplets. The other 80% originates directly from the original egg cortex, when the latter is progressively covered, after the germ layer formation has occured, by the proliferating blastoderm. As to the morphogenetic role of the yolk syncytium, it can be reasonably considered as an excellent physical and nutritive substrate for the overlying proliferating and differentiating cells. It also allows a coordinated cellular arrangement during wound closure in embryos at early organogenetic stages (5, 8). The relationship of the yolk syncytium with the cellular blastoderm we have termed in our determination model as "the first primary interaction system" (5). We are very conscious of the significance of the yolk syncytium in morphogenesis. However, we feel unable to support the hypotheses suggesting that the yolk syncytium (the egg cortex) would have an organ-specific inductive function (12-15 ; see also, 5, 8).

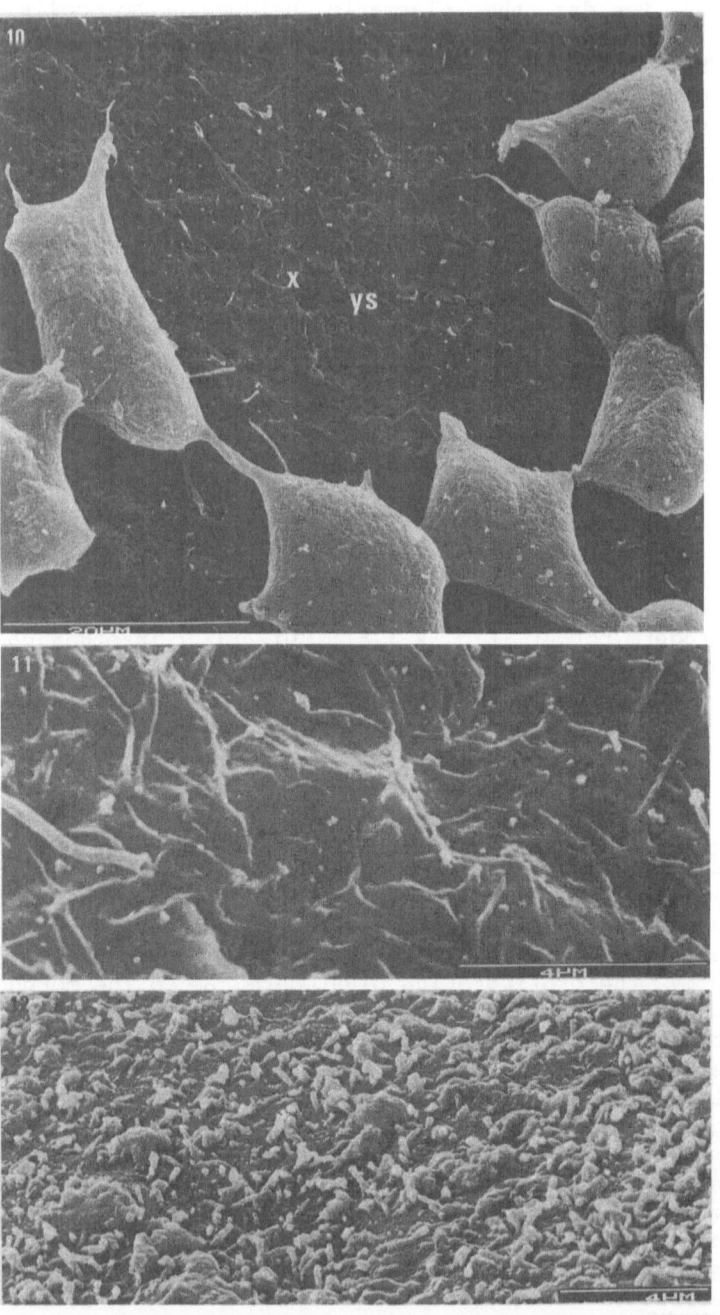

In the living embryo and at low magnification, the surface of the yolk syncytium looks like a rather homogenous and structureless sheet. On SEM-pictures however, it appears that it has considerable structure and that regional differences exist which correspond roughly to a) the embryonic and b) the extra-embryonic area (Fig. 2: 6, 8).

a) The embryonic area. Figure 10 and 11 present a portion of the yolk syncytium close to the "transition zone" of the embryonic to the extra-embryonic area. This surface structure is typical of the area of interconnected cells. Despite the fact, that it no longer resembles to the egg cortex (Fig. 12), it is certain, that it has been directly derived from the latter. For as yet unknown reasons, the cortical surface becomes modified, when it becomes covered by cells. The yolk syncytium of the embryonic area is characterized by folds and ribs, delimiting flatter portions, and dispersed microvilli. Neither a particular order nor an orientation of the ribs is recognizable. Beneath the intercellular spaces within the posterior parts of the mes-endodermal complex, the same structures are seen (Fig. 8, 9). One may assume, that they are also identical underneath the mes-endodermal cells. The original egg cortex (Fig. 12) is characterized by many short villi and irregularly intermingled bulbous surface protrusions.

b) The extra-embryonic area. The typical surface structures of this area is shown in Figs. 13 and 14 (see also, Fig. 2 : 8). It appears literally "thrown up like a stormy sea"; a surface formed by villi, ribs, bulbs and variuos other protrusions. It differs

Fig. 10. Detail n^{0}.6 from Fig. 2. Mesodermal cells in the peripheral rostral region of the mes-endodermal cell complex (approximately the region of the future arms). ys : yolk syncytium ; x : see Fig. 11. Bar : 20 μm/1.630 x.

Fig. 11. Detail from Fig. 10. Surface of the yolk syncytium. Various types of villi, "ribs" and folds (arrows) are irregularly arranged. This type of substrate is only observed in the embryonic part differing from that in the extra-embryonic part (Figs. 13, 14). Bar : 4 μm/8.100 x.

Fig. 12. Detail n^{0}.7 from Fig. 2. Surface of the egg cortex. Short microvilli and various bulbous protrusions are evident. Bar : 4 μm/4.700 x.

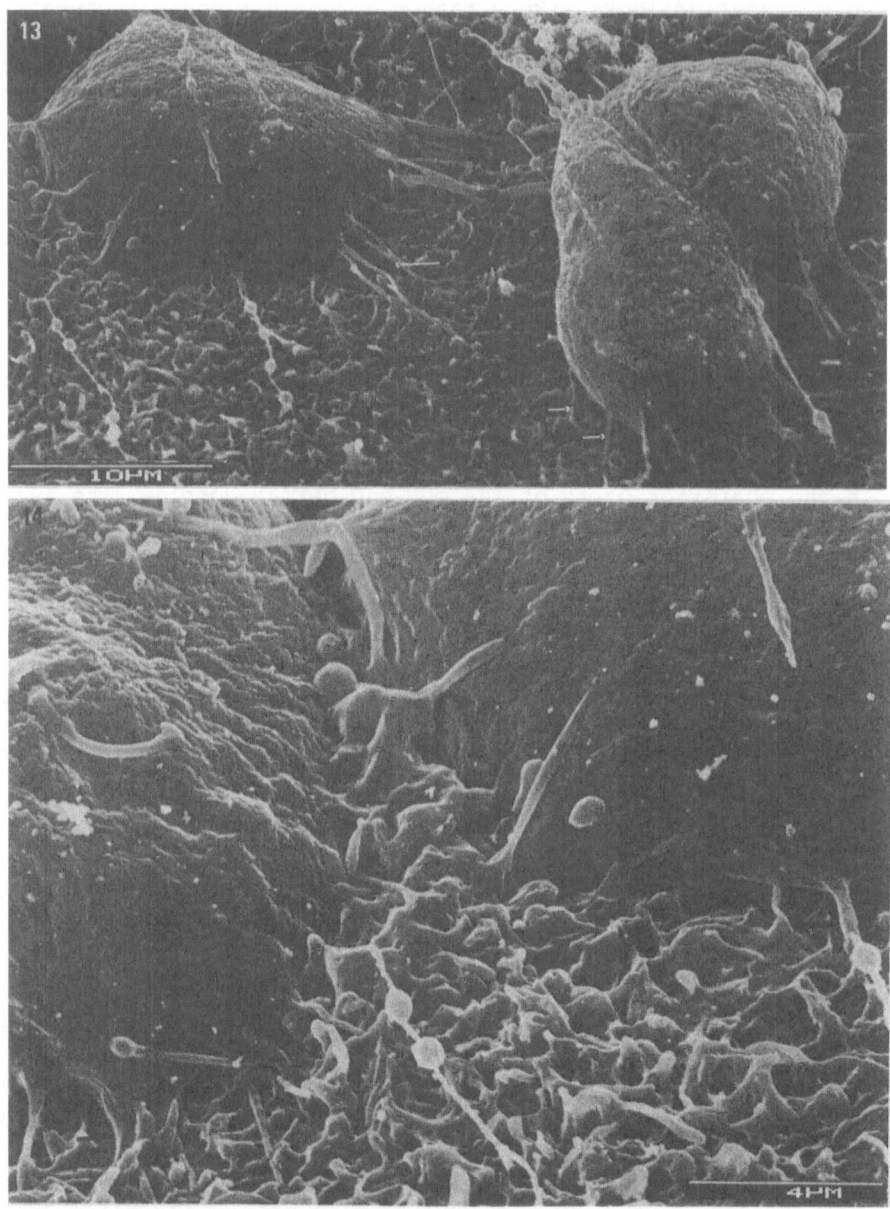

Fig. 13. Detail nº.8 from Fig. 2. Individual mesodermal cells
         in the extra-embryonic area. The cells are anchored
         in the yolk syncytium (arrows), on and between the
         surface protrusions of the yolk syncytium.
         Bar : 10 μm/2.200 x.

from that in the embryonic region as well as from that of the egg
cortex (Fig. 6 ; Fig. 12). In fact, one is tempted, to conclude
from these morphological findings that it reveals a region-specific
yolk digesting activity.

The interaction between the individual cells and the yolk
syncytium is structurally expressed. On the one hand, each cell is
anchored by filopodia to the syncytial substrate (Fig. 13). The
tips of the filopodia reach the substrate on the tops of the
protrusions as well as in the grooves. On the other hand, it
appears that not only the cell itself makes actively contact with
the substrate, but that the latter extends, on its turn, long
protrusions towards the cells and over the cell surfaces (Fig. 14).
Within the relationship cell-syncytial substrate, both partners are
likely to be equally active. (This is, of course, a fundamental
difference to an in vitro cell-to-substrate interaction.)

Interdermal material as described above is found in any region
of the yolk syncytium.

B) Tissues and cells of an embryo at the early organogenetic
   stage VII

The first "real" organogenetic stage is termed "stage VII"
(Fig. 1). Bilateral, dorsoventral and antero-posterior symetry is
morphologically clearly expressed. First the ectodermal organ
primordia become prominent : the shell gland, the optic vesicles,
the arm buds etc... From histological work it is known, that the
mes-endodermal complex also proceeds towards organogenetic
differentiation (16-18). Within the chorionic space the embryo
rotates slowly, thus indicating fully functional cilia grown on the
epithelial cells of the extraembryonic ectoderm. A small portion of
the original egg cortex at the vegetal pole is still seen.

On the living embryo, when mechanically separating the
ectoderm from the underlying mes-endoderm, it becomes evident that
many cells and smaller cell groups of the latter adhere so strongly

---

Fig. 14. Detail from Fig. 13. The appearance of the yolk syn-
         cytium is different from that in the embryonic area
         (Fig. 11), as well as from that of the egg cortex
         (Fig. 12) from which it originates. Dense surface
         protrusions and folds are present. The cell is
         anchored not only in the yolk syncytium (arrows),
         the substrate itself also occasionally extends over
         the cell surfaces (arrow heads).
         Bar : 4 μm/5.400 x.

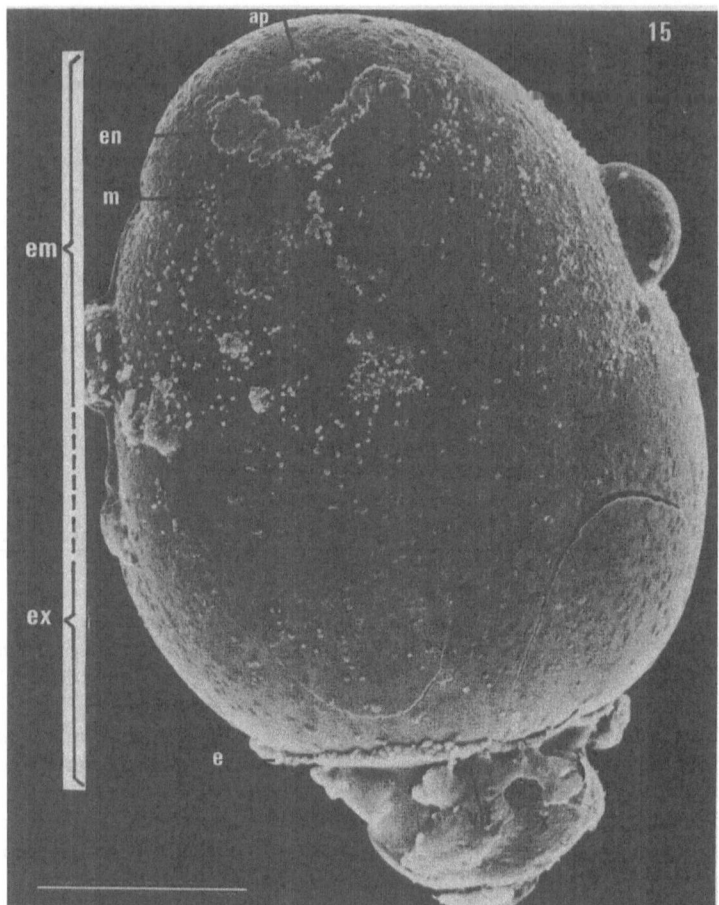

Fig. 15. Ventral view on an operated embryo, at about stage
        VII (slightly more advanced than that in Fig. 1).
        The organogenesis is well advanced. The ectodermal
        cell layer and the mes-endodermal complex adhere
        increasingly closely ; therefore, when removing the
        ectoderm, numerous cells of the mesoderm are also
        removed. The remaining cells still provide a good
        insight into the cellular arrangment, cell shapes,
        sizes and interconnections, as well as the
        appearance of the syncytial substrate. The endoderm
        (1: en) becomes distinct from the mesodermal cell
        material. Individual cells are observed in the
        extra-embryonic area. ap : animal pole ; vp :
        vegetal pole ; m : mesoderm ; x : artefacts (out-
        leaking yolk) ; arrow : migrating single cells.

to the former, that they are removed together with it. Compared to the situation described for stage VI, the increase in germ layer affinity at stage VII is drastic. Since larger parts of the internal cell population are therefore detached with the ectoderm, a study of the surface morphology of the mes-endoderm and the yolk syncytium has consequently to be performed on an incomplete specimen. However, the remaining tissue fragments and cells, laid out on the yolk syncytium still give good insight into the actual morphological situation.

Analysis of the SEM-pictures (Fig. 15-25)

Figure 15 gives an overview on the ventral face of the mes-endoderm and the yolk syncytium of an embryo at stage VII. In fact, comparing this figure with Fig. 2 (stage VI), the morphological aspect has considerably changed. Specific cell and organ differentiation has advanced. The interconnected cell complex has broken down into placodes of cells (Fig. 15 : 1), into groups of round cells with each cell in contact with others (Fig. 15 : 2) and into groups of spread but still interconnected cells (Fig. 15 : 4). In the extra-embryonic area the number of individual cells of various forms has increased (Fig. 15 : 4-6). The yolk syncytium has also been modified, and, most importantly, the mes-endoderm has been segregated into an actual endoderm and a mesoderm.

The endoderm

In the posterior region of the embryonic area, one distinguishes a solid, nearly bilaterally symetrical placode (Fig. 15 : en). From its position and its form (16, 19-21), no doubt seems possible, that this placode corresponds to the medioventral and partially lateral endoderm, that is, the common anlage for hind- and midgut (19-21). Whereas the median part corresponds to the hindgut and ink gland rudiment, the enlarged lateral parts (incomplete on one side) comprize the paired anlagen for caecum, digestive gland and stomach. Figure 16 gives more insight into the cells' shape, size and connections. The cells are closely packed, presenting many surface protrusions and have a somewhat "angular" shape. The average diameter of a cell is 15 µm. Those cells in direct contact with the substrate are anchored to it by filopodia.

1: see Fig. 16 ; 2: see Figs. 17 and 18 ; 3: see Fig. 19 ; 4: see Figs. 20-23 ; 5: see Fig. 24 ; 6: see Fig. 25. Bar : 0.5 mm / 39 x.

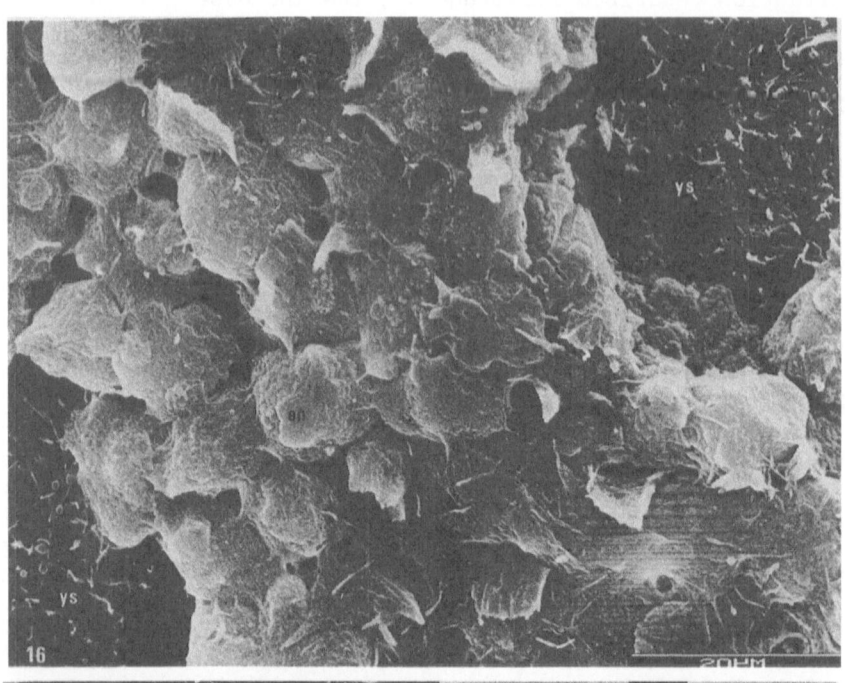

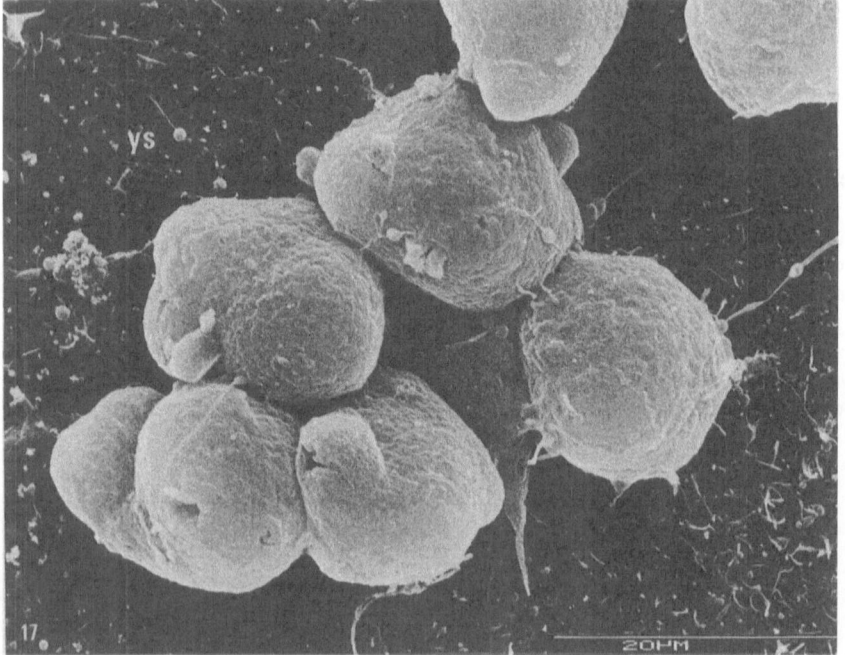

The mesoderm

     We also might assume that the mesodermal component too should
reveal   a   morphologically   recognizable   initial state of specific
cell and organ differentiation. However, in SEM-pictures of stage
VII   embryos   no well defined organ anlage is seen yet, but various
region-specific cell types have to be noted.

     - In the embryonic area. The cell population of the posterior
part is mainly composed of round or oval cells (Fig. 15 : 2), often
grouped in small   clusters   (Fig.   17).   They   present   a   "smooth"
surface,   have   various surface protrusions (i.e., Fig. 18) and are
poorly anchored on the yolk syncytium.

     - Towards the equatorial egg region (transition   zone   between
the embryonic and extra-embryonic area). Groups of round as well as
of fusiform and flattened cells are found (Fig. 15 : 3,   4 ;   Fig.
19-21).   Also   few single cells are seen. (The groups of spread and
flattened cells are   likely   to   be   "incorporated"   into   the   arm
anlagen.) Mainly   the   border   cells   of   these   clusters are well
anchored by filopodia within the yolk syncytium. From their spread
and   stretched   forms,   on might assume that such cell groups might
occasionally move as a whole.

     - In the extra-embryonic area. The individual mesodermal cells
have   proved   particularly   interesting.   In fact, this is the only
cell type, which migrates individually and actively and for which
origin and final fate are known. The migratory behaviour was first
recognized by analyzing SEM-pictures of stage V-VI embryos (1 :
p.232).   It   was   clearly   confirmed   by   in situ recording using
time-lapse   micro-cinematography   (23).   A   more   detailed

---

     Fig. 16. Detail n⁰.1 from Fig. 15. Differentiating endoderm,
              forming the midgut and hindgut rudiment. Cell shapes
              and sizes differ from those of mesodermal cell
              groups. en : endoderm ; ys : yolk syncytium.
              Bar : 20 μm/1.340 x.

     Fig. 17. Detail n⁰.2 from Fig. 15. Small group of differen-
              tiating   mesodermal   cells   (precise   final fate
              unknown, presumably incorporated in the branchial
              region). Loose contact between the large cells. The
              aspects of the yolk syncytium (ys) resembles that of
              earlier   stages   (Fig.   11), but the amount of
              interdermal material has increased.
              Bar : 20μm/1.770 x.

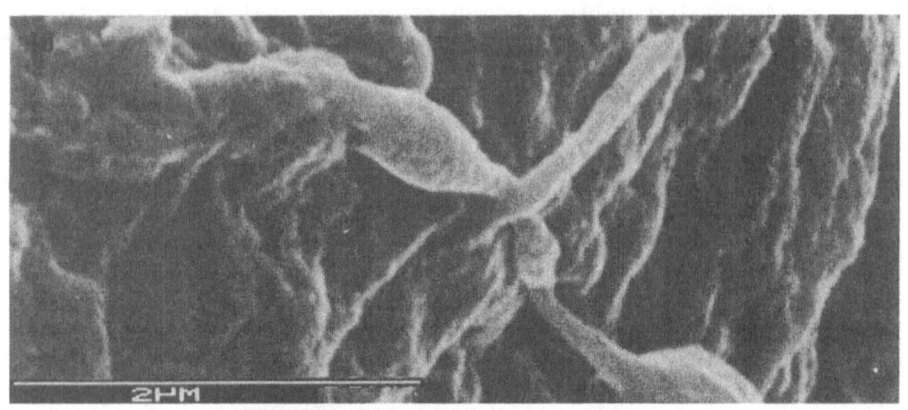

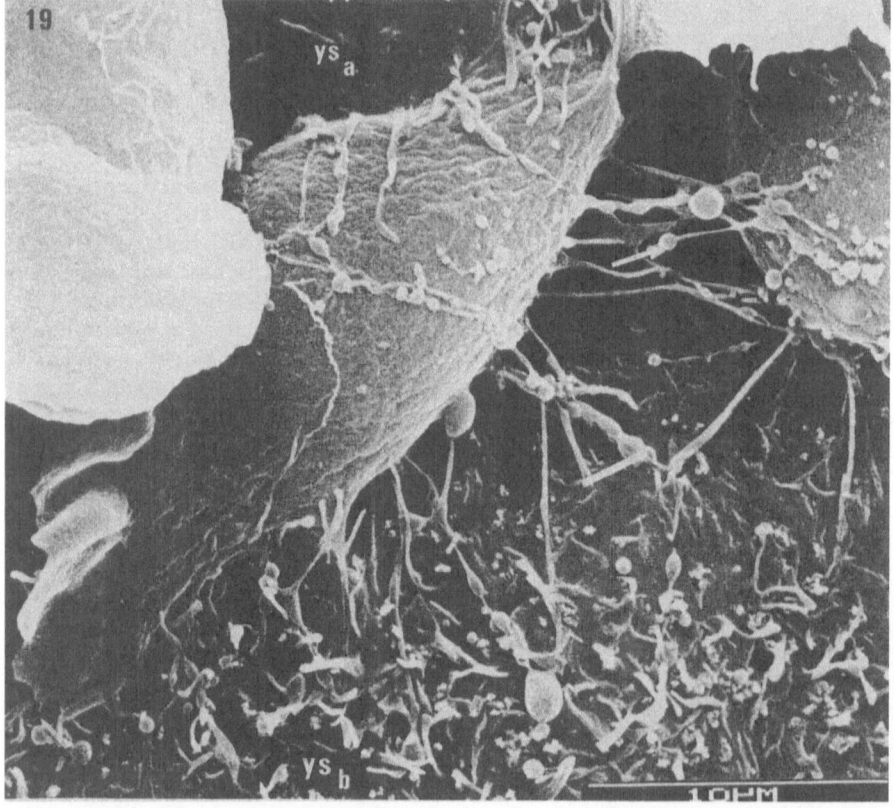

Fig. 18. Detail from Fig. 17. Pronounced surface protrusions of a mesodermal cell. Bar : 2 μm/10.000 x.

Fig. 19. Detail n°.3 from Fig. 15. Mesodermal cells close to the equatorial egg region. Transition zone of two regions of the yolk syncytium (ys$_a$: embryonic area and ys$_b$: extra-embryonic area). Large amounts of interdermal or extracellular material (arrows) are present. Bar : 10 μm/2.600 x.

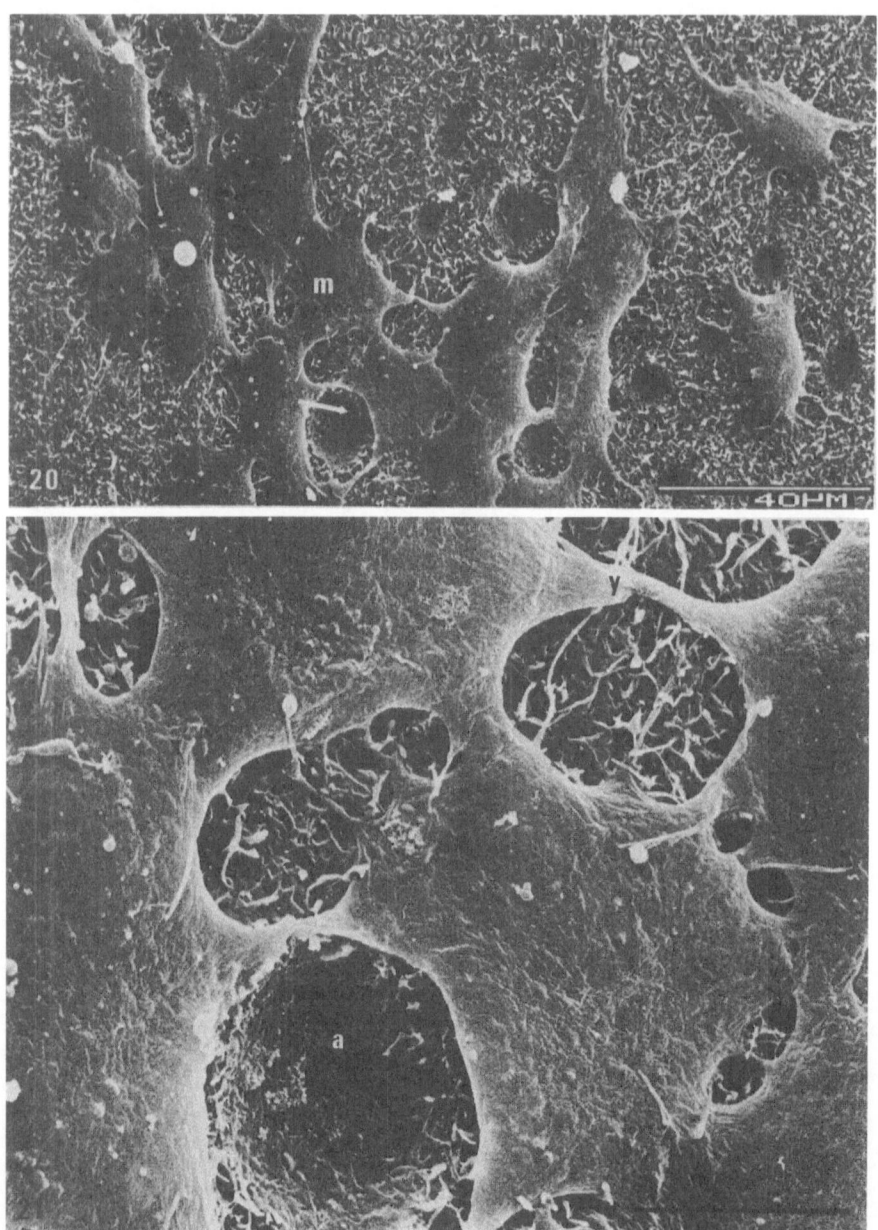

Fig. 20. Detail n⁰.4 from Fig. 15. Interconnected (m) as well
as individual mesodermal cells in the equatorial region. The surface of the yolk syncytium is composed
of villi and large protrusions. The arrow indicates
a depression in the syncytial substrate, where a
cell was presumably removed during the operation
procedure. Bar : 40 μm/480 x.

Fig. 21. Detail from Fig. 20. Interconnected stretched mesodermal cells on the yolk substrate. In places where
cells have been accidentally detached (a), the
villosity of the substrate appears less pronounced.
x : cytoplasmic bridges ; y : membrane contact between cells. Bar : 10 μm/1.800 x.

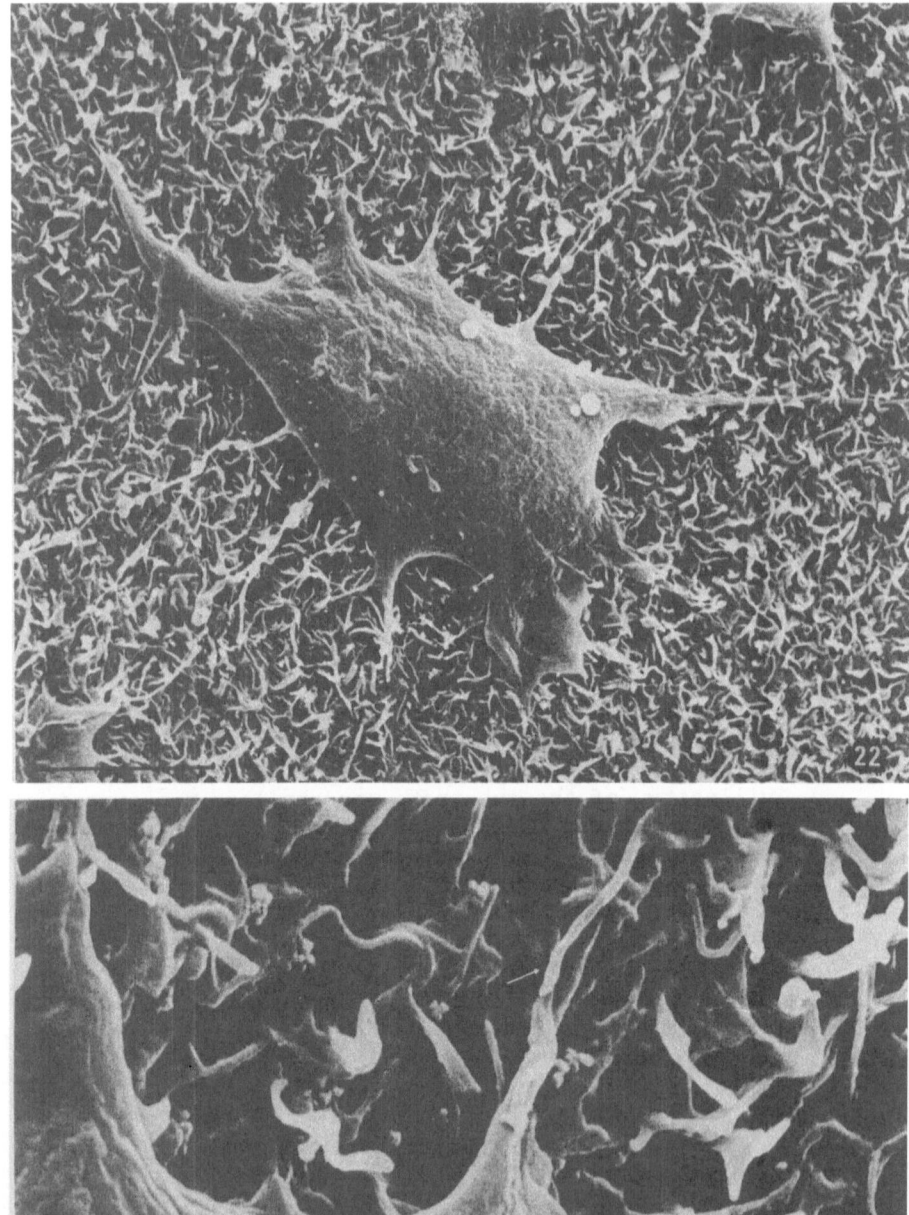

Fig. 22. Detail n⁰.4 from Fig. 15. Individual cell migrating
        on the villious yolk syncytium. The anchoring points
        of the filopodia on the substrate are essentially
        between and not on the microvilli and surface
        protrusions (arows). Bar : 10 µm/1.800 x.

Fig. 23. Detail from Fig. 22. Folds and ribs of the surface
        of the yolk syncytium. The attachment point of the
        cell to the syncytial substrate is marked with an
        arrow. Bar : 2 µm/9.600 x.

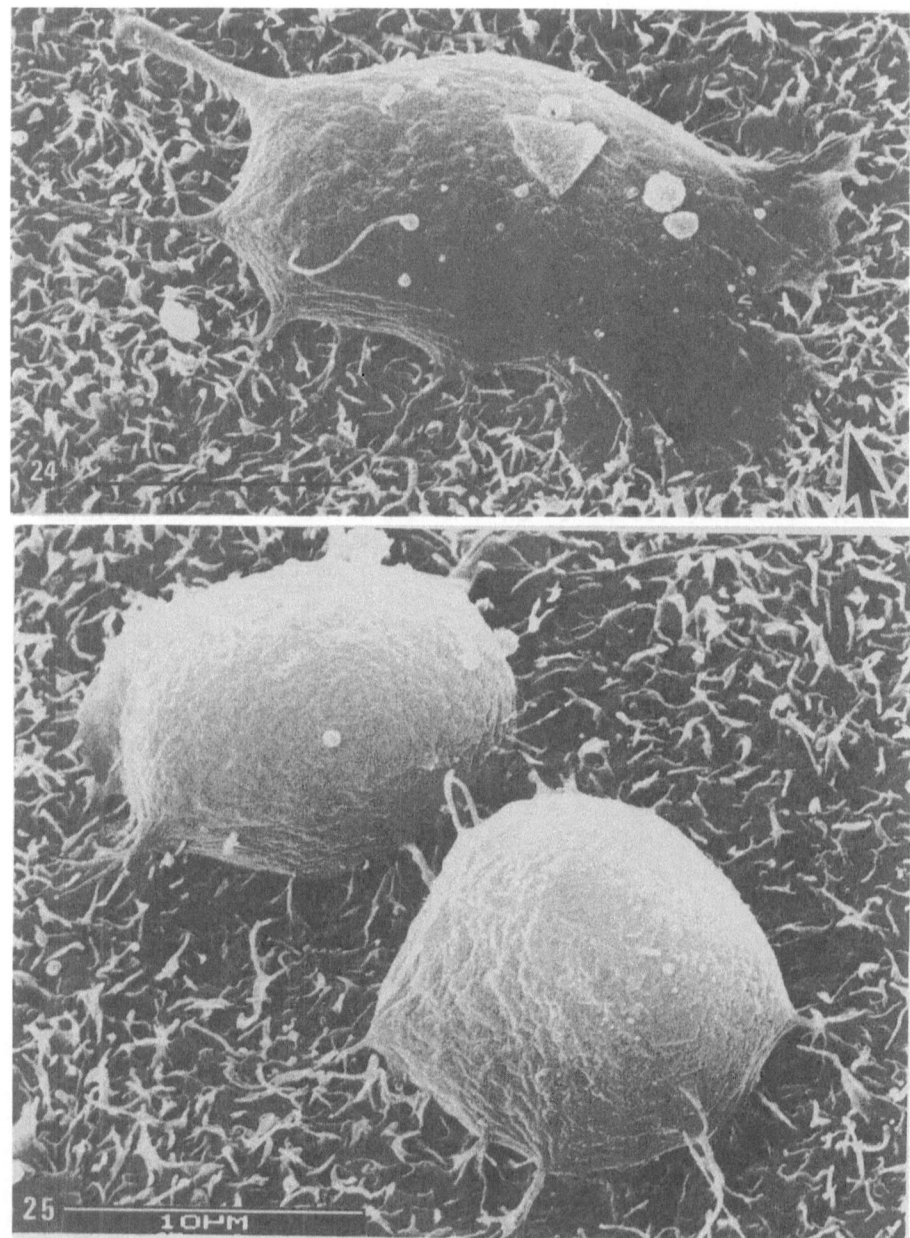

Fig. 24. Detail n⁰.5 from Fig. 15. Migrating cell with dis-
         tinct lamellipodium-like leading edge (arrow).
         Adhesion sites between cells and yolk substrate are
         mainly between the folds and villi.
         Bar : 10 μm/2.200 x.

Fig. 25. Detail n⁰.6 from Fig. 15. Two cells, their rounded
         shape indicating that they have divided shortly
         before. The round cells, less well anchored in the
         syncytium are often detached during the removal of
         the overlying ectoderm from a living embryo.
         Bar : 10 μm/2.600 x.

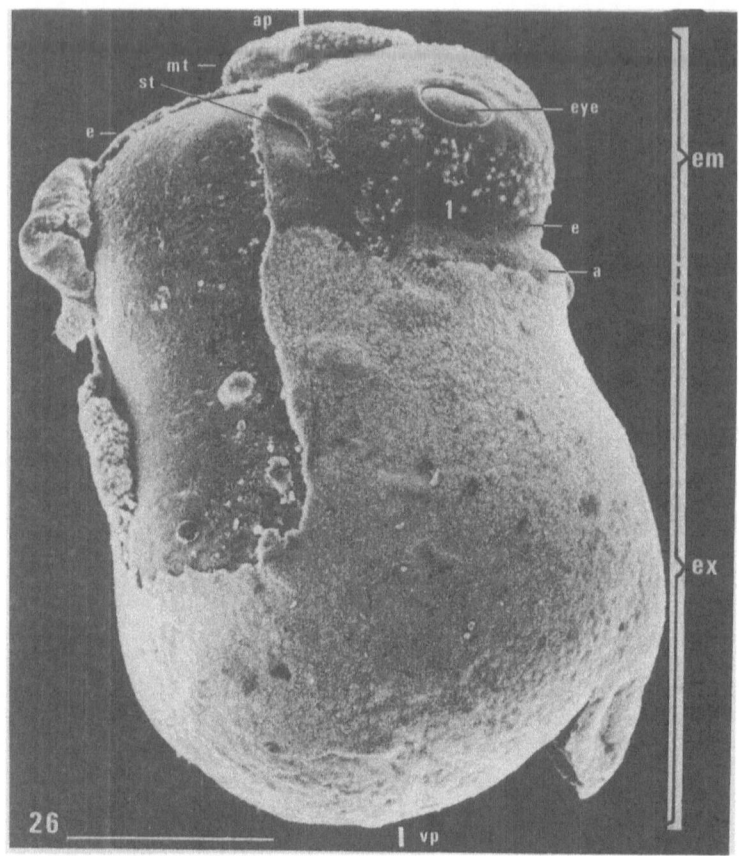

Fig. 26. Dorsolateral view on an operated embryo of stage VIII
         -IX. The ectoderm of the dorsolateral side has been
         removed giving access to the inner cell layer(s) and
         groups and the yolk syncytium. Organogenesis is well
         advanced and several organ rudiments can be dis-
         tinguished. a : arm rudiment ; e : ectoderm ; m :
         mesoderm ; mt : mantle rudiment ; eye : rudiment of
         the optic vesicle (central part : retina, peripheral
         wall : lentigenic material and iris) ; em : embryonic
         part ; ex : extra-embryonic part = outer yolk sac ;
         st : invagination of the stomodeum ; ap : animal pole
         vp : vegetal pole.
         1: see Figs. 27 and 28 ; 2: see Fig. 29 ; 3: see Fig.
         30 ; 4: see Figs. 31 and 32. Bar : 0,5 mm / 40 x.

cinematographic study is well advanced. The SEM-pictures shown here
(Fig. 22, 24) provide further details as to the morphology of the
migrating cell and its relation to the substrate. The cells have a
"typical migrating" aspect. They are slightly elongated, and they
are in contact with the syncytial substrate with long and thin
filopodia. The tips of the filopodia reach to the substrate mainly
underneath the microvilli (see also Fig. 23). The migration
direction can be deduced from the position of a lamellipodium-like
"leading edge". Although the lamellipodium extends essentially over
the microvilli, it protrudes locally between them. The substrate is
not modified under the influence of a migrating cell ; the
microvilli are not crushed. Since migrating cells forming
lamellipodia generally are found in in vitro but not in
multicellular in vivo systems, this (intermediate) case of an in
vitro-behaviour in an in vivo-system, certainly merits particular
attention. The more since the substrate is a living, active and
microvillous syncytial substrate.

Besides "migrating cell forms", one also observes in the
extra-embryonic area numerous cells with round shapes (Fig. 25).
From time-lapse recordings we know that migrating cells
occasionally stop and divide (23 ; and unpublished observations) ;
the daughter cells then continue migrating. We therefore assume
that these round cells are such cells immediately before or after
cell division. However, at this moment, we can not entirely
exclude, that different mesodermal cell types differentiate within
the extra-embryonic area.

The yolk syncytium

The yolk syncytium of the embryonic area in stage VII embryos
looks similar to that described before (Fig. 16, 17 : ys). However,
in the transition zone to the extra-embryonic area as well as in
the proper extra-embryonic area, it has developed into a dense
"carpet" of microvilli (Fig. 19-25). In any region of the yolk
syncytium small depressions are found (i.e., Fig. 15 : 6 ; Fig. 20
and Fig. 21 : a). These depressions are considered as "imprint"
left by round cells. (These cells, less well anchored to the
substrate are particularly easely detached from the substrate when
the ectoderm is removed.) If "round cells" are cells close to or
during mitosis, this could mean that such cells locally modify the
syncytial substrate. Not only mechanically by "crushing" the
microvilli and surface protrusions, but physiologically.

Finally, the interdermal material is locally accumulated in
considerable amount (i.e., Fig. 19).

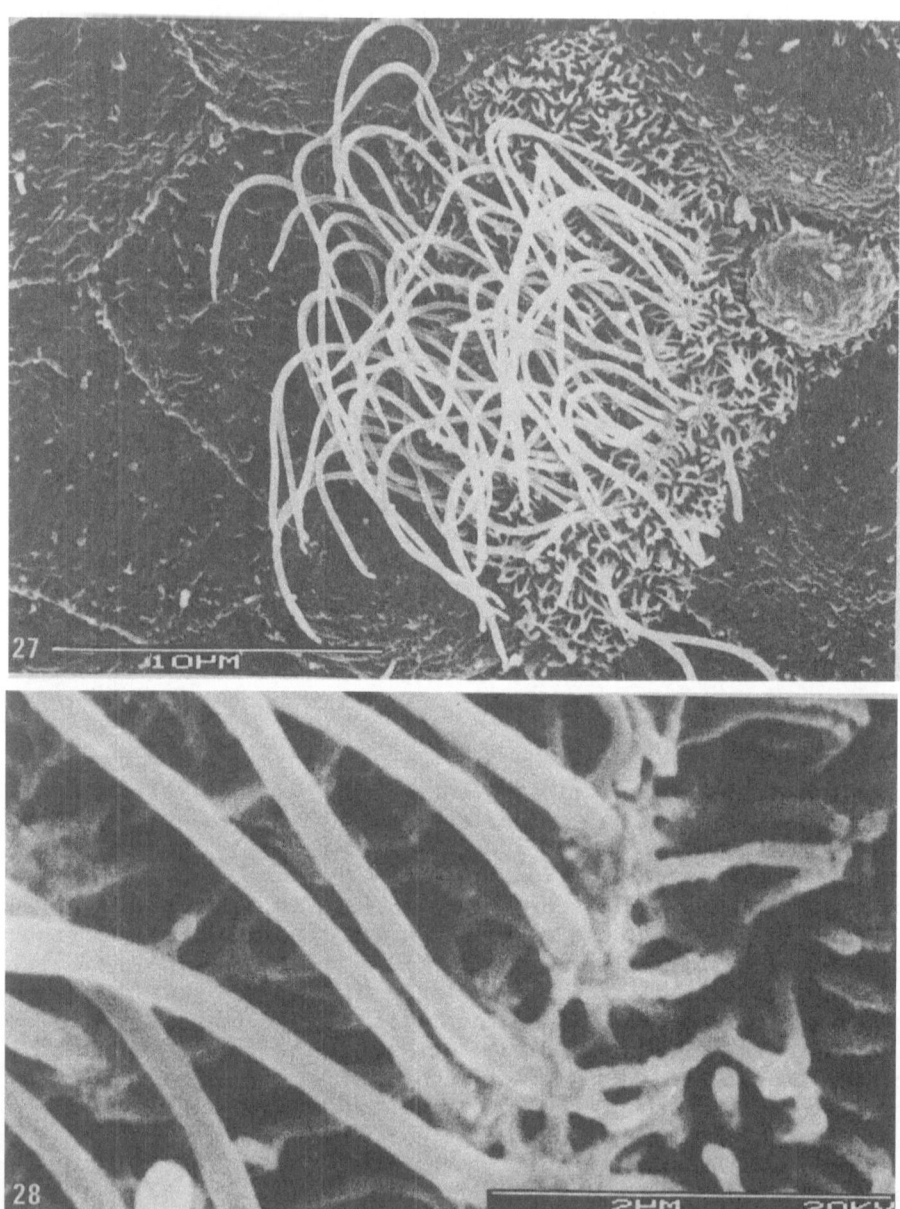

Fig. 27. Detail nº .1 from Fig. 26. Groups as well as indivi-
        dual ciliated cells are seen within the embryonic
        ectoderm. Compare Fig. 2 and 28.
        Bar : 10 μm/2.900 x.

Fig. 28. Detail from Fig. 27. Bases of cilia within the mi-
        crovilli "carpet". Bar : 2 μm/17.700 x.

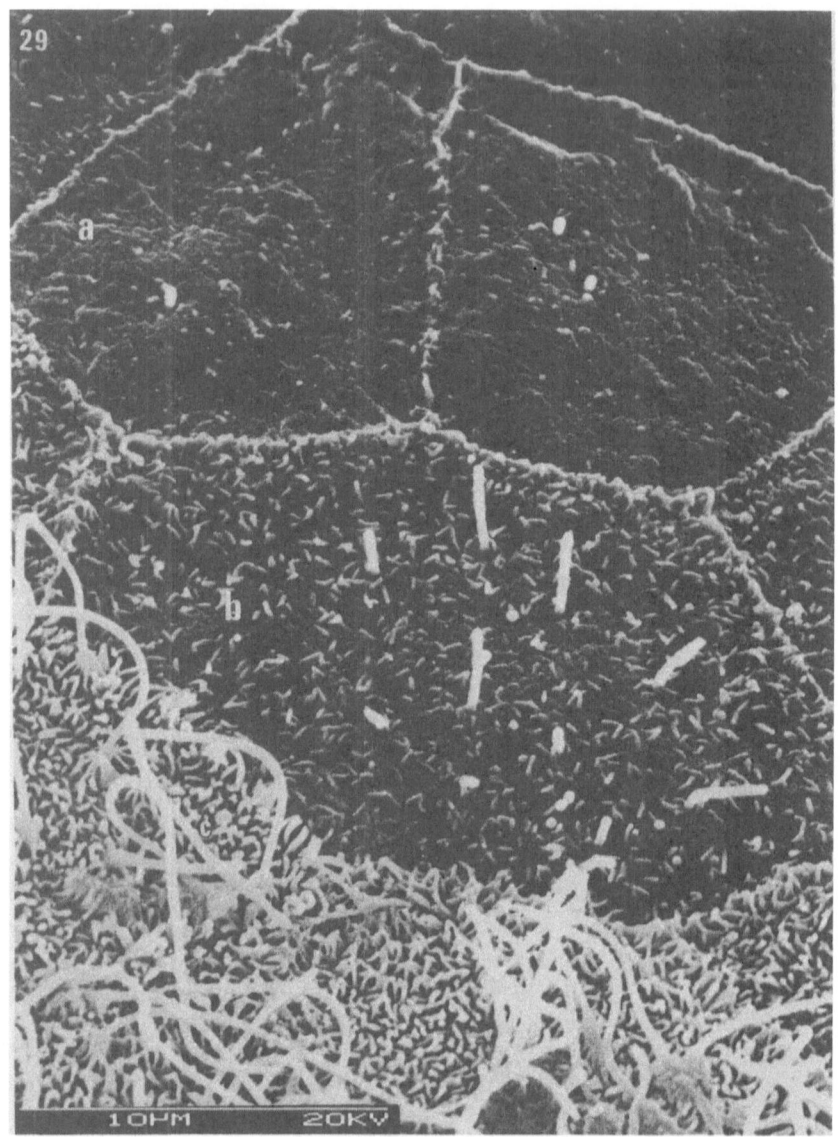

Fig. 29. Detail n⁰ .2 from Fig. 26. Transition zone between the
         ectoderm of the embryonic part (a) and that of the
         outer yolk sac envelope (c). The cells of this zone
         appears to be in a intermediate stage   showing
         microvilli and few growing cilia (b).
         Bar 10 μm/3.500 x.

C) Tissues and cells of an embryo at stage VIII-IX (Fig. 26-33)

    Once the anlagen of the shell gland, the mantle, the eyes, the
statocysts, the stomodeum,   the   tentacles   etc...   become   clearly
distinct as tissue placodes, walls and invaginations, organogenesis
becomes rapidly advanced and precise staging criteria can be used.
An (operated) embryo at stage VIII-IX is shown in Fig. 26 (i.e.,
one of the staging criteria : half-closed eye vesicle).   At this
stage, the   outer   yolk sac pulsates rhythmically and is therefore
the first fully functional "organ". Whereas in the embryogenic part
organogenesis precedes histogenesis (24), in the outer yolk sac the
order is reversed. It is still possible to remove parts of the
ectoderm, but large parts of the mesoderm, strongly adhering to the
ectoderm and forming in common   various   organ   anlagen,   are   also
removed. The yolk syncytium appears to be very fragile and when
touching it with needles, it easely breaks. Analyzing the different
cell and tissue structures, one realizes now that "squid-specific"
differentiation phenomena relieve the more common   cell   biological
phenomena as seen at the stages before.

Analysis of the SEM-pictures (Fig. 26-33)

    - The ectoderm : As one could already expect at stage VI, in
stage VIII-IX embryos (Fig. 26 : 1, 2), the ectodermal cover of the
embryo proper and of the outer yolk sac looks different. Whereas in
the former two cell types are found (Fig. 27, 29 : a), only one
constitutes the epithelium of the latter (Fig. 29 : c).

    For its larger portion, the embryonic ectoderm is composed of
"primary cilia" cells as described earlier. It appears, that these
"quiescent" cells perform the morphogenetic movements necessary for
spatial organ formation (i.e., see invagination of the stomodeum).
Irregularly intermingled individual cells are found. The surface of
these cells is formed primarily by densely packed microvilli, but
also contains many long cilia. The base of an individual cilium is
maintained by a wreath of microvilli (Fig. 28).

    The ectodermal cover of the outer yolk sac is an epithelium of
ciliated cells. The cells are large and carry on their surfaces
densely arranged microvilli and long cilia (Fig. 26 : 2, Fig. 29
:b). Finally, at the transition zone cells are found carrying a few
microvilli and a few short cilia (Fig. 29 : b).

    - The mesoderm : The description of the mesodermal component
is fragmentary and limited to some cells adhering to the yolk
syncytium (Fig. 26 : 3 and 4). Underneath the stomodeum cells are
interconnected by numerous ramifications (Fig. 30). These cells are
also anchored on the yolk syncytium. They may be muscle cells. A
cell group is shown in the outer yolk sac area in Fig. 31. These
cells are stretched on   the   yolk   syncytium   and   strongly

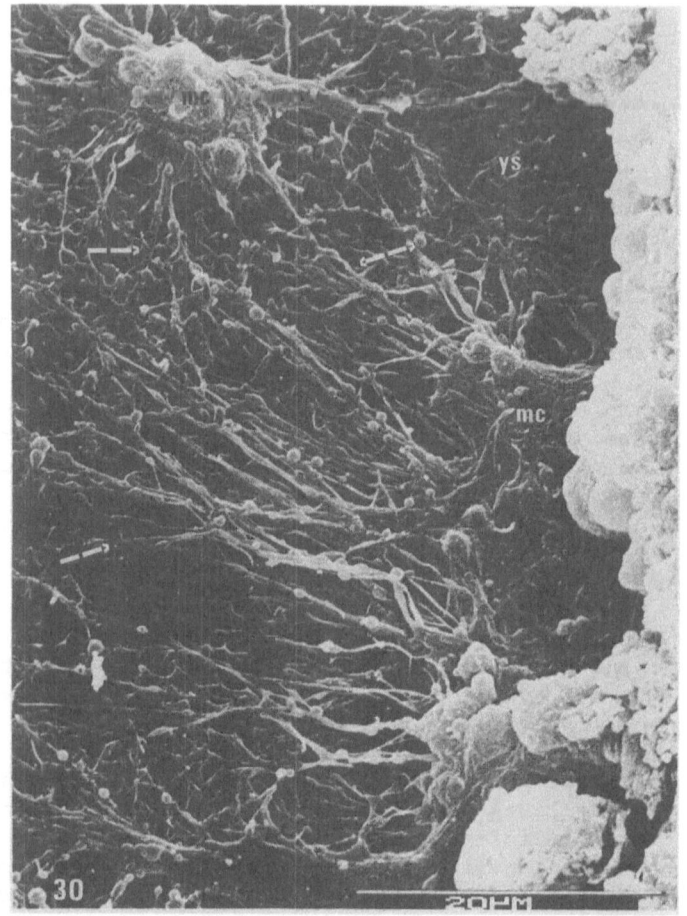

Fig. 30. Detail n⁰.3 from Fig. 26. Mesodermal cells (presumably muscle cells (mc)), interconnecting (arrows)
and anchoring (dotted arrows) on the yolk syncytium
(ys). Bar : 20 µm/1.460 x.

interconnected (see also Fig. 33). In this case, there can be not
doubt, that these cells are functional muscle cells, producing the
rhythmic pulsation of the outer yolk sac. They represent the final
differentiation state of the migrating cells observed at earlier
stages.

    - The yolk syncytium : The surface of the yolk syncytium
appears to be modified again by stage VII. The microvilli have
disappeared and, in the outer yolk sac area, the surface is
extremely broken up (Fig. 31 and 32). This is less the case in the

embryonic area. The "interdermal material" (granules, droplets, filaments) has considerably increased. It is dispersed all over the yolk syncytium surface and one might imagine, that it is produced by the latter too and not entirely by the ectoderm as supposed at stage VI.

CONCLUDING REMARKS

The "principle of Krogh" means : "For a large number of problems there will be some animal of choice, on which it can be most conviently studied" (25, 26). With this idea in mind, we are convinced, that the Cephalopod embryo, at pre-organogenetic and early organogenetic stages, is a system of choice, on which many actual problems found in cell-to-cell and cell-to-substrate interactions can be conviently studied. In presenting here a short overview on various cellular, acellular and syncytial structures in embryos of the squid, Loligo vulgaris, near to and during early organogenesis, we want to illustrate and support the usefulness of this (Invertebrate) system for future, detailed studies on problems such as cells' adhesivness, motility, positioning, recognition and (selective) communication. These studies may focus on either structural, physiological, cell behavioural or cell molecular aspects, the well illustrated description given here may serve as an initial morphological basis.

Throughout this paper the presentation of the tissues, the cells and the syncytial substrate has been accompanied by various additional comments. We do not want to repeat them. However, we only wish to re-emphasize two points, which we consider as essential for promising future studies :

- An embryo at stage VI (based on the cellular arrangement of the inner germ layer ; i.e., Fig. 2, Fig. 8) is likely to be a valuable model system for studying direct transfer and exchange of "developmental information" between cells.

- An embryo at stage VI and stage VII offers interesting possibilities for studying in situ various migration phenomena of cells interacting with a living, active and microvillious substrate (i.e., Fig. 15, Fig. 22, 24). Also the main topic of this course, the mechanisms of control during the formation and the functioning of cell-to-cell and cell-to-substrate interactions, may be recognized as far as they are expressed on a cell structural, functional and behavioural level. There is no reason to doubt that this material should finally be suitable to methodologies used in Molecular Biology as well. Therefore, it is our open aim to "promote" the Cephalopod embryo and to claim that it may progressively find its place among the more "classical", generally

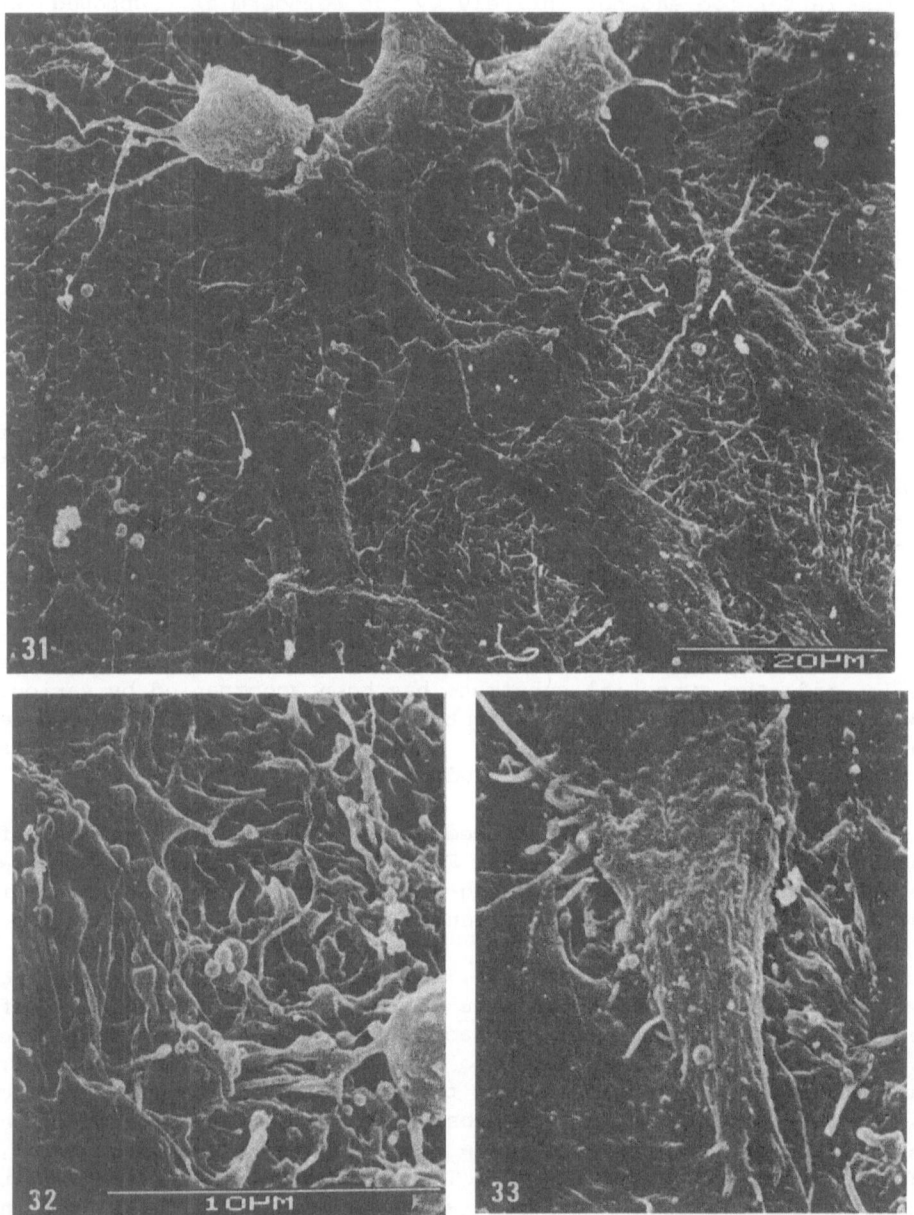

Fig. 31. Detail n°.4 from Fig. 26. Muscle cells in the outer
         yolk sac region. These cells are functional muscle
         cells originating from the individual migrating
         cells as shown in earlier stages (Fig. 15). The
         cells are interconnected as well as anchored in the
         yolk syncytium (and possibly also on the inner
         surface of the ectoderm). Their functioning ensures
         the rhythmic pulsation of the outer yolk sac.
         Bar : 20 µm/950 x.

Fig. 32. Detail from Fig. 31. The surface of the yolk syncy-
         tium is extremly folded and a great amount of
         "extracellular material" is observed.
         Bar : 10 µm/2.900 x.

Fig. 33. Detail from Fig. 26. Zone of contact between two mus-
         cle cells. Bar : 10 µm/2.300 x.

better   known,   in   vivo systems currently studied in Developmental
Biology.

Of course, the morphological presentation could now   be
extensively discussed from a "squid-specific" embryological point
of view. For instance, the structures found could be interpreted in
the context of our determination hypothesis on the existence of two
primary cell and tissue interaction systems   (1, 5) ; one might
speculate on how the induction-reaction system ("the second primary
interaction   system"   (5,   26)),   works   in   the   presence   of   an
interdermal material ; one might argue for the importance of purely
physical factors, also essential for specific cell differentiation
(anchoring of the blastoderm border (5) ; temporary stretching of
cells, phase-specific blocking of cells in well defined positions)
and one could further discuss, how individual cells detach first
from a solid complex for a migratory phase and again link up in an
other generation period. These points and many others have and will
be considered with the necessary detail in specific studies.   This
paper is aimed primarily at the specialists in cell interaction
phenomena, interested in a complementary or alternative "research
material" and secondarily to the (few) specialists in experimental
Cephalopod embryology.

ACKNOWLEDGEMENTS

We thank Prof. W.J. Gehring, Biozentrum, Basel (CH) for
hospitality,   stimulating   discussions   and   for offering the
SEM-facilities of his Department. The technical assistance of M.
Düggelin, SEM-Laboratory, University of Basel (CH), is greatly
aknowledged. We also thank Dr. E.L. Benedetti, Institut J. Monod,
Paris, for many encouraging discussions and advice. Dr. R. O'Dor,
University of Halifax (Canada) and R. Tait (Banyuls-sur-mer) for
having red the manuscript.

REFERENCES

1. H.-J. Marthy, Progress in Clin. and Biol. Res., In: "Embryonic
    Dev.", Part B., M.M. Burger and R. Weber, eds., Allan R. Lyss,
    N.Y., 85 : 223-233 (1982).
2. A. Naef, Fauna Flora del Golfo di Napoli, V-IX, 35 (2) : 1-357
    (1928).
3. H.-J. Marthy, J.E.E.M., 33 : 75-83 (1975).
4. H.-J. Marthy, Thèse d'Etat nr. 12 426, 55 p. 7 pl., Université
    Paris VI (1976).
5. H.-J. Marthy, Vie et Milieu, 28/29 (1) : 121-142 (1978/79).
6. G. Albrecht-Bühler and A. Bushnell, Exp. Cell Res., 126 :
    427-438 (1980).

7. M.L. Vialleton, Thèse ser. A no. 113 638, Université Paris, 1-116 (1888).
8. H.-J. Marthy, J.E.E.M., 29 (2) : 347-361 (1973).
9. J.M. Arnold, Diff., 2 : 335-341 (1974).
10. J.R. Cartwright, Jr. and J.M. Arnold, Cell Mot., 1 : 455-468 (1981).
11. J.R. Cartwright, Jr. and J.M. Arnold, J.E.E.M., 57 : 3-24 (1980).
12. J.M. Arnold, Biol. Bull., 129 : 72-78 (1965).
13. J.M. Arnold, Dev. Biol., 18 : 180-197 (1968).
14. J.M. Arnold and L.D. Williams-Arnold, J.E.E.M., 31 (1) : 1-25 (1974).
15. J.M. Arnold and L.D. Williams-Arnold, Repr. Mar. Invert. IV, ch. 5, Academic Press, N.Y., 243-290 (1977).
16. G. Meister, Zool. Jb. Anat., 89 : 247-300 (1972).
17. P. Fioroni and G. Meister, Gr. Zool. Prakt 16c/2, R. Siewing ed., G. Fischer Verlag, Stuttgart (1974).
18. P. Fioroni, Cephalopoda, In: "Morph. Tiere", F. Seidel ed., 1: R. Lief, 2: G_5-I, VEB G. Fischer, Jena (1978).
19. S.v. Boletzky, Rev. Suisse Zool., 74 : 555-562 (1967).
20. S.v. Boletzky, Experientia, 26 : 880-881 (1970).
21. G. Meister and P. Fioroni, Zool. Jb. Anat., 96 : 394-419 (1976).
22. S.v. Boletzky, Rev. Suisse Zool., 85 : 379-380 (1978).
23. M. Segmüller and H.-J. Marthy, Experientia, 40 : 636 (1984).
24. S. Ranzi, Bull. Soc. Ital. Sperimentale, 6 (11) : 1-2 (1931).
25. A. Krogh, Am. J. Physiol., 90 : 245-251 (1929).
26. H.A. Krebs, J. exp. Zool., 194 : 221-226 (1975).
27. H.-J. Marthy, C. r. hebd. Séanc. Acad. Sci. Paris, 278 : 1345-1348 (1978).

FIBRONECTIN, LAMININ AND PROTEOLYSIS IN CELLULAR INTERACTIONS

A. Vaheri

Department of Virology
University of Helsinki
SF-00290 Helsinki, Finland

Distinct changes occur in the extracellular matrix glycoproteins during the early embryonic development and during tissue morphogenesis such as kidney differentiation. In the adult organism all division-competent cells are associated with components of the extracellular matrix (1, 2).

Normal adherent cells deposit in culture a pericellular matrix, in which noncollagenous (fibronectin, laminin) and collagenous glycoproteins and proteoglycans and hyaluronic acid interact with each other and with other cell surface molecules (3, 4).

Fibronectin (5) is a macromolecular dimer of disulfide-bonded subunits each with Mr 250 000. It is found in soluble form in blood and other body fluids and in insoluble form in interstitial connective tissues and in or close to basement membranes. Other distinctive features of fibronectin include its domain structure, its susceptibility to proteolytic fragmentation and its multiple molecular and biological interactions, thought to be involved in cell migration and anchorage, organization of the extracellular matrix, chemotaxis and opsonization.

Laminin (6, 7) is a Mr 900 000 matrix glycoprotein, confined in the adult organism to basement membranes. It is produced by cells of epithelial origin and also by normal astrocytes and is used by neurons for attachment and spreading in culture (8, 9, 10). This is a function of laminin apparently distinct from its structural role in basement membranes. After injury laminin is found in vivo in reactive brain astrocytes and is also found in the embryo in association with certain glial cells.

199

The cell-binding site in fibronectin has been defined as a tetrapeptide sequence in the molecule (11), but the cell surface receptor for fibronectin remains to be determined. In contrast, the laminin receptor has been characterized as a Mr 70 000 membrane protein (12).

Invasiveness and metastasis, distinguishing properties of malignant cells, involve penetration through components of the extracellular matrix. Enzymatic degradation of matrix components is involved in these phenomena (13).

Defined gelatin-binding fragments of fibronection have transformation-promoting activity in experimental conditions (14). This activity is shared by t-PA (15). Interestingly, t-PA (but not u-PA) has been reported to have partial structural homology with fibronectin (16) and it is bound by immobilized fibronectin and immobilized laminin (17, 18). Fibronectin fragments detected in body fluids of tumor patients may serve as markers for tumor progression (19).

Pericellular proteolysis is regulated in part through α_2-macroglobulin (α_2M), plasminogen activators (u-PA and t-PA) and their inhibitors, as detected in cultures of various types of normal and malignant human cells (20). Macrophages seem to direct pericellular proteolysis by regulation at several levels (u-PA, its Mr 65 000 inhibitor and α_2M) (21). Proteolytic targets of the pericellular matrices of cells in culture include fibronectin and an Mr 66 000 matrix-associated protein (22).

The pericellular matrices, and in particular their cell-associated glycoprotein components, fibronectin and laminin are essential for the integrity of differentiated cellular phenotypes. Regulation of pericellular proteolysis is one of the ways the cell controls the cell surface-matrix interactions.

REFERENCES

1. J. Wartiovaara and A. Vaheri, In: "Development of Mammals",
       Vol. 4, M.H. Johnson ed., North-Holland Publishing Company,
       Amsterdam, pp. 233-269 (1980).
2. K. Alitalo and A. Vaheri, Adv. Cancer Res., 37 : 111-158
       (1982).
3. K. Hedman and A. Vaheri, In: "Fibronectin", E. Mosher ed.,
       Academic Press, (1984).
4. K. Yamada, Ann. Rev. Biochem., 52 : 761-799 (1983).
5. T. Vartio and A. Vaheri, TIBS, 8 : 442-444 (1983).
6. R. Timpl, P. Rohde, P. Robey, S. Rennard, J.-M. Foidart and G.
       Martin, J. Biol. Chem., 254 : 9933-9937 (1979).
7. R. Timpl, H. Rohde, L. Risteli, U. Ott, P. Robey and G. Martin,

Meth. Enzymol., 82 : 831-838 (1982).

8. P. Liesi, D. Dahl and A. Vaheri, J. Cell Biol., 96 : 920-924
    (1983).

9. P. Liesi, D. Dahl and A. Vaheri, EMBO J., 3 : 683-686 (1984).

10. P. Liesi, D. Dahl and A. Vaheri, J. Neurosci. Res., 11 :
    241-251 (1984).

11. M.D. Pierschbacher and E. Ruoslahti, Nature, 309 : 30-33
    (1984).

12. H. Lesot, U. Kühl and K. von der Mark, EMBO J., 2 : 861-865
    (1983)

13. T. Vartio, A. Vaheri, G. De Petro and S. Barlati, Invasion and
    Metast., 3 : 125-138 (1983).

14. G. De Petro, S. Barlati, T. Vartio and A. Vaheri, Proc. Natl.
    Acad. Sci. USA, 78 : 4965-4969 (1981).

15. G. De Petro, T. Vartio, E.-M. Salonen, A. Vaheri and S.
    Barlati, Int. J. Cancer, 33 : 563-567 (1984).

16. L. Banyai, A. Varadi and L. Patthy, FEBS Lett., 163 : 37-41
    (1983).

17. E.-M. Salonen, A. Zitting and A. Vaheri, FEBS Lett., 172 :
    29-32 (1984).

18. E.-M. Salonen, O. Saksela, T. Vartio, A. Vaheri, L. Nielsen and
    J. Zeuthen, J. Biol. Chem., in press (1985).

19. G. De Petro, S. Barlati, T. Vartio and A. Vaheri, Int. J.
    Cancer, 31 : 157-162 (1983).

20. O. Saksela, A. Vaheri, W.-D. Schleuning, P. Mignatti and S.
    Barlatti, Int. J. Cancer, 33 : 609-616 (1984).

21. O. Saksela, T. Hovi and A. Vaheri, J. cell Physiol., 122 :
    125-132 (1985).

22. J. Keski-Oja and A. Vaheri, Biochem. Biophys. Acta, 720 :
    1142-1146 (1982).

CELL-CELL AND CELL-SUBSTRATE CONTACTS : INVESTIGATION BY

FUNCTIONALLY ACTIVE MONOCLONAL ANTIBODIES

Beat A. Imhof[1], Joachim Krieg[2], Jürgen Behrens,
H. Peter Vollmers and Walter Birchmeier

Friedrich-Miescher-Laboratorium
der Max-Planck-Gesellschaft
D-7400 Tübingen, Germany

SUMMARY

It has previously been shown that the anti-FC-1 antibody prevents cell-substrate adhesion of fibroblasts, and the FC-1 antigen has been characterized as a focal contact protein of cells in tissue culture (Oesch and Birchmeier, Cell, 31 : 671-679 (1982)). In the present study, we show that the enrichment of the FC-1 antigen in the focal contacts depends on the nature of the substrate; it is present in the contacts when cells are grown on components of the basement membrane such as laminin, collagen IV, or fibronectin, but absent on collagen I or Concanavalin A. Basement membrane components seem also to influence the distribution of the FC-1 antigen in vivo; immunolocalization on frozen sections of various chicken tissues revealed that the antigen is concentrated at the basement membrane but not the junctional complex of different epithelia.

In contrast, the monoclonal antibody anti-Arc-1 interferes with cell-cell contacts and changes the morphology of MDCK (canine kidney) epithelial cells in vitro (Imhof et al., Cell, 35 : 667-675 (1983)). In this communication we demonstrate that the Arc-1 antigen is enriched at the cell-cell contacts both of MDCK cells in tissue culture and of cells in various canine tissues. In the

Present addresses
1 Institut d'Embryologie du CNRS, 94130 Nogent-Sur-Marne, France.
2 Friedrich-Miescher-Institut, Ciba-Geigy,
  4000 Basel, Switzerland.

intestinal epithelium the antigen could be localized at the region of the junctional complex. It is therefore plausible that the anti-Arc-1 antibody acts by direct binding to components of the intercellular junctions. We further examined possible mechanisms by which the antibody might produce its cell dissociating effect. Antibody action was inhibited in the presence of colchicine but not cytochalasin B, and when the cellular cAMP level was raised.

INTRODUCTION

    In order to function in integrated tissues living cells form complex adhesion systems with their adjoining cells and with acellular structures (i.e. they form cell-cell and cell-substrate contacts, for review see 1, 2). These two types of cell adhesion systems are either organized in organelle-like structures (e.g. junctional complexes and hemidesmosomes of epithelial cells) or they are uniformally distributed over cell surfaces. On the molecular level different surface components seem to be responsible for these two types of adhesions. For instance, defined adhesion proteins such as N-CAM and uvomorulin (L-CAM, cadherin, CAM 120/80) are involved in specific cell-cell contacts (3-6), and furthermore, distinct proteins have been found in cell-cell junctions such as gap junctions or desmosomes (7, 8). A variety of components have been implicated in specific cell-substrate adhesion. For instance, a 120-160 kd membrane protein complex (defined by poly- and monoclonal antibodies) seems to be involved (9-11), as are a variety of other cellular (12-14) and extracellular components (fibronectin, laminin, collagens, ref. 15-17).

    Adhesion molecules were analyzed by mainly two procedures. Contact organelles were isolated by cell-fractionation leading to electron-microscopically pure structures (7, 8), and the molecules were then analyzed. In another line of research, functionally active antibodies were produced, which interfered with either cell-cell or cell-substrate contacts. By this latter procedure a whole series of molecules involved in both types of adhesions were characterized (3-6, 9-11, 18-20).

    Our laboratory has recently selected monoclonal antibodies which prevent the adhesion of fibroblasts to tissue culture dishes (anti-FC-1, 19), or interfere with epithelial cell-cell contacts (anti-Arc-1, 20). The FC-1 antigen was identified as a component concentrated at the fibroblast focal contacts, organelles thought to be involved in cell-substrate adhesion. The cell-cell contact antigen Arc-1, on the other hand, was found to be located at epithelial cell boundaries where it might function in the maintenance of cell-cell contacts. It has recently been realized, however, that antibodies can also be used to influence cells by indirect means; for instance, antibodies can replace hormones and growth factors by reacting with corresponding surface receptors (21-24). It is therefore possible that certain anti-adhesion antibodies might induce their effects by indirect interference.

In the present study, we compare the localization of the FC-1 and Arc-1 antigen in frozen sections of epithelial cells of intestinal tissue. We also describe how the distribution of the FC-1 antigen in focal contacts is modulated by different components of the extracellular matrix, and we use our anti-Arc-1 monoclonal antibody to examine the possible involvement of cytoskeletal - and metabolism - affecting factors on the dissociation process.

EXPERIMENTAL PROCEDURE

Screening of hybridomas

BALB/C mice were immunized with chicken embryo fibroblasts or MDCK epithelial cells. The spleen cells of the immunized mice were fused with FO or NS-1 myeloma cells, and the hybridomas cultured (19, 20). The hybridoma supernatants were then tested in two functional assays with living cells : A) For the attachment assay, trypsinized cells (400) were plated in 50 µl hybridoma supernatants or myeloma control medium on tissue culture plates, the plates incubated at 37°C for 20 min, and the attached cells counted. B) For the morphology assay, MDCK cells were plated on Terasaki plates (3000 per well), and the medium replaced by the different supernatants or controls. After 1-4 hrs the plates were fixed and the wells microscopically scanned for cell dissociation.

Immunofluorescence of cells and frozen sections

Chicken embryo fibroblasts and MDCK cells were cultured on glass cover slips over night, fixed for 15 min with 3% formaldehyde in phosphate-buffered saline (pH 8.3), and permeabilized with Triton X-100 for 5 min. They were stained with purified anti-FC-1 or anti-Arc-1 antibody followed by appropriately diluted fluorescein-labelled rabbit anti-mouse immunoglobulin. All steps were carried out at room temperature. Specimens were mounted in Moviol 4-88 containing 1 mg/ml p-phenylene diamine (to reduce fluorescence quenching) and examined on a Leitz Orthoplan photomicroscope.

Cryostat sections of various organs were cut to thicknesses between 0.5-7 µm on a Reichert-Jung Frigocut mod. 2700 or a Sorvall ultramicrotome MT-2B equiped with a frozen thin sectioning system (25). Sections were dried on gelatine-coated glass cover slips and prepared for immunofluorescence as above.

Coating of surfaces with components of the extracellular matrix
_____

Chicken embryo fibroblasts were plated for 5-10 hrs on siliconized glass cover slips which were precoated with fibronectin, laminin, collagen IV, collagen I, gelatine, or Concanavalin A (50 µg/ml). Cells did not adhere to uncoated cover slips. (Laminin and collagen IV were generous gifts of Drs. Timpl, von der Mark, and Hühl (Munich).) Cells were prepared for immunofluorescence as described above.

For immunolocalization in the electron microscope, cells were cultured on fibronectin-coated Petriperm dishes (Heraeus), extracted with 0.2% saponin as described (26), fixed with 3% formaldehyde for 10 min, permeabilized with 0.1% Triton X-100, and immunolabelled for vinculin (using a rabbit anti-vinculin IgG followed by 5 nm Protein A-coated colloidal gold) and for FC-1 (using our monoclonal antibody followed by 5 nm goat anti-mouse IgG1-coated colloidal gold). Specimens were Epon-embedded, sectioned, and inspected in a Siemens electron microscope.

Drug experiments
_____

MDCK cells grown on glass cover slips to 70% confluency were incubated with medium containing 50 µg/ml anti-Arc-1 antibody plus the various drugs at concentrations that alone did not markedly affect cell morphology : Colchicine at 1 µg/ml, cytochalasin B at 0.05 µg/ml, 3-isobutyl 1-methylxanthine at 50 µM, forskolin at 50 µM, N^6O^2-dibutyryl cAMP at 10 µM, A 23187 at 10 µM, calmidazolium at 15 µM, and cycloheximide at 10 µg/ml.

RESULTS

Effect of two types of monoclonal antibodies which perturb
cell-cell and cell-substrate adhesion
_____

The monoclonal antibodies anti-FC-1 and anti-Arc-1 were obtained after immunization of BALB/C mice with intact tissue culture cells (see refs. 19, 20). They were chosen out of hundreds of hybridomas by functional assays as shown in Fig. 1. The anti-adhesion antibody anti-FC-1 prevents chicken embryo fibroblasts from attaching to tissue culture dishes (compare Fig. 1 a and b). The anti-Arc-1 antibody perturbs cell-cell contacts of MDCK (canine kidney) epithelial cells. It also induces a shape change of the cells, i.e. leads to a spindle-like morphology (compare Fig. 1 c and d). Both antibodies are unique in their activity : anti-Arc-1 does not effect cell-attachment of MDCK cells and anti-FC-1 does not perturb the shape of chicken fibroblasts once the cells are attached.

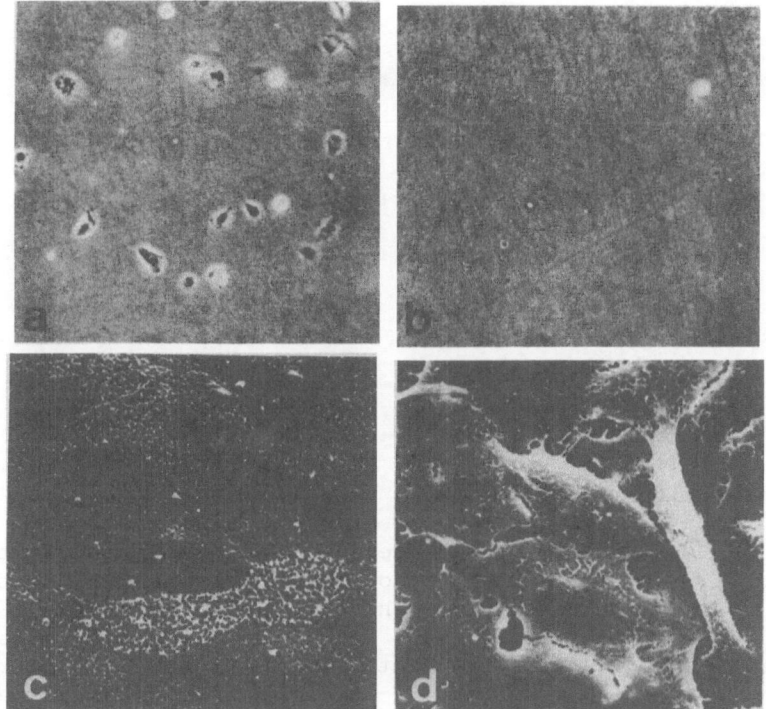

Fig. 1. Effect of the anti-FC-1 monoclonal antibody on the
adhesion of chicken embryo fibroblasts to tissue
culture dishes (compare b with the control in a),
and of anti-Arc-1 on the cell-cell contacts of MDCK
(canine kidney) epithelial cells (compare d with the
control in c).

Localization of the FC-1 and Arc-1 antigens on cultured cells and on various tissue sections

The fact that the antibodies discriminated between cell-cell
(anti-Arc-1) and cell-substrate (anti-FC-1) adhesion in vitro
prompted us to examine whether these functional effects correlate
with a particular location of the antigen. In the immunofluorescence we found the FC-1 antigen to be enriched in the focal
adhesion plaques of chicken embryo fibroblasts, i.e. at sites where
intimate cell-substrate contact occurs (Fig. 2 a, b). On the other
hand, the Arc-1 antigen was enriched at the cell-cell boundaries of
MDCK cells (Fig. 2 c, d). Often, the staining showed a punctuate
pattern.

Both types of contacts could also be demonstrated in
epithelial tissues. Here, the cells express well characterized

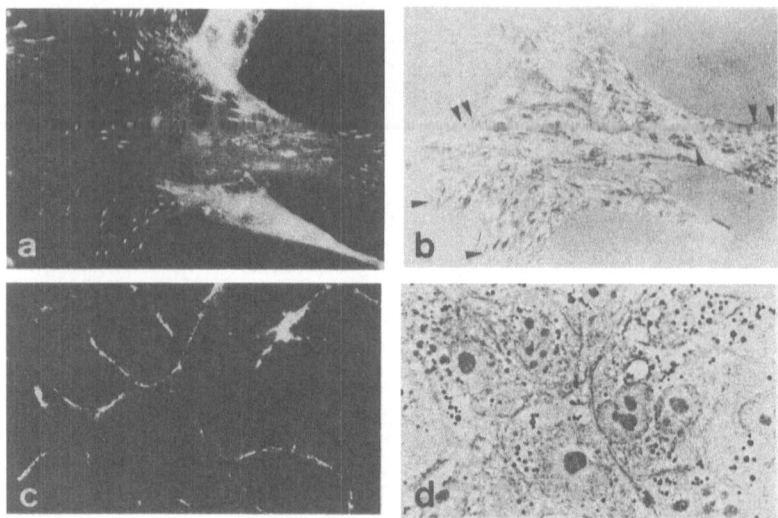

Fig. 2.   Localization of antigens on cultured cells by the
          two   functional   monoclonal   antibodies.   In   the
          immunofluorescence,   the   FC-1   antigen   (a)   is
          concentrated   at   the   focal contacts of fibroblasts
          (b, identified by   interference   reflection   micro-
          scopy, see arrows). The Arc-1 antigen (c) was en-
          riched at the cell-cell contacts of MDCK cells   (d,
          identified by phase contrast microscopy).

cell-cell   junctions   and   are   ventrally   attached to an acellular
basal lamina. On frozen sections of chicken   intestinal   villi   we
found   the   FC-1 antigen to be markedly enriched at the base of the
epithelial cells (Fig. 3 a, b). In contrast, the Arc-1 antigen   was
restricted to the region of the epithelial junctional complex (Fig.
3 c, d).

     We have also examined the location of the   antigens   in   other
epithelia. In the stratified epithelium of the chicken cornea, FC-1
was confined to a thin layer close to the basement membrane (Fig. 4
A, a-d). In   the   underlying   substantia   propria the antigen was
absent except in slender fibroblastic cells (keratocytes). The FC-1
antigen was also detected at the base of the pigment epithelium of
the eye (Fig. 4 A, e-h), the epithelium of the   gall   bladder,   and
the epithelia of the renal tubules of the kidney (data not shown).
The Arc-1 antigen, on the other hand, was identified in the
epithelial   cell-cell   contacts of the distal tubules of the canine
kidney (Fig. 4 B, a and b). It was absent from the proximal tubules
and the glomeruli. We could also detect the Arc-1 antigen in
sections of the canine epidermis (Fig. 4 B, c), i.e. it was present
in an intermediate layer between the fully keratinized and the

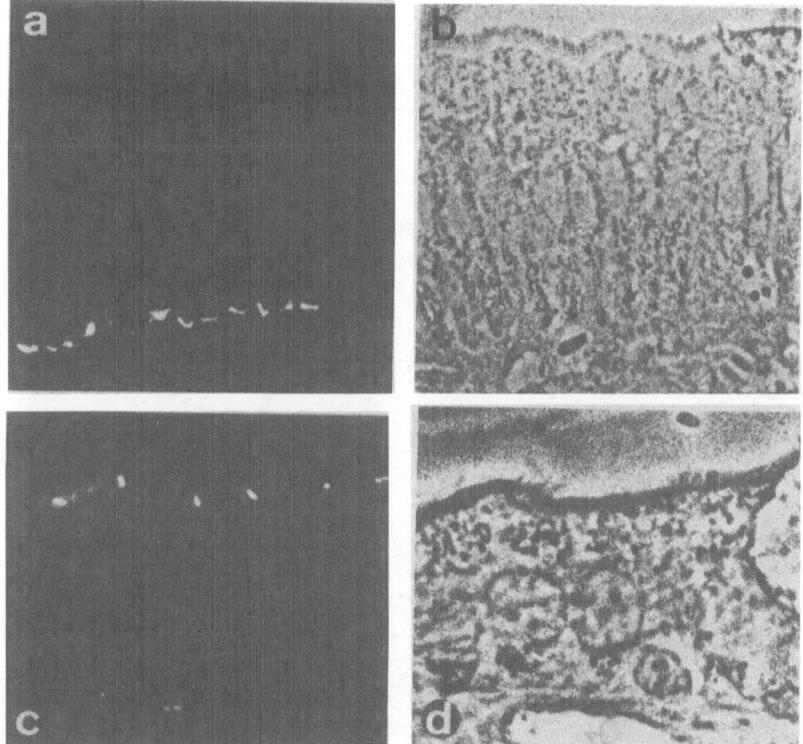

Fig. 3.   Immunolocalization of the FC-1 and Arc-1 antigens on
          intestinal tissue sections. (a) Immunofluorescence
          of   FC-1   on   the   chicken   small   intestine,   (b)
          corresponding phase contrast image, (c)   immunofluo-
          rescence   of   Arc-1   in   the   canine   intestinal
          epithelium, (d) corresponding phase contrast image.

regenerating regions of the   tissue.   In   the   hair   follicle,   the
antigen was found in the external root sheath (Fig. 4 B, d). In all
tissues (including the liver parenchyme, Fig. 4 B,   e),   Arc-1   was
clearly enriched at the cell-cell contacts.

Substrate-dependent expression of the FC-1 antigen

    In tissue culture fibroblasts can adhere and grow on different
extracellular matrix components. It seems that   the   cells   express
specific   receptor   molecules for these substances. Here we studied
the expression of the FC-1   antigen   in   focal   contacts   of   cells
cultured   on   such   matrix   components.   Interestingly,   FC-1   was
detected in focal contacts of   cells   grown   only   on   fibronectin,

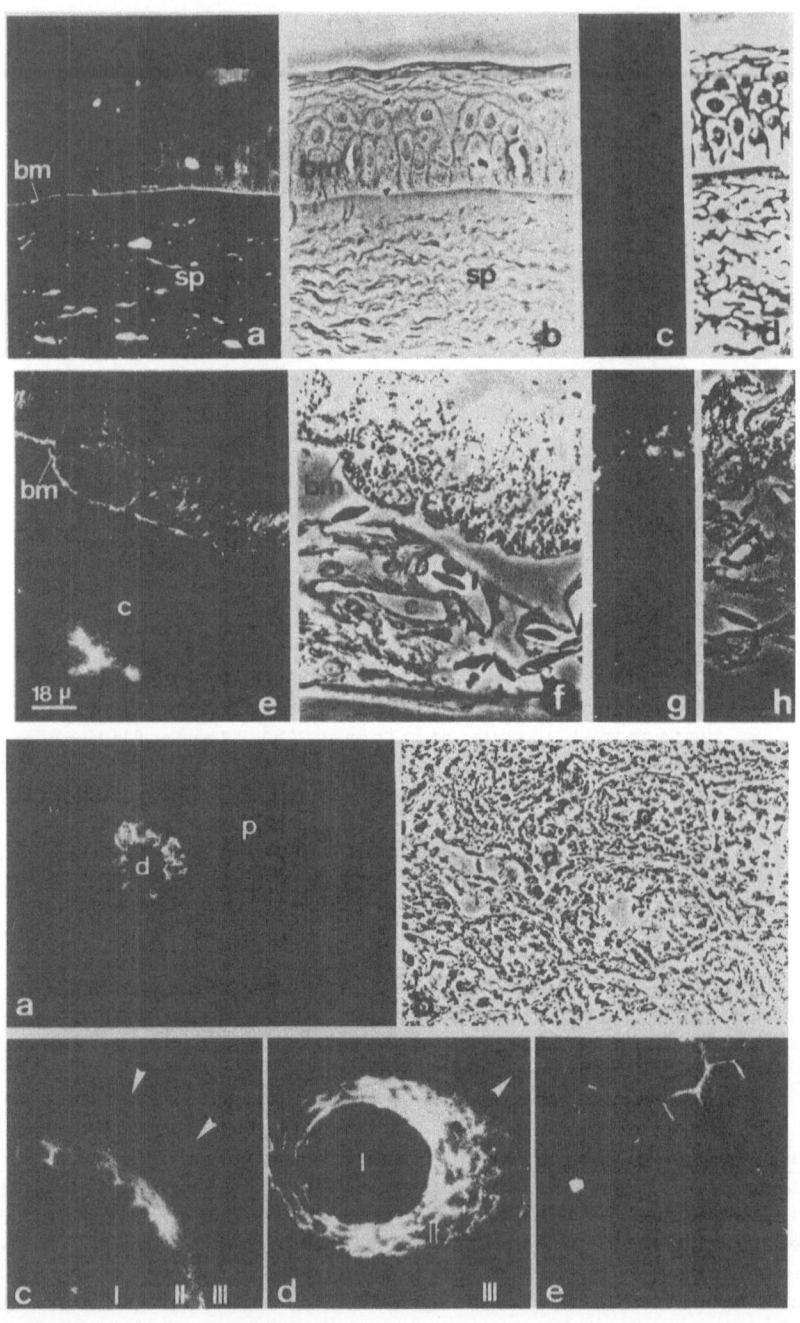

laminin, and collagen IV (Fig. 5 a-c), but not on collagen I (or gelatine) or on Concanavalin A (Fig. 5 d, e). However, on all substrates the cells produced vinculin-containing focal contacts (as seen by immunofluorescence and interference reflection microscopy; shown for gelatine and Con A in Fig. 5 f-h). It was possible to convert cells between these two states : for instance, FC-1 appeared in the contacts when cells grown on Con A or gelatine were later incubated with soluble collagen IV or laminin (shown for the pair Con A and collagen IV in Fig. 5 i). Conversely, FC-1 disappeared from focal contacts when soluble gelatine was added to cells on laminin or collagen IV (data not shown).

Immunolocalization of the FC-1 antigen was performed on saponin-treated cell models of chicken embryo fibroblasts. It has been shown by others (26) that such models are enriched for focal contacts and have removed most cytoplasmic structures. In these preparations FC-1 antigen could clearly be visualized at the outside of the ventral cell surface (Fig. 6 a, b). In contrast, vinculin was unequivocally located on the cytoplasmic face together with microfilamentous structures (Fig. 6 c, d).

Fig. 4.   Immunolocalization of FC-1 and Arc-1 in various tissues.
A) Frozen sections of chicken tissues were stained for FC-1 : (a-d) Crossection of the corneal epithelium, a) immunofluorescence, b) corresponding phase contrast image, (f, g) control where the anti-FC-1 antibody was omitted. (e-h) Pigment epithelium of the eye, e) immunofluorescence, (f) phase contrast, (g, h) controls. (bm) basement membrane, (sp) substantia propria.
B) Frozen sections of canine tissues were stained for Arc-1 : (a) Immunofluorescence of the kidney cortex, (b) corresponding phase contrast picture. (d) Shows the distal, (p) the proximal tubule. Glomeruli (not on this section) are not stained. (c) Immunofluorescence of the epidermis, (I) marks the stratum basale, (II) the stratum spinosum and stratum granulosum, and (III) the stratum corneum. Arrows point to the surface of the epidermis. (d) Immunofluorescence of a hair follicle, (I) hair shaft and internal root sheath, (II) external root sheath, (III) dermal root sheath. The arrow indicates the border of the follicle. (e) Immuno-fluorescence of liver tissue.

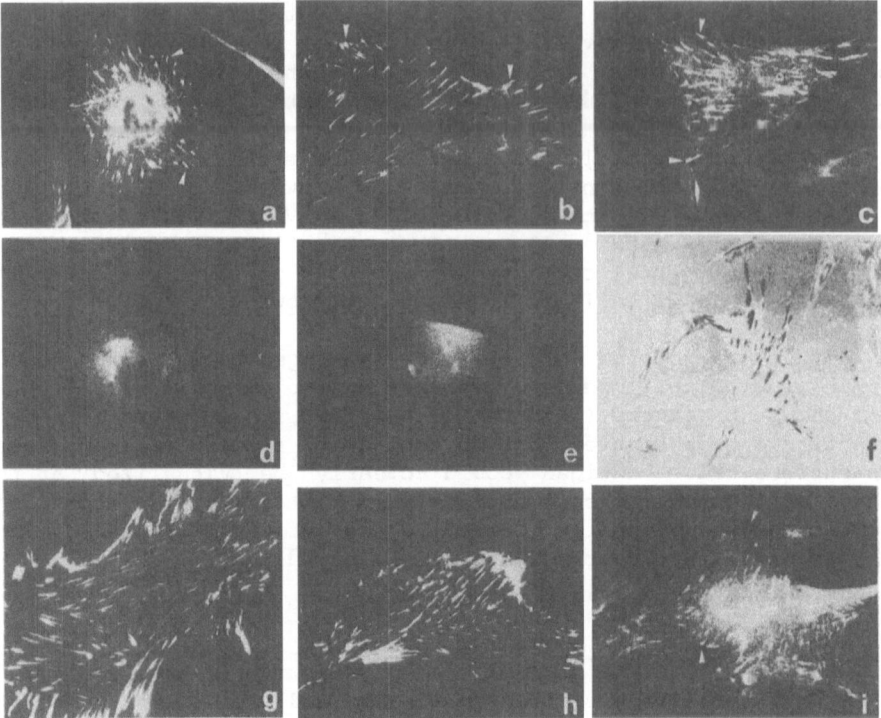

Fig. 5.   Modulation of the   FC-1 antigen in cells grown on
          different substrates. Cells were cultured on glass
          cover slips coated with fibronectin (a), laminin
          (b), collagen IV (c), gelatine (d), and Concanavalin
          A (e), and were examined for the FC-1 antigen by
          immunofluorescence. Focal contacts are marked by
          arrowheads. Collagen I produced the same pattern as
          gelatine. On all substrates focal contacts were
          formed as judged by interference reflection
          microscopy (shown for gelatine in f) or immunofluo-
          rescence for vinculin (shown for gelatine in g and
          Con A in h). In (i) cells grown on Con A were sub-
          sequently incubated with 50 µg/ml collagen IV.

Modulation of anti-Arc-1 action by cytoskeleton –
and cAMP – affecting drugs

     To examine the possible involvement of cytoskeletal components
in the process of cell dissociation by anti-Arc-1, the effects of
colchicine and cytochalasin were tested. In the presence of
colchicine, disruption of the MDCK monolayer by anti-Arc-1 was
clearly inhibited (Fig. 7 c, compare with a, b). In contrast,
cytochalasin B rather promoted antibody action (Fig. 7 d).

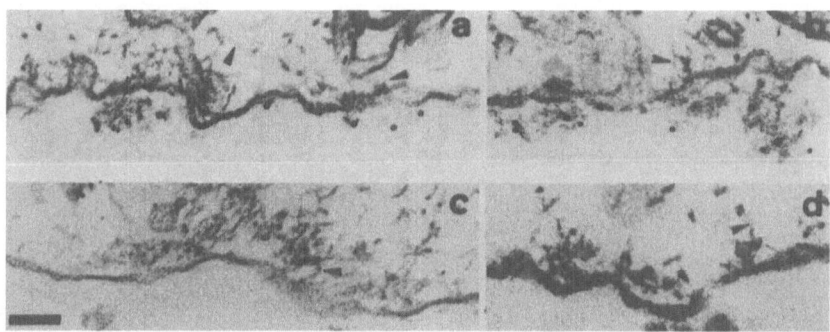

Fig. 6. Localization of the FC-1 antigen by immuno-electron
         microscopy of saponin-treated cell models. Cell
         models of chicken embryo fibroblasts were prepared
         and labelled as described. Regions of the ventral
         cell surface are shown here. (a, b) Labelling for
         FC-1, (c, d) staining for vinculin. Arrowheads point
         to actin filaments on the cytoplasmic side. The bar
         represents 100 nm.

Apparently, an intact microtubular network is necessary for proper
antibody action to occur, but the microfilamentous network is not.

We also tested the effect of various drugs which influence
cell metabolism. Antibody action was prevented when cellular cAMP
was raised (Fig. 7 e, f). This was accomplished directly by
addition of dibutyryl cAMP, or indirectly by addition of isobutyl
methylxanthine which inhibits cyclic nucleotide phosphodiesterase
or forskolin which activates adenylate cyclase (27, 28). Compounds
affecting other metabolic processes such as cycloheximide, serum
$(0-20\%)$, calmidazolium (an inhibitor of calmodulin action) and the
Ca^{2+} ionophore A 23187 did not interfere with the antibody effect
(data not shown).

DISCUSSION

In the present study we compare the functional monoclonal
antibody anti-FC-1 which affects cell-substrate adhesion with the
anti-Arc-1 antibody which perturbs cell-cell contacts. It has
previously been shown (19) that the FC-1 antigen is concentrated at
fibroblasts focal contacts, i.e. at cell-substrate contact
organelles. We show here that the Arc-1 antigen (20) is enriched at
cell-cell contacts of cultured epithelial cells. In the intestinal
epithelium the different location of the two antigens was even more
conspicuous : antibody anti-FC-1 stained the cell-basement membrane

214   B.A. IMHOF ET AL.

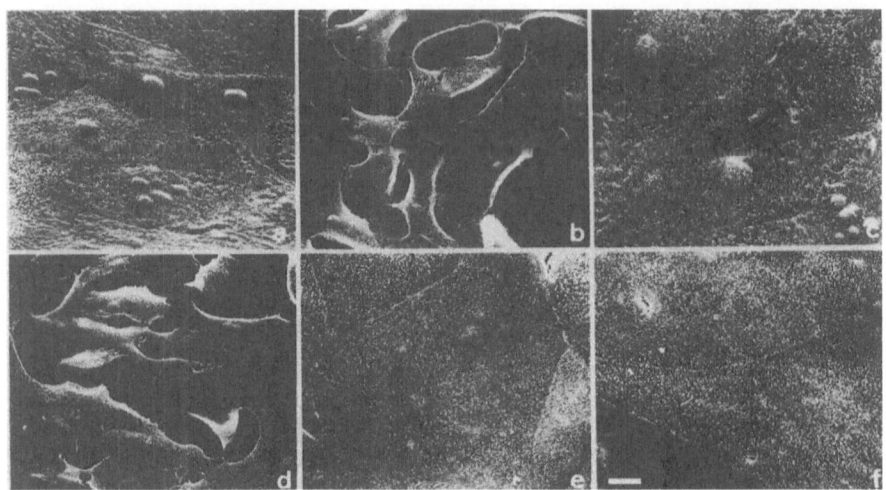

Fig. 7. Effect of the anti-Arc-1 antibody on cell disso-
         ciation : Modulation by various drugs. MDCK mono-
         layers were incubated for 3 hrs with anti-Arc-1
         antibody in the presence and absence of the drugs as
         described in Experimental Procedures. (a) Cells
         without antibody and drugs, (b) with anti-Arc-1
         antibody alone, (c) anti-Arc-1 plus colchicine, (d)
         anti-Arc-1 plus cytochalasin B, (e) anti-Arc-1 plus
         3-isobutyl methylxanthine, and (f) anti-Arc-1 plus
         forskolin. Cells were processed for scanning
         electron microscopy as described (5). Bar represents
         10 μm.

contacts whereas Arc-1 was confined to the region of the junctional
complex.

     It is well established that epithelial cells are polarized
with respect to certain surface components. For instance,
hydrolytic enzymes of the intestine are exclusively present in the
apical membrane of the brush border (29), whereas Na^+K^+-ATPase
accumulates at the basolateral surface (30). Moreover, epithelial
cells are polarized with respect to the budding of certain viruses:
Influenza virus leaves MDCK epithelial cells through the upper cell
surface whereas vesicular stomatitis virus uses the basolateral
surface (31). In the cases of FC-1 and Arc-1, polarization of the
epithelial membrane is even more restricted. The cell-substrate
contact molecule FC-1 is present only at the basal but not the
lateral membrane. The present data suggest that this type of
epithelial polarization is induced by the components of the
underlying acellular basement membrane. In the case of the
cell-cell contact molecule Arc-1, a preferential enrichment in the

junctional complex region of epithelia is seen. Arc-1 might be located ventrally to the tight junctions since in intact MDCK epithelial cells the antibody seems to reach its target from the basolateral rather than from the apical side (32).

It is quite remarkable that the FC-1 antigen is induced in focal contacts by collagen IV (the basement membrane component) but disappears from focal contacts in the presence of collagen I (the mesenchymal component). In contrast, the adhesion complex identified by another monoclonal antibody (CSAT, ref. 10) also seems to be involved in cell-collagen I interaction. Thus, different types of focal contacts seem to exist in cells depending on the nature of the substrate. To our knowledge, FC-1 is the first protein which monitors such differencies. The intracellular components of the focal contacts show a similar behaviour : for instance, vinculin is present both in focal contacts plus in certain cell-cell junctions (33), whereas the protein talin is enriched specifically in cell-substrate contacts (34). Thus, each type of adhesive contact seems to contain a characteristic set of both extra- and intracellular components, some being unique to and others being common to different types of contacts.

The question arises how the anti-Arc-1 antibody might interfere with the adhesion process and how the cytoskeleton - and metabolism - interfering drugs might interplay with the antibody ? In our previous study (20) we detected some Arc-1 antigen at the apical surface of MDCK cells by sensitive immunoscanning electron microscopy. Accordingly, we considered an indirect signaling mechanism for anti-Arc-1 action. However, immunolocalization of Arc-1 on fixed cells after detergent extraction clearly revealed high antigen concentration at the cell-cell borders. This rather favours a direct mechanism of antibody action. Conceivably, the antibody (approaching its target from below the tight junctions) might first act on the junctional complex and then induce a cascade of events, e.g. a local influx of Ca^{2+} followed by closure of the gap junctions (35). The cells, having now broken off communications with their neighbours, might then change into a fibroblastic morphology, a change requiring cytoskeletal rearrangement. In the unperturbed epithelia MDCK cells are highly communicative, develop strong intercellular junctions and express an elaborated vertical polarity (20, 36, 37). The spindle-shaped state after the antibody action would rather reflect the morphology of fibroblastic cells which exhibit little communication, few intercellular junctions, and a strong anterior-posterior polarity (38). We could hypothesize that any modulator which shifts the cells between these two extreme states might also affect the degree of anti-Arc-1 action. For instance, it is known that cAMP can increase gap junctional communication in certain systems (39) and colchicine has been shown to destroy anterior-posterior polarity of fibroblasts (40). Thus, under the influence of both these effectors the epithelial form might be favoured. By contrast, cytochalasin B itself can induce spindle-shaped MDCK cells at higher concentration (e.g. at 10 µg/ml), and might therefore potentiate anti-Arc-1 action.

In the present and the preceding studies (19, 20, 25, 32) we have illustrated that functionally active monoclonal antibodies allow the identification of new and minor cell surface antigens exhibiting a particular location and possibly a particular function. It is expected that this approach will lead to the identification of other antigens involved in both cell-substrate and cell-cell adhesion.

ACKNOWLEDGEMENTS

We thank Dr. S.L. Goodman for helpful discussions, Annette Hiesel, Suzanne Braun, and Marianne Kohtz for excellent technical assistance, Regine Braun and Inge Zimmermann for embedding and sectioning, Dr. H. Frank for advice with the electron microscopes, and Margrit Hipp for secretarial help. This work was partly supported by the Dr. Mildred-Scheel-Stiftung/Deutsche Krebshilfe e. V.

REFERENCES

1. F. Grinnel, Int. Rev. Cytol., 53 : 65-144 (1978).
2. L.A. Staehelin, Int. Rev. Cytol., 39 : 191-284 (1974).
3. G. Edelman, Science, 49 : 450-457 (1983).
4. C. Yoshida and M. Takeichi, Cell, 28 : 217-224.
5. C.H. Damsky, J. Richa, D. Solter, K. Knudsen and C. Buck, Cell, 34 : 455-466 (1983).
6. F. Hyafil, D. Morello, C. Babinet and F. Jacob, Cell, 21 : 927-934 (1980).
7. W.W. Franke, R. Moll, H. Mueller, E. Schmid, C. Ruhn, R. Krepler, U. Artlieb and H. Denk, Proc. Natl. Acad. Sci. USA, 80 : 543-547 (1983).
8. E.L. Hertzberg and N.B. Gilula, J. Biol. Chem., 254 : 2138-2147 (1979).
9. C.H. Damsky, K.A. Knudsen, J.R. Dorio and C.A. Buck, J. Cell Biol., 89 : 173-184 (1981).
10. N.T. Neff, C. Lowrey, C. Decker, A. Tovar, C. Damsky, C. Buck and A.F. Horwitz, J. Cell Biol., 95 : 654-666 (1982).
11. J.M. Greve and D.I. Gottlieb, J. Cell Biochem., 18 : 221-230 (1982).
12. G. Tarone, G. Galetto, H. Prat and P.H. Comoglio, J. Cell Biol., 94 : 179-186 (1982).
13. N. Oppenheimer-Marks and F. Grinnel, Exp. Cell Res., 152 : 467-475 (1984).
14. J.D. Aplin and R.C. Hughes, J. Cell Science, 50 : 89-103 (1981).

15. M.D. Pierschbacher and E. Ruoslathi, Nature, 309 : 30-33
    (1984).
16. R. Timpl, J. Engel and G.R. Martin, Trends Biochem. Sci., 8 :
    207-209 (1983).
17. H.K. Kleinmann, R.J. Klebe and G.R. Martin, J. Cell Biol., 88 :
    473-485 (1981).
18. C. Ocklind, P. Odin and B. Obrink, Exp. Cell Research, 151 :
    29-45 (1984).
19. B. Oesch and W. Birchmeier, Cell, 31 : 671-679 (1982).
20. B.A. Imhof, H.P. Vollmers, S.L. Goodman and W. Birchmeier,
    Cell, 35 : 667-675 (1983).
21. A.B. Schreiber, I. Lax, Y. Yarden, Z. Eshhar and J.
    Schlessinger, Proc. Natl. Acad. Sci. USA, 78 : 7535-7539
    (1981).
22. D. Baldwin, S. Terris and D.F. Steiner, J. Biol. Chem., 255 :
    4028-4034 (1980).
23. D. Leiber, S. Harbon, J.G. Guillet, C. André and A.D.
    Strosberg, Proc. Natl. Acad. Sci. USA, 81 : 4331-4334
    (1981).
24. A.F. Lopez and M.A. Vadas, Proc. Natl. Acad. Sci. USA, 81 :
    1818-1821 (1984).
25. B.A. Imhof, J. Krieg, H. Reggio and W. Birchmeier, Proc. Natl.
    Acad. Sci. USA, (in press) (1985).
26. A.A. Neyfakh and T.M. Svitkina, Exp. Cell Research, 149 :
    582-586 (1983).
27. W. Montague and J.R. Cook, Biochem. J., 120 : 9 p. (1970).
28. F.J. Darfler, M.D. Muellen and P.A. Insel, Biochem. Biophys.
    Acta, 803 : 203-209 (1984).
29. D. Louvard, Proc. Natl. Acad. Sci. USA, 77 : 4132-4136 (1980).
30. J. Kyte, J. Cell Biol., 68 : 287-303 (1975).
31. E.R. Boulan and D.D. Sabatini, Proc. Natl. Acad. Sci. USA, 75 :
    5071-5075 (1978).
32. J. Behrens, W. Birchmeier, S.L. Goodman and B.A. Imhof,
    submitted, (1985).
33. B. Geiger, A.H. Dutton, K.T. Tokuyasu and S.J. Singer, J. Cell
    Biol., 91 : 614-628 (1981).
34. K. Burridge and L. Connell, J. Cell Biol., 97 : 359-367
    (1983).
35. P.N.T. Unwin and G. Zampighi, Nature, 283 : 545-549 (1980).
36. D.A. Herzlinger and G.K. Ojakian, J. Cell Biol., 98 :1777-1787
    (1984).
37. M. Certijido, E. Robbins, D.D. Sabatini and E. Stefani, J.
    Membrane Biol., 81 : 41-48 (1984).
38. W.T. Chen, J. Cell Biol., 81 : 684-691 (1979).
39. E.C. Wiener and W.R. Loewenstein, Nature, 305 : 433-435
    (1983).
40. J.M. Vasiliev and I.M. Gelfand, In: "Cell motility" R. Goldman,
    T. Pollard and J. Rosenbaum eds., Cold Spring Harbor Press,
    pp. 279-304 (1976).

MECHANISMS OF EMBRYONIC LIMB BUD INTERCELLULAR ADHESION :

KINETIC ANALYSES AND CHARACTERIZATION OF THE MOLECULAR MECHANISM

James A. Bee[1] and Klaus von der Mark

Max-Planck-Institut fur Biochemie
D-8033 Martinsried, Federal Republic of Germany

ABSTRACT

When cultured in suspension, dissociated early limb bud cells segregate into aggregating and non-aggregating populations. While the non-aggregating cells die, resulting aggregates differentiate exclusively as cartilage. We are investigating the role of the cell surface in pre-chondrogenic condensation by analyzing the mechanism of aggregation in suspension culture. Immediately after their tryptic dissociation, limb bud cells undergo calcium-independent aggregation . This mechanism is completely and reversibly sensitive to cycloheximide, although this drug does not inhibit the re-appearance of all cell surface proteins. Fab' fragments prepared from antisera directed against the surface of these cells fail to inhibit their aggregation. In contrast, after 16 h recovery equivalent cells exhibit a different, calcium-dependent aggregation mechanism and demonstrate distinct sensitivity to cycloheximide. This calcium-dependent aggregation mechanism can be maintained upon cells by the addition of exogenous calcium during trypsin dissociation. Fab fragments prepared from an antiserum directed against these recovered cells inhibit only calcium-dependent aggregation. A surface glycoprotein with an approximate molecular weight of 8.5 kD can be isolated directly from total solubilized limb bud membrane proteins by affinity chromatography against equivalent proteins covalently coupled to Sepharose in the presence of calcium. This same protein is recognized by the aggregation-inhibiting antiserum. We present evidence that this homophilic protein is the major limb bud cell adhesion molecule.

[1] To whom all correspondence should be addressed : Department of Anatomy, The Royal Veterinary College, University of London, Royal College Street, London, NWL OTU.

INTRODUCTION

The morphological onset of skeletal development in the avian
limb is characterized by a localized increase in cell density
within prospective chondrogenic regions of the mesoderm. This phase
of condensation (Fell, 1925 ; Fell and Canti, 1935) is established
as mesodermal cells assume a rounded profile, decrease their
filopodial projections, and become intimately associated to
distinguish them from surrounding loosely packed, stellate
mesodermal cells (Thorogood and Hinchliffe, 1975). This transient
event appears not to be due to selective proliferation of
pre-chondrogenic cells (Summerbell and Wolpert, 1972) and has been
attributed to a novel enhancement of their mutual cell-cell
contacts (Elmer, 1982). Condensation terminates as definitive
chondrocytes secrete a thick pericellular matrix between one
another (Dessau et al., 1980).

Although the significance of condensation is not known,
considerable evidence implicates the role of intimate cell-cell
association in ultimate chondrogenic expression by early limb bud
mesoderm. When cultured over agar, dissociated stage 24 limb bud
mesodermal cells segregate into a non-aggregating population unable
to survive suspension culture conditions, and an aggregating
population which differentiates exclusively as cartilage (von der
Mark and von der Mark, 1977). The most compelling evidence for the
role of homotypic cell contact in establishing limb bud
chondrogenesis is the inhibition of cartilage nodule formation by
stage 24 mesoderm when co-cultured with numbers of stage 19 cells
sufficient to prevent contact among the older cells (Solursh and
Reiter, 1980).

The methodology pioneered by Gerisch and his co-workers in
their demonstration of a cell adhesion mechanism in Dictyostelium
has served as the paradigm by which a variety of cell adhesion
mechanisms have subsequently been characterized. Such an approach
utilizes the ability of monovalent (Fab) fragments derived from
antisera against the cell surface to inhibit their aggregation. The
cell surface is then screened for components capable of
neutralizing this inhibition (reviewed by Gerisch, 1980).
Accordingly, cell adhesion molecules have now been isolated from
neural retina (Rutishauser et al., 1976 ; Thiery et al., 1977),
brain (Grumet and Edelman, 1984), liver (Bertolotti et al., 1980 ;
Nielsen et al., 1981 ; Grady et al., 1982 ; Ocklind et al., 1984),
and teratocarcinoma cells (Yoshida and Takeichi, 1982 ; Grabel et
al., 1983 ; Yoshida-Noro et al., 1984). Although investigation has
focussed upon the isolation of cell adhesion molecules, in certain
cases their role in tissue architecture (Rutishauser et al., 1978 ;
Buskirk et al., 1980) and morphological cell interactions
(Rutishauser and Edelman, 1980 ; Shirayoshi et al., 1983) have been
also demonstrated.

To initiate an investigation into the role of intercellular
contact in cartilage differentiation we have analyzed the mechanism
of cell adhesion in the embryonic chick limb bud. Employing cell
aggregation in suspension culture as an index of adhesivity we have
demonstrated two distinct mechanisms : (i) Divalent cation-
independent, exhibited immediately after cell isolation with
trypsin in the absence of exogenous calcium, and (ii) Divalent
cation-dependent, expressed only either after complete cell surface
recovery or cell isolation with trypsin in the presence of
exogenous calcium. Fab fragments prepared from an antiserum
directed against the surface of recovered (type ii) cells inhibit
their aggregation but not that of freshly isolated (type i) cells.
This antiserum recognizes a number of surface proteins including
the one with an approximate moleculare weight of 8.5 kD isolated by
limb bud plasma membrane protein affinity chromatography against
equivalent proteins coupled to Sepharose. Here we review the
kinetics of limb bud cell aggregate formation and, from both
immunological and biochemical data, indicate an 8.5 kD
calcium-binding surface glycoprotein to be the major limb bud cell
adhesion molecule.

AGGREGATION

(i) The initial mechanism

Immediatly after their dissociation with trypsin-collagenase
in the absence of divalent cations, stage 24 limb bud cells exhibit
extensive aggregation over a 5 h period (Fig. 1 a). The majority of
cells are aggregation competent while only minimal cell death
occurs during the assay. This would not be predicted given that
aggregation in suspension culture is selective for the chondrogenic
lineage and the relatively small contribution of limb bud
mesodermal cells to cartilage development in situ. Since it is only
partially affected by the addition of 5 mM EDTA (see later), this
initial aggregation occurs largely independent of divalent cations
(Fig. 1 a). Fab' fragments prepared from numerous rabbit antisera
directed against the surface of these cells at each of the assay
time points are unable to disrupt this aggregation (not shown).

Aggregation is completely inhibited by the addition, during
both dissociation and the assay, of 25 µg/ml cycloheximide (Fig. 1
b). This inhibition is completely reversible : Upon the removal of
cycloheximide after 2.5 h exposure, cells aggregate spontaneously
in a manner identical to control (Fig. 1 b). On the assumption that
cycloheximide inhibits protein synthesis, and thus recovery of the
cell surface from proteolysis, employing the lactoper-
oxidase-glucose oxidase technique (Schneider et al., 1982) we
$[^{131}I]$-iodinated the surface of these cells at each time point
during the assay, in either the presence or absence of drug. To our

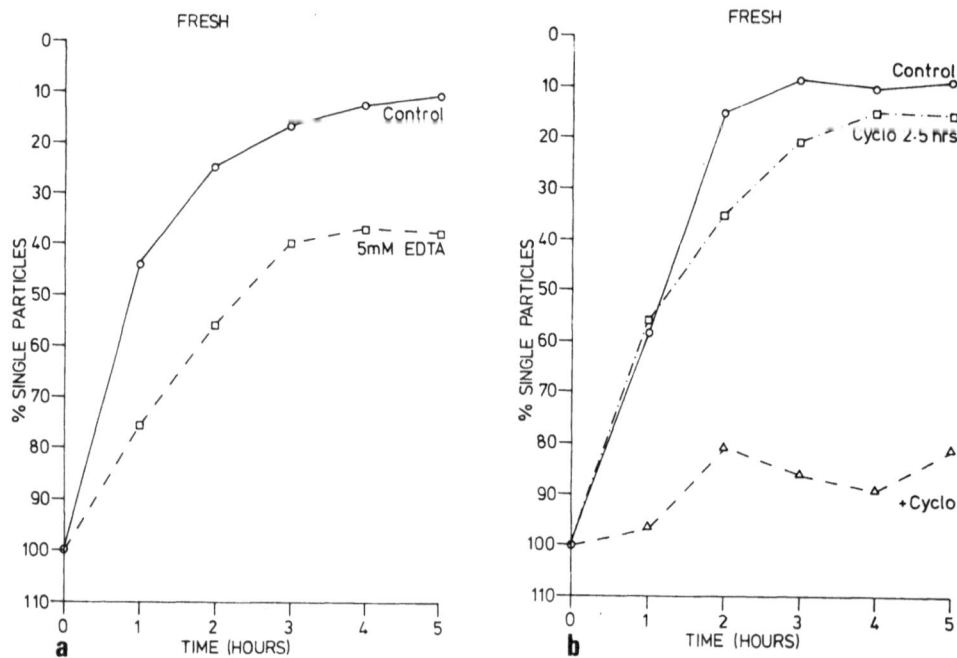

Fig.   1 a and 1 b. Limb bud cell aggregation immediately after
        trypsin dissociation in the absence of calcium.
        For each time point, the aggregation ability of 1.5 x
        10^6 limb bud cells suspended in 200 μl Ham's F12
        culture medium in a 1.5 ml Eppendorf tube was
        monitored under constant agitation at 37°C, 5% CO_2.
        Aggregation is presented as decrease in single
        particle number with reference to the 100% control
        starting value.
        (a) Aggregation is divalent cation-independent. Under
        standard (control) conditions, cells undergo regular
        and sequential aggregation to achieve a plateau after
        5 h. This type of aggregation is largely unaffected by
        the addition of 5 mM EDTA : the rate of aggregation is
        reduced and the plateau value is approximately 30%
        below that of the control. Nevertheless, this partial
        inhibition is reflected as a decrease in the size of
        individual aggregates and not an increase in the
        number of single cells.
        (b) Reversible sensitivity to cycloheximide. The addi-
        tion of 25 μg/ml cycloheximide to the assay medium
        almost totally abolishes aggregation (cyclo). Immedia-
        tely upon its removal following 2.5 h exposure to the
        drug (cyclo 2.5 hrs), cells spontaneously exhibit ag-
        gregation equivalent to that demonstrated by control.

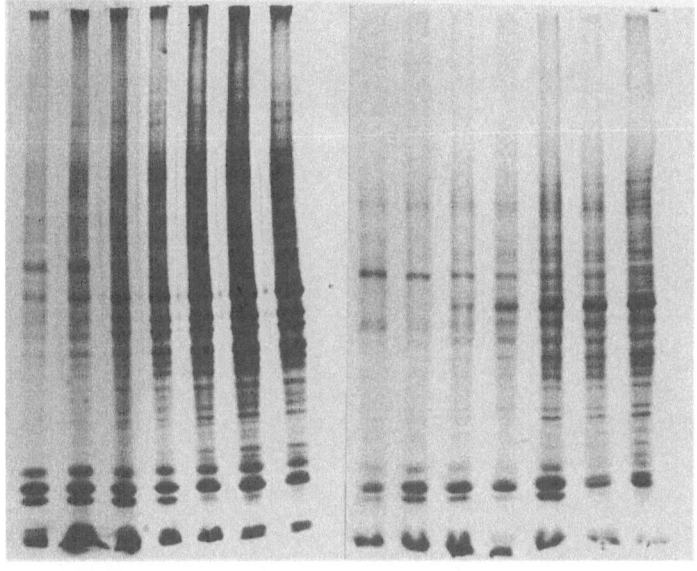

a   **b**

Fig. 2. Limb bud cell surfaces were [131 I]-iodinated at each
         of the aggregation assay time points in either the
         absence (A) or presence (B) of cycloheximide.
         Resultant radio-ionated proteins were analyzed by
         polyacrylamide electrophoresis on 5 - 18% gradient
         gels followed by autoradiogaphy. Although this drug
         completely inhibits aggregation and protein synthesis
         by these cells it appears not to have a major
         qualitative effect on surface recovery. The proteins
         re-appearing during exposure to cycloheximide (B) are
         presumed to derive from an intra-cellular pool.

surprise, recovery of the cell surface in either the presence or
absence of cycloheximide was extremely similar (Fig. 2 A and 2 B).
Although cycloheximide may quantitatively inhibit cell surface
recovery , the ultimate repertoire of proteins is identical. Since
cycloheximide completely inhibits protein synthesis in this system
(not shown) we conclude that proteins appearing on the surface of
these cells are derived from an intra-cellular pool and reflect
membrane cycling. Cycloheximide has been reported to reduce cell
surface protein lateral mobility (Edidin and Weiss, 1972). When
incubated with anti-limb bud cell surface rabbit sera and
fluorescein-conjugated anti-rabbit immunoglobulin at 4°C, limb bud
cells in either the presence or absence of cycloheximide exhibit

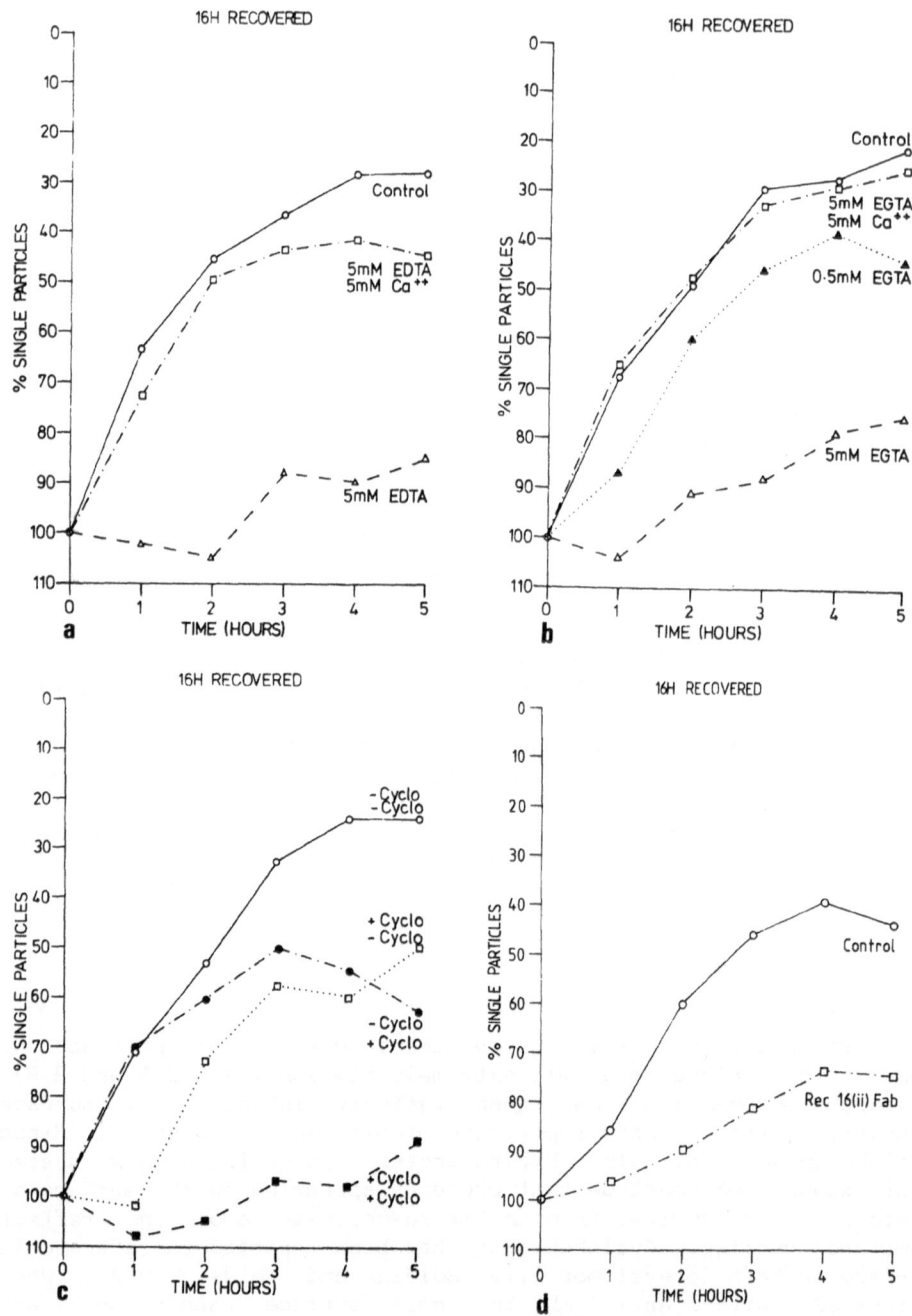

Fig.   3   a, 3 b, 3 c and 3 d. The aggregation of isolated limb
           bud cells recovered for 16 h after dissociation.
           Aggregation was assayed as described in Fig. 1.
           (a) and (b) Dependency upon divalent cations.
           Recovered cells exhibit a regular pattern of aggre-
           gation which is completely inhibited by the inclusion
           of 5 mM EDTA (a). The effect of EDTA is reversed by
           the simultaneous addition of 5 mM calcium chloride
           (a). Similarly, aggregation is partially inhibited by
           0.5 mM EGTA and completely inhibited by 5 mM EGTA
           (b). The addition of equimolar amounts of calcium
           chloride relieves the inhibition of 5 mM EGTA (b).
           (c) Sensitivity to cycloheximide. Control values
           (--o--, -cyclo/-cyclo) were derived from cells both
           recovered and assayed in the absence of 25 µg/ml
           cycloheximide. When present throughout the recovery
           and   assay   period   (-- ■ --,   +cyclo/+cyclo)
           cycloheximide   completely   abolishes   aggregation.
           Recovery in the absence and assay in the presence of
           cycloheximide (-˙-●˙-˙, -cyclo/+ cyclo) results in an
           initial phase of aggregation equivalent to control
           which   subsequently reduces in rate, achieves a
           maximum after 3 h, and is followed by partial
           dissociation. Recovery in the presence and assay in
           the   absence   of   cycloheximide   (˙˙˙ □ ˙˙˙,
           +cyclo/-cyclo) demonstrates a lag period followed by
           sequential recovery of the aggregation mechanism.
           (d) Inhibition by Fab fragments derived from rabbit
           antisera directed against the limb bud cell surface.
           Antisera directed against recovered limb bud cells
           were   prepared   in   rabbits.   Immunoglobulins were
           isolated by DEAE Affi Gel Blue Chromatography,
           converted to monovalent Fab fragments by papain
           digestion (Palmer et al., 1962), and   the   Fab
           fragments isolated by chromatography on Staph protein
           A sepharose (Mollenhauer et al., 1984). When present
           at a final concentration of 8 mg/ml, one of these
           antisera (rec[16](ii)) demonstrated an inhibition of
           aggregation when compared to that of an equal
           concentration of pre-immune rabbit Fab fragments
           (control).

patching at 21 °C and capping and internalization of antigen–antibody complexes at 37°C (not shown). Although this type of aggregation is stimulated by the addition of mixed ganglioside micelles we have not analyzed the effect of cycloheximide on surface ganglioside composition (see Harmer, 1978 for review).

(ii) The recovered mechanism

When recovered for 16 h in dilute rotation culture under conditions which prevent their aggregation, approximately 60% of the limb bud cells die. The remaining 40% of this initial starting population exhibit 70% aggregation (Fig. 3 a). Thus, aggregation is now restricted to 30% of the total dissociated cells. In contrast to the initial mechanism, aggregation is completely inhibited by the inclusion of 5 mM EDTA and this inhibition is relieved by equimolar amounts of calcium chloride (Fig. 3 a). Similarly, aggregation is inhibited partially by 0.5 mM EGTA and completely by 5 mM EGTA (Fig. 3 b). Like EDTA, inhibition of aggregation by 5 mM EGTA is completely reversed by the addition of 5 mM calcium chloride (Fig. 3 b). Consequently, this type of cell aggregation is dependent upon divalent cations.

The susceptibility of the recovered aggregation mechanism to cycloheximide was examined in four ways : cells were both recovered for 16 h and assayed in either the presence or absence of 25 µg/ml cycloheximide. Control cells (i), recovered and assayed in the absence of drug, demonstrate typical (reference) aggregation (Fig. 3 c, -cyclo/-cyclo). In contrast, when both recovered and assayed in the presence of drug (ii), aggregation is completely abolished (Fig. 3 c, +cyclo/+cyclo). Following recovery in its absence and assay in the presence of cycloheximide (iii), aggregation capability is restricted to an initial 3 h period after which dissociation begins to occur (Fig. 3 c, -cyclo/+cyclo). Similarly, recovery in the presence and assay in the absence of drug (iv) demonstrates an initial lag period followed by sequential recovery of the aggregation mechanism (Fig. 3 c, +cyclo/-cyclo). Thus, the type of aggregation demonstrated by the recovered cells is both inhibited (ii) and disrupted (iii) by cycloheximide and these effects are reversible (iv).

Fab fragments prepared from a rabbit antiserum directed against the surface of recovered cells, rec[16] (ii), demonstrate a marked inhibition of aggregation (Fig. 3 d). Equivalent concentrations of Fab fragments prepared from pre-immune sera (Fig. 3 d, control) or antisera directed against cells exhibiting the initial aggregation mechanism (not shown) have no effect on recovered cell aggregation. Similarly, rec[16] (ii) Fab fragments do not disrupt the aggregation of freshly isolated cells.

(iii) Protection of the calcium-dependent mechanism during
     proteolysis

     Immediately   after   trysin-collagenase dissociation in the
presence of 5 mM calcium chloride, limb bud cells demonstrate
aggregation which is inhibited by the inclusion of 5 mM EGTA into
the assay medium (Fig. 4 a). This inhibition is relieved by 5 mM
calcium   chloride but not 5 mM magnesium chloride. Thus, a
calcium-dependent aggregation typical of recovered, and not freshly
isolated cells is demonstrated. Dissociation in the presence of
EGTA during the initial phases of the assay is due to disruption of
the large number of cell pairs generated by this dissociation
procedure. The ability of exogenous calcium to preserve the
calcium-dependent limb bud cell aggregation mechanism is further
revealed by the inability of 25 µg/ml cycloheximide to completely
inhibit aggregation (Fig. 4 b). Presumably, recovery of the
mechanism is not required (Fig. 4 b, cyclo). However,   the
simultaneous presence of both 25 µg/ml cycloheximide and 5 mM EGTA
completely abolishes aggregation.

     Although exogenous calcium retains the calcium-dependent limb
bud   cell   adhesion mechanism, it does not protect it from
proteolysis : when cells are trypsin-collagenase dissociated in the
presence of calcium, briefly incubated with 5 mM EGTA, and washed,
the resultant aggregation is unaffected by the further inclusion of
5 mM (Fig. 4 c). They thus spontaneously exhibit the type of
aggregation characteristic of cells immediately after   their
trypsin-collagenase isolation in the absence of calcium - having
previously possessed the mechanism typically expressed by cells
after 16 h recovery. It should be noted that EGTA does not release
the cell adhesion mechanism from recovered cells and does not
convert them to divalent cation-independent aggregation. When the
material released by EGTA from cells dissociated in the presence of
calcium is dialyzed to remove the EGTA, concentrated, and added to
recovered cells, it partially inhibits their aggregation (Fig. 4
d).   Therefore, this EGTA extract apparently contains the
solubilized domain of the calcium-dependent cell adhesion
mechanism.

(iv) Characterization of the limb bud cell adhesion molecule

     Following   their   separation   by   polyacrylamide   gel
electrophoresis according to Laemmli (1970), total plasma membrane
proteins isolated by sucrose density gradient centrifugation
(Mollenhauer and von der Mark, 1983) were transferred on to
nitrocellulose paper (Towbin et al., 1979). Strips were incubated
with the antiserum rec[16] (ii), Fab fragments derived from which
inhibit the calcium-dependent (recovered) aggregation mechanism
(Fig. 3 d), followed by peroxidase-conjugated goat anti-rabbit
immunoglobulin and a peroxidase substrate (Dziadek et al., 1983 ;

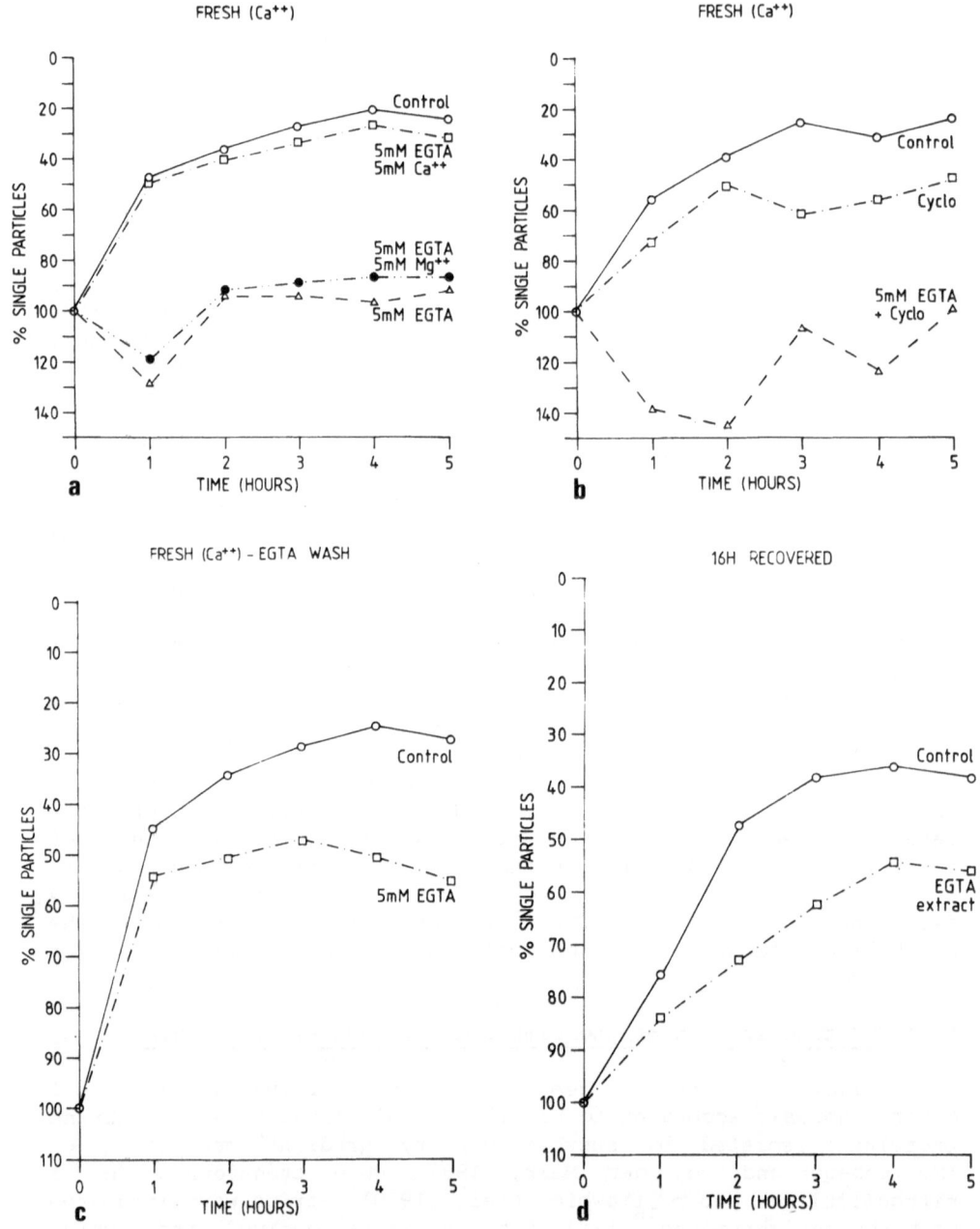

Fig. 4 a, b, c and d. The aggregation of limb bud cells trypsin dissociated in the presence of calcium. Aggregation was assayed as in Fig. 1.

(a) Protection of the calcium-dependent aggregation mechanism during proteolysis. In contrast to cells dissociated in the absence of calcium (Fig. 1 a) aggregation is now completely inhibited by the inclusion of 5 mM EGTA. The inhibitory effects of this chelator are relieved, as in the naturally recovered mechanism (Fig. 3 b) by the further addition of 5 mM calcium chloride but not 5 mM magnesium chloride.

(b) Lack of inhibition of cell aggregation by cycloheximide. In comparison to control, the addition of 25 µg/ml cycloheximide during both dissociation and assay (cyclo) only partially inhibits cell aggregation. Such an effect of cycloheximide is distinct from its complete inhibition of aggregation by cells immediately after their dissociation in the absence of calcium (Fig. 1 b). It is very similar to the progressive inhibition demonstrated by this drug when equivalent cells are first recovered in its absence and then assayed in its presence (Fig. 3 c). The addition of 5 mM EGTA together with cycloheximide (5 mM EGTA + cyclo) completely abolishes aggregation.

(c) Spontaneous conversion from calcium-dependent to -independent aggregation. Immediately after their dissociation, cells were incubated for 15 min in the presence of 5 mM EGTA, washed, and assayed for their ability to aggregate. Aggregation now occurs in the presence of 5 mM EGTA and achieves a plateau value approximately 30% lower than control. It is thus almost indistinguishable from the aggregation of cells trypsin dissociated in the absence of calcium (Fig. 1 a).

(d) The EGTA-extractable material inhibits calcium-dependent aggregation. Material released from limb bud cells during incubation with 5 mM EGTA (c) was dialyzed free of EGTA and concentrated by Amicon ultrafiltration. When added to equivalent cells recovered for 16 h, this material partially inhibits their aggregation (EGTA extract).

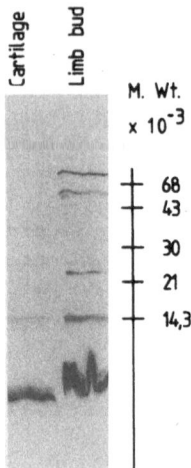

Fig. 5. Total plasma membrane proteins from either limb bud
         cartilage were separated by electrophoresis on linear
         18% polyacrylamide gels and transferred onto nitro-
         cellulose. Strips were incubated with the antiserum
         rec[16] (ii), Fab fragments derived from which inhibit
         calcium-dependent limb bud cell aggregation, followed
         by a peroxidase-conjugated goat anti-rabbit antibody
         and a peroxidase substrate. A cell surface component
         with a molecular weight of 8.5 kD is a prominent
         antigen in both preparations.

Lesot et al., 1983). The resultant pattern demonstrates those
plasma membrane proteins recognized by this antiserum (Fig. 5) in
both limb bud and cartilage. A particularly prominent band with an
apparent molecular weight of approximately 8.5 kD is revealed in
both preparations although it is clearly more pronounced in the
stage 24 limb bud.

     In an attempt to isolate the limb bud cell adhesion molecule
directly, purified limb bud plasma membranes were solubilized with
0.1% octyl glucoside and the proteins coupled to Sepharose 4B with
cyanogen bromide (Cuatrecasas et al., 1968). An equivalent plasma
membrane protein preparation was [^{125}I]-iodinated by the chloramine
T method (Greenwood et al., 1963), solubilized in 0.1% octyl
glucoside and applied to the limb bud membrane protein-sepharose
affinity column in the presence of 5 mM calcium chloride. Following
the removal of unbound material, a 0 - 1 M sodium chloride gradient
was applied to the column, and the retained fraction eluted as a
single peak at 0.6 M sodium chloride. Polyacrylamide gel
electrophoresis and autoradiography of the salt-eluted material
reveals it to be a single protein with an apparent molecular weight
of 8.5 kD (Fig. 6). It is extremely pertinent that in the absence
of exogenous calcium no material is retained by the column.

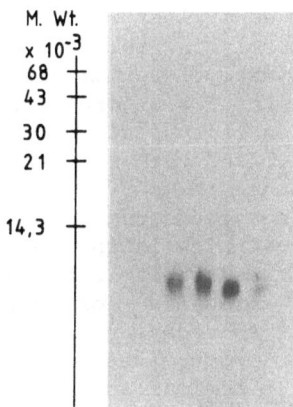

Fig. 6. Characterization of the material eluted at 0.6 M sodium chloride from the limb bud membrane protein
affinity column : the salt-eluted peak of radioiodinated material was subjected to polyacrylamide
electrophoresis in the presence of sodium dodecyl
sulfate on linear 18% gels. Autoradiography reveals
it to be a single protein with a molecular weight of
approximately 8.5 kD. Each lane represents a different region of the salt-eluted peak.

CONCLUSION

Utilizing their ability to aggregate in suspension culture,
two distinct mechanisms governing limb bud cell adhesion are
demonstrated. The first of these is exhibited by cells immediately
after their trypsin-collagenase dissociation in the absence of
exogenous calcium, occurs independently of divalent cations and,
although it is exquisitely sensitive to cycloheximide, it has not
been possible to resolve its molecular basis : even in the presence
of cycloheximide, the majority of proteins re-appear on the cell
surface after proteolysis. The immediate recovery of aggregation
capability after cycloheximide exposure indicates an atypical
action of this drug on intercellular adhesion. Despite repeated
attempts, we have been unable to generate an antiserum against the
surface of these cells, Fab fragments from which will inhibit their
aggregation. Nevertheless, each of the antisera agglutinate the
cells and recognize a variety of surface components by
immunoblotting. It should be remembered that this type of
aggregation is exhibited by cells immediately after their enzymic
dissociation. We therefore suggest that this type of aggregation is
the artefactual result of extensive surface proteolysis.

In contrast, when equivalent cells are first allowed to recover from surface proteolysis their aggregation is divalent cation-dependent as well as cycloheximide sensitive. Recovery from exposure to cycloheximide follows a predictable lag period more indicative of an inhibition of surface protein synthesis. This type of aggregation can be inhibited with Fab fragments prepared from an antiserum directed against the surface of these recovered cells. The aggregation-inhibiting antiserum recognizes a variety of surface proteins - one of which is the 8.5 kD protein isolated by affinity chromatography. The affinity purified protein binds to the column only in the presence of calcium, reflecting the calcium-dependency of cell aggregation, and elutes at 0.6 M sodium chloride. It apparently binds to itself with the high affinity and specificity predicted for a homophilic cell adhesion molecule. The 8.5 kD protein is also the major limb bud cell surface calcium-binding protein (Bee and von der Mark, unpublished).

Whether calcium functions as a bridge in limb bud cell adhesion remains unclear. Evidence is presented for the role of calcium bound to the limb bud cell adhesion molecule in both its function and retention during proteolysis. It is plausible to predict that internal calcium maintains a functional architecture of the 8.5 kD protein : following proteolysis in the presence of calcium and subsequent exposure to a calcium chelator, a soluble fraction is released which is itself capable to inhibiting aggregation. This extract apparently contains the functional domain of the 8.5 kD protein distinct from the intact, insoluble cell adhesion molecule.

It is now pertinent to localize the limb bud cell adhesion molecule within the developing limb and to test our hypothesis that specific cell adhesion mechanisms underlie condensation and cartilage differentiation.

ACKNOWLEDGEMENTS

We are indebted to Prof. Klaus Kuhn for his support of the work described here and to our colleagues for their moral encouragement. Special thanks to Jenny Woods for her assistance with the photography. This work was completed while JAB was a fellow of the Max Planck Gesellschaft.

REFERENCES

Bertolotti R., Rutishauser U., and Edelman G.M., 1980, A cell
    surface molecule involved in aggregation of embryonic liver
    cells, Proc. Natl. Acad. Sci. USA, 77 : 4831-4835.
Buskirk D.R., Thiery J.P., Rutishauser U., and Edelman G.M., 1980,

Antibodies to a neural cell adhesion molecule disrupt
    histogenesis in cultured chick retinae, Nature, 285 :
    488-489.

Cuatrecasas P., Wilchek M., and Anfinsen C.B., 1968, Selective
    enzyme purification by affinity chromatography, Proc. Natl.
    Acad. Sci. USA, 61 : 636-643.

Dessau W., von der Mark H., von der Mark K., and Fischer S., 1980,
    Changes in the patterns of collagens and fibronectin during
    limb-bud chondrogenesis, J. Embryol. exp. Morph., 57 :
    51-60.

Dziadek M., Richter H., Schachner M., and Timpl R., 1983,
    Monoclonal antibodies used as probes for the structural
    organization of the central region of fibronectin, Fed. Eur.
    Biol. Soc. Lett., 155 : 321-325.

Edidin M., and Weiss A., 1972, Antigen cap formation in cultured
    fibroblasts : A reflection of membrane fluidity and of cell
    motility, Proc. Natl. Acad. Sci. USA, 69 : 2456-2459.

Elmer W.A., 1982, Developmental cues in limb bud chondrogenesis,
    Collagen Rel. Res., 2 : 257-279.

Fell H.B., 1925, The histogenesis of cartilage and bone in the long
    bones of the embryonic fowl, J. Morph., 40 : 417-451.

Fell H.B., and Canti R.G., 1935, Experiments on the development in
    vitro of the avian knee joint, Proc. R. Soc., B 166 :
    316-351.

Gerisch G., 1980, Univalent antibody fragments as tools for the
    analysis of cell interactions in Dictyostelium, Curr. Top.
    Dev. Biol., 14 (2) : 243-270.

Grabel L.B., Singer M.S., Martin G.R., and Rosen S.D., 1983,
    Teratocarcinoma stem cell adhesion : The role of divalent
    cations and a cell surface lectin, J. Cell Biol., 96 :
    1532-1537.

Grady S.R., Nielsen L.D., and McGuire E.J., 1982, Organ and class
    specificity of cell adhesion blocking antisera, Exp. Cell
    Res., 142 : 169-180.

Greenwood F.C., Hunter W.M., and Clover J.S., 1963, The preparation
    of [^{131}I]-labelled human growth hormone of high specific
    radioactivity, Biochem. J., 89 : 114-123.

Grumet M., and Edelman G.M., 1984, Heterotypic binding between
    neuronal membrane vesicles and glial cells is mediated by a
    specific cell adhesion molecule, J. Cell Biol., 98 :
    1746-1756.

Harman R.E, 1978, ed. "Cell surface carbohydrate chemistry",
    Academic Press, New York.

Laemmli U.K., 1970, Cleavage of structural proteins during the
    assembly of the head of bacteriophage T4, Nature, 227 :
    680-685.

Lesot H., Kuhl U., and von der Mark K., 1983, Isolation of a
    laminin binding protein from muscle cell membranes, EMBO J.,
    2 : 861-865.

Mollenhauer J., Bee J.A., Lizarbe M.A., and von der Mark K., 1984,

Role of anchorin CII, a 31,000-mol-wt membrane protein, in the interaction of chondrocytes with type II collagen, J. Cell Biol., 90 : 1572-1578.

Mollenhauer J., and von der Mark K., 1983, Isolation and characterization of a collagen binding glycoprotein from chondrocyte membranes, EMBO J., 2 : 45-50.

Nielsen L.D., Pitts M, Grady S.R., and McGuire E.J., 1981, Cell-cell adhesion in the embryonic chick : partial purification of liver adhesion molecules from liver membranes, Develop. Biol., 86 : 315-326.

Ocklind C., Odin P., and Obrink B., 1984, Two different cell adhesion molecules - cell-CAM 105 and a calcium-dependent protein - occur on the surface of rat hepatocytes, Exp. Cell Res., 151 : 29-45.

Palmer J.L., Mandy W.J., and Nisonoff A., 1962, Heterogeneity of rabbit antibody and its subunits, Proc. Natl. Acad. Sci. USA, 48 : 49-53.

Rutishauser U., and Edelman G.M., 1980, Effects of fasciculation on the outgrowth of neurites from spinal ganglia in culture, J. Cell Biol., 87 : 370-378.

Rutishauser U., Thiery J.P., Brackenbury R., and Edelman G.M., 1978, Adhesion among neural cells of the chick embryo. III. Relationship of the surface molecule CAM to cell adhesion and the development of histotypic patterns, J. Cell Biol., 79 : 371-381.

Rutishauser U., Thiery J.P., Brackenbury R., Sela B.A., and Edelman G.M., 1976, Mechanisms of adhesion among cells from neural tissues of the chick embryo, Proc. Natl. Acad. Sci. USA, 73 : 577-581.

Schneider C., Sutherland R., Newman R., and Greaves M., 1982, Structural features of the cell surface receptor for transferrin that is recognized by the monoclonal antibody OKT9, J. Biol. Chem., 257 : 8516-8522.

Shirayoshi Y., Okada T.S., and Takeichi M., 1983, The calcium-dependent cell-cell adhesion system regulates inner cell mass formation and cell surface polarization in early mouse development, Cell, 35 : 631-638.

Solursh M., and Reiter R.S., 1980, Evidence for histogenic interaction during in vitro limb chondrogenesis, Develop. Biol., 78 : 141-150.

Summerbell D., and Wolpert L., 1972, Cell density and cell division in early morphogenesis of the chick wing, Nature, 239 : 24-26.

Thiery J.P., Brackenbury R., Rutishauser U., and Edelman G.M., 1977, Adhesion among neural cells of the chick embryo. II. Purification and characterization of a cell adhesion molecule from neural retina, J. Biol. Chem., 252 : 6841-6845.

Thorogood P.V., and Hinchliffe J.R., 1975, An analysis of the condensation process during condensation in the embryonic

        chick hind limb, J. Embryol. exp. Morph., 33 : 581-606.
Towbin H., Staehelin T., and Gordon J., 1979, Electrophoretic
        transfer of proteins from polyacrylamide gels to
        nitrocellulose sheets : procedure and some applications,
        Proc. Natl. Acad. Sci. USA, 76 : 4350-4354.
von der Mark K., and von der Mark H., 1977, Immunological and
        biochemical studies of collagen type transitions during in
        vitro chondrogenesis of chick limb mesodermal cells, J. Cell
        Biol., 73 : 736-747.
Yoshida C., and Takeichi M., 1982, Teratocarcinoma cell adhesion :
        identification of a cell-surface protein involved in
        calcium-dependent cell aggregation, Cell, 28 : 217-224.
Yoshida-Noro C., Suzuki N., and Takeichi M., 1984, Molecular nature
        of the calcium-dependent cell-cell adhesion system in mouse
        teratocarcinoma and embryonic cells studied with a
        monoclonal antibody, Develop. Biol., 101 : 19-27.

SECTION III

ACETYLCHOLINE RECEPTOR CLUSTERING AND THE FORMATION

OF THE NEUROMUSCULAR JUNCTION

Robert J. Bloch, David W. Pumplin[1]
and Manfred Baetscher

Departments of Physiology and Anatomy[1]
University of Maryland School of Medicine
Baltimore, Md. 21201

INTRODUCTION

The molecular mechanisms involved in the formation of the nervous system are not yet understood, and experimental approaches to study them are not easy to come by. Several problems face the researcher in this area. For example, in any particular part of the brain there are several different types of neurons, which are difficult to distinguish while alive. These neurons are involved in several different types of synapses, utilizing different neurotransmitters and neurotransmitter receptors. Molecular markers for the different cell types and for their pre- and postsynaptic connections are useful only for autoradiographic studies of whole brain slices. Without markers for use at the cellular level, a detailed description of the assembly of synapses and neuronal circuits will not be possible.

One synapse for which we do have such markers is the synapse formed in the peripheral nervous system between motor neurons and skeletal muscle fibers -- the neuromuscular junction. The breakthrough in this field was provided by the discovery of α-bungaro-toxin (BT), a polypeptide from the venom of the krait, Bungarus multicinctus, which binds specifically and essentially irreversibly to the nicotinic acetylcholine receptor, the neurotransmitter receptor in the postsynaptic membrane of the neuromuscular junction. Thanks largely to the experiments made possible by BT, we now know quite a bit about the stages of development of this synapse, and have identified several proteins which participate in its formation. From our knowledge of the roles these proteins play in other cells and in model systems for the

239

postsynaptic membrane, we can begin to propose possible functions for them during synaptic development. This is a first step in understanding the cellular and molecular events involved in synapse formation.

Our first aim in this article is to summarize the four stages of development of the neuromuscular junction and to discuss some of the processes that may be involved. Many investigators have done important research in this area. We refer to their results whenever possible, but lack of space precludes listing a complete bibliography. For this purpose, the reader should consult several recent reviews (1-4). Our second aim is to outline the approaches we have been using to study the initial stages of junction formation. Our research has concentrated on a model system for the postsynaptic region of the rat neuromuscular junction.

FORMATION OF THE NEUROMUSCULAR JUNCTION

Muscle cells in the anterior region of the rat embryo start to differentiate on embryonic day 13 (E13). At this time, processes of motor neurons are already present in the muscle mass where they are likely to influence muscle differentiation.

The first stage of neuromuscular junction formation occurs as newly formed myotubes start to interact with the neurons. In the rat, this period lasts from approximately E13 to E15. Newly formed myotubes synthesize acetylcholine receptors (AChR) and insert them into their cell membrane, where they become distributed diffusely over the surface. These receptors interact with ACh released from the nerve terminal, generating small unitary depolarizations, called miniature endplate potentials or "mepps". These early mepps are characterized by slow rising and falling phases typical of a situation in which the postsynaptic receptor density is low. Nevertheless, even at this early stage it is clear that synaptic interactions are occurring (5). This is confirmed by numerous electron microscopic studies which show that limited regions of the nerve and muscle cells are very closely apposed to one another (e.g., 6).

The second stage of junction formation involves the accumulation of AChR in the region of the muscle membrane near the nerve. This stage may begin on E15.5 and continue through E18. It is characterized electrophysiologically by mepps showing faster rising and falling phases. Morphologically, AChR concentrated at the postsynaptic region become detectable using fluorescent or radioiodinated BT derivatives. In the light microscope, postsynaptic AChR aggregates or clusters appear mottled or speckled, with areas of receptor enrichment interspersed with areas lacking high receptor densities (7, 8). At the ultrastructural level, AChR aggregates also appear discontinuous, and are associated with electron-dense regions of the postsynaptic region near, but not always under, regions of nerve contact (6). We refer to these early AChR clusters as "embryonic clusters".

The <u>third stage</u> of neuromuscular junction formation is associated with a condensation of the postsynaptic AChR aggregate into a more uniformly organized "plaque". Receptor-free areas within the postsynaptic region are not seen in plaques, and overall receptor densities are therefore considerably higher than those at stage 2 (6). In the anterior muscles of the rat, plaque-formation is usually complete by birth. We refer to these structures as "neonatal clusters".

The <u>fourth</u>, and final stage of neuromuscular junction formation starts during the first postnatal week, and is completed by 2 weeks after birth, when the postsynaptic AChR aggregate resembles in all respects except size the structure found at the mature motor endplate. We therefore refer to the AChR organization at junctions which are two weeks old, or older, as an "adult cluster". Such structures show a very high density of AChRs which closely follows the lines of the nerve terminus (6-8). To reach this stage, the "neonatal clusters", or plaques, probably undergo a "sculpting", or removal of AChRs from regions not directly contacted by nerve, while innervated strips of membrane retain their high density of AChRs.

Some of the properties of the receptors in embryonic, neonatal and adult clusters are presented in Table 1. Several points become clear from this table. First, the properties of the AChRs of embryonic clusters are essentially identical to those of AChRs found at non-junctional membrane of newly innervated, embryonic myofibers. In addition, these properties are similar, if not identical, to those of the extrajunctional receptors which appear throughout adult muscle after denervation or blockade of neurotransmission. Second, the properties of AChRs at developing neuromuscular junctions change gradually, in at least two distinct steps. The first step, from embryonic to neonatal clusters, involves stabilization of the receptors, but no changes in antigenicity or channel properties. The second step, from neonatal to adult, involves changes in the latter, but not the former, properties. The molecular mechanisms responsible for these transitions are not yet understood. Their elucidation should reveal many of the ways in which a nerve cell interacts with and alters its target.

FORMATION OF AChR CLUSTERS

The mechanisms contributing to the formation of embryonic

Table 1. Properties of AChR at the developing
rat neuromuscular junction and in
clusters of cultured rat myotubes.

Stage	Appearance	Property Density	Stability	Open-time	Antigen
Embryonic	Speckled	∿2,000/ sq.mic.	Turns over rapidly. Dispersable	∿5 msec	Extra-junctional
Neonatal	Plaque	∿10,000/ sq.mic.	Turns over slowly. Not dispersable	∿5 msec	Extra-junctional
Adult	Follows nerve terminus	∿10,000/ sq.mic.	Turns over slowly. Not dispersable	∿1 msec	junctional
Myotube Cluster	Speckled	∿2,000/ sq.mic.	Turns over rapidly. dispersable	Not known	Not known

(References: 6-8, 17   6, 9   10-14   15   16)

clusters have been investigated in considerable detail. Understanding how AChRs cluster in muscle membrane involves answering a number of questions. (i) Where do the AChRs in the clusters come from ? (ii) How does the muscle cell maintain a cluster once it has begun to form ? (iii) What is the effect exerted by the nerve on the muscle which promotes localized aggregation of receptors ? These questions will be considered briefly in the following paragraphs.

The similarities between diffuse (i.e., unclustered) receptors and those in embryonic clusters suggest that some of the AChrs in embryonic clusters are recruited from diffuse receptor pools in the sarcolemma. Alternatively, the innervated muscle membrane may become a site for selective insertion of new receptors, creating a high local concentration. Current evidence strongly indicates that recruitment of diffuse receptors is involved in the formation of embryonic clusters (18-20), but there is also evidence that at least some receptors are selectively inserted into clusters from

intracellular pools (21). The relative importance of these two
sources of AChRs in embryonic clusters has not yet been determined.
Different molecular mechanisms will probably be involved in
generating clusters if the receptors arize from intercellular or
sarcolemmal pools.

Whatever the source of the AChRs, the muscle must employ
special means for maintaining a cluster once it has begun to form.
Two kinds of macromolecular arrays have been found at AChR-rich
membrane sites. On the outer membrane surface of cultured or
embryonic muscle cells, basal lamina is often localized at sites of
AChR aggregation. On the inner membrane surface, microfilaments and
associated cytoskeletal proteins underly many AChR clusters. There
is now considerable evidence linking both of these intracellular
and extracellular structures to the process of AChR cluster
formation and stabilization. For example, drugs which disrupt the
cystoskeleton disrupt or disorganize pre-existing AChR clusters,
and prevent the formation of new clusters (12, 13, 22). Drugs which
increase or decrease the levels of collagen synthesized in muscle
cultures also increase or decrease the number of AChR clusters in
the cultures (23). Similarly, collagenase treatment of intact
myotubes decreases the number of AChR clusters (24). A molecule
isolated from the extracellular matrix of Torpedo electric organ is
also able to induce AChR clustering in cultured myotubes (25).
Immunocytochemical methods reveal that a number of intracellular
and extracellular proteins are enriched at AChR clusters. These are
listed in Table 2. In the final section of this article, we will
consider the possible roles of some of these proteins, as
established in other "model" systems.

Finally, to induce an embryonic AChR cluster at the site of
innervation and not elsewhere on the muscle fiber surface, the
nerve must exert an highly localized effect. There are two models
which can account for this effect of nerve. The nerve secretes
soluble factors, among which there may be a "trophic" factor
capable of interacting with the muscle cell surface and inducing
the local accumulation of AChRs. The nerve can also provide a
structure to which the muscle can adhere, thereby distinguishing
the membrane attached to the nerve terminal as special. In fact,
there is now considerable evidence that both "trophic", or
chemical, and "adherent", or physical, factors can induce the
formation of AchR clusters. "Trophic" substances of high and low
molecular weight have been found in extracts from brain and Torpedo
electric organ, and in medium conditioned by neuronal cells (23,
25, 38-40). Substances to which muscle cells attach and
subsequently produce AChR clusters include substrates of glass,
plastic and collagen, latex beads, and, if they can be considered
in this category, elaborations of the extracellular matrix (17, 29,
32, 41-45). Examples are presented in Fig. 1 of AChR clusters
induced by the presumably physical interaction of a rat myotube

Table 2. Macromolecules associated with AChR-
rich membrane in vivo and in vitro.

Molecule	Cluster in vitro	Synapse in vivo	Ref.
Intracellular			
Actin	+	+	26, 27
Vincullin	+	+	17, 28
α-Actinin	+	+	28, 29
Filamin	+	+	28, 29
51 Kd	Not known	+	30
43 Kd	Not known	+	31
Extracellular			
Heparan Sulfate Proteoglycan	+	+	32
Acetylcholin-esterase	+	+	33, 34
Collagen	+	+	23, 35
Laminin	+ (partial)	+	35-37
Fibronectin	+ (partial)	+	35, 37

with a glass fiber and a glass cover slip. These clusters induced
in vitro resemble in many ways the embryonic clusters which form at
the developing neuromuscular junction.

AChR CLUSTERING IN RAT MYOTUBES

Rat myotubes can be cultured from dissociated hind limb muscle
of embryonic or neonatal animals. The myogenic cells in the
dissociated preparations plate down onto a variety of substrates,
grow, and fuse to form myotubes. The myotubes produce a number of
muscle-specific proteins ; some of these, including AChRs, are
inserted into the surface membrane. On glass or collagen, about 10%
of the AChRs spontaneously cluster at the myotube-substrate
interface. (e.g., Fig. 1). The properties of these
substrate-apposed AChR clusters are summarized in Table 1.

It is clear that they greatly resemble the embryonic clusters
described earlier in this report. They may therefore be an
appropriate model system with which to study the processes of
cluster induction and stabilization under easily controlled
conditions. Our approach to these problems has been to identify the

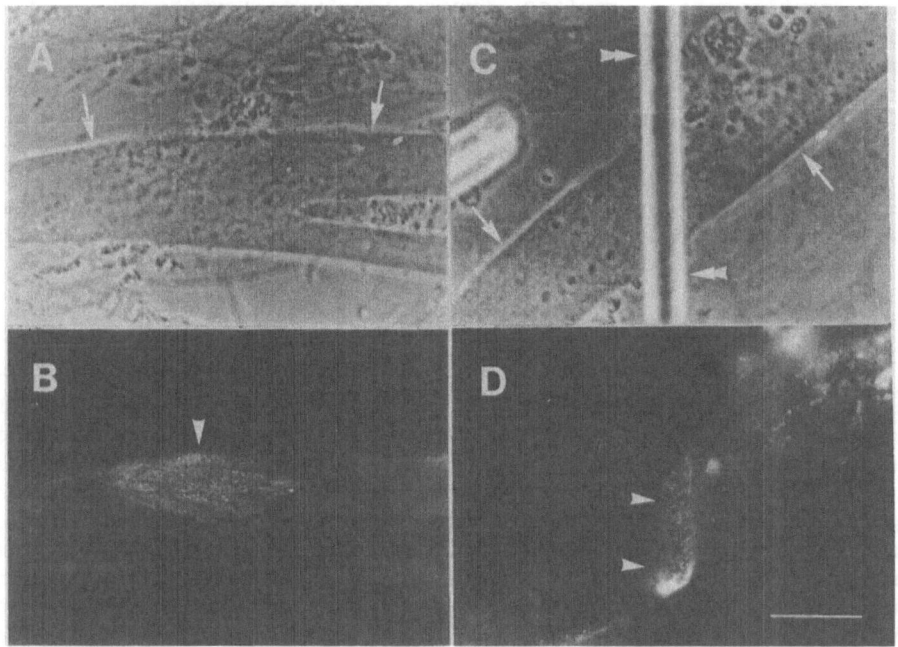

Fig. 1. AChR clusters in cultured rat myotubes form in
response to interaction with glass.
Rat myotubes cultured on glass cover slips were
either stained with tetramethylrhodamine-BT (R-BT)
(panels A and B), or incubated overnight with glass
fibers, and then stained (panels C and D). Regions of
enhancement of R-BT label, representing AChR clusters, are associated with regions of the membrane
contact by the cover slip (B), or by the glass fiber
(D). Arrowheads point to the AChR clusters, visualized in fluorescence. Small arrows point to the
myotube, and double arrowheads point to the glass
fiber, seen in phase. The bar represents 10 microns.

molecules asspciated with clusters of AChRs using intact cells and
cell fragments prepared so as to contain some, or essentially no,
cytoplasmic contamination. Whenever possible, we have also tried to
compare the clusters formed in vitro to the nerve-induced,
embryonic clusters formed in vivo.

The correlation seen in rat myotubes of clustering sites to
sites of cell-substrate attachment has had several important
implications. (i) The clusters, and the structures closely
associated with them, always appear in a single plane of focus when

observed in the light microscope. (ii) The clusters can be studied using complementary replica freeze-fracture techniques. (iii) Substrate-attached material tends to be stable when cultures are sheared or exposed to the cholesterol-specific detergent, saponin. With such manipulations, clusters can be separated from the bulk of the cytoplasm. (iv) More generally, their close correlation makes us wonder what it is about attachment sites that promotes the clustering of AChR nearby. In the following paragraphs, we will consider some of these points, using the information we and others have obtained to present a model for the organization of AChR clusters of rat myotubes.

The "Domain Structure" of AChR clusters

AChR clusters are associated preferentially with regions of myotube-substrate attachment, but these regions are usually heterogeneous in appearance. When visualized with interference reflection microscopy, they often appear to be organized in strips. Some strips show a grey interference color, indicating close membrane-to-glass apposition, while the intervening membrane strips show no interference color, indicating greater membrane-to-glass distances. A close comparison of this organization with that of clustered AChRs, which are also organized in AChR-rich and AChR-poor strips, shows that the AChR are concentrated in the regions farther from the glass, and essentially absent from the regions of close membrane-to-glass contact (17). This interdigitating pattern has given rise to the idea that AChR clusters are organized in distinctive membrane domains, termed "AChR domains", rich in receptors, and "contact domains", poor in receptors but closer to the substrate. This distinctive organization allows us to differentiate between molecules or structures which are preferentially associated with AChR or with regions involved in attaching the myotube to the substrate.

Using this criterion, we have been able to assign a number of components of AChR clusters to one membrane domain or the other. For example, vinculin, a protein known to be enriched at the intracellular face of the membrane at the focal contacts of mononucleate cells (46), is present at the contact domains, but not the AChR domains of AChR clusters. This can be seen in intact myotubes after mild treatment with neutral detergent to permeabilize the membrane, or, more clearly, in fragments isolated from myotubes by gently shearing the culture (17). Cholesterol, on the other hand, is relatively enriched in the AChR domains, but depleted in the contact domains (47).

Isolation of AChR clusters

Although cholesterol is present in AChR domains, it is not as concentrated there as it is in regions of the muscle membrane outside the clusters. The greater amounts of cholesterol outside of clusters probably accounts for the observation that in the presence of saponin, a cholesterol-specific detergent, the bulk of the myotube is destabilized while the clusters remain intact. After brief extraction with saponin, clusters are found still attached to the substrate, while > 99% of the cellular protein, and ∿ 90% of the AChR, are shed into the saponin solution. The clusters which remain on the substrate are essentially unchanged from those seen in intact cells (48). AChR domains and contact domains are clearly distinguishable. Vinculin, however, is missing from these isolated AChR clusters. This observation suggests that vinculin is not required either for the continued integrity of clusters which have already formed, or for the continued attachment of the myotube membrane to the substrate.

Presumably, the molecules which are responsible for these two processes are retained in AChR clusters isolated by saponin extraction. In fact, there are relatively few proteins visible in SDS-PAGE of these preparations (48), so it should be possible to assign a role in substrate attachment or receptor clustering to several of them. The major aim of our recent research has been to identify and characterize such molecules. In the following few paragraphs, we discuss some of the evidence which suggests that an actin-like molecule is involved in maintaining the organization of AChR in isolated clusters. We then present evidence which suggests that glycoproteins capable of binding concanavalin A are enriched at the contact domains of isolated clusters.

Actin involved in AChR clustering

Isolated AChR clusters occupy a relatively large expanse of membrane, and have distinct domains, making them an excellent preparation with which to study the interaction of membrane constituents with the cytoskeleton. The cytoplasmic face of the cluster membrane is completely accessible to the bathing solution, and it can be acted upon by enzymes or other modifications of this solution. After fixation, it can be reacted with antibodies or other specific macromolecular probes for cytoskeletal proteins.

Upon staining of isolated clusters with a monoclonal antibody specific for actin (mAb HP249 : kindly provided by Dr. R. Anthony, Department of Pathology, University of Maryland School of Medicine), followed by a fluoresceinated second antibody, a clear pattern of labelling is obtained (Fig. 2 B). Comparison of this pattern with the pattern seen with R-BT-labelled AChR (Fig. 2 A) shows them to be essentially identical. This suggests that an actin-like molecule is present at the AChR domains of isolated clusters.

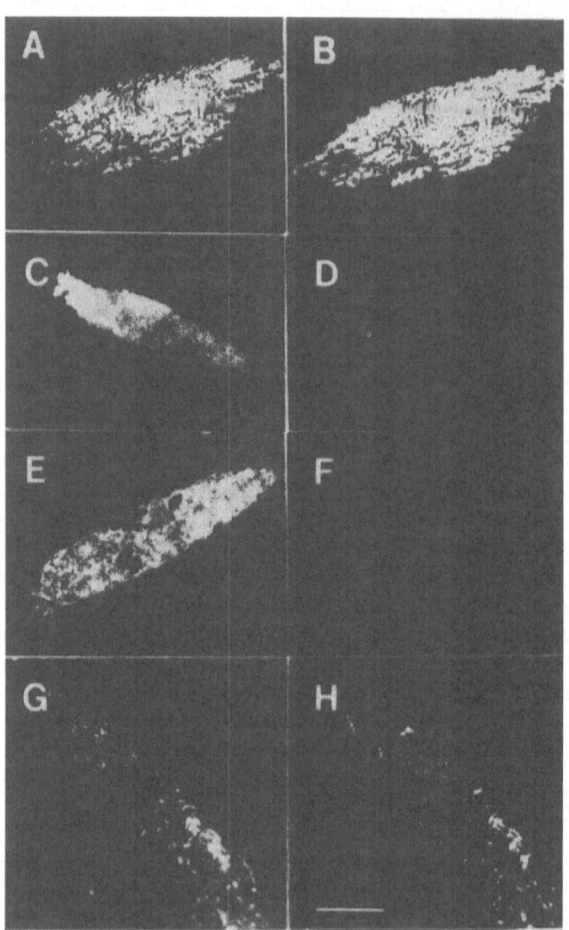

Fig. 2.   Actin is involved in maintaining the organization of
           AChR clusters isolated by saponin extraction.
           AChR clusters were labelled with R-BT and isolated
           from myotube cultures by saponin extraction. R-BT
           label is represented in the left-hand panels, A,   C,
           E and G. Isolated clusters were then fixed, or
           treated and fixed, and then stained with HP249
           antiactin antibody, followed by FITC-labelled goat
           antimouse antibody (fGAM). FGAM label is shown in
           the right-hand panels, B, D, F, and H. A, B :
           Control clusters showing the   typical   streaked
           appearance of AChR clusters, and the colabelling of
           the AChR domains with antiactin and fGAM.   C, D :
           Clusters extracted at low ionic strength before
           fixation and staining with antibody. Actin is no

How similar is the actin-like molecule of AChR clusters to other forms of actin ? We have not yet characterized it chemically, but several additional experiments suggest that the material stained with the monoclonal antiactin antibody is truly actin. Other monoclonal antiactins obtained from different investigators also stain the AChR domains of isolated receptor clusters. After SDS-PAGE and blotting onto nitrocellulose paper, actin isolated from rabbit skeletal muscle is labeled with the HP249 antibody. Of the polypeptide chains present in preparations of isolated AChR clusters, only a band at ∿ 42 kd reacts with this antiactin antibody under the same conditions. Finally, the isolated clusters can be decorated with filaments of purified smooth muscle myosin. These filaments label the AChR domains preferentially. Like myosin-actin interactions in other systems, labelling by the filaments is inhibited by 5 mM MgATP. All these experiments suggest that a form of actin is present at the AChR domains of isolated receptor clusters.

Does the presence of actin have anything to do with the distribution of AChRs in isolated clusters ? The answer seems to be, "yes". After extraction of actin with solutions of low ionic strength, isolated clusters no longer stain with antiactin antibody (Fig. 2 D). Although these structures, depleted of actin, retain nearly all their AChR, the receptors undergo a significant rearrangement. Instead of the distinctive linear organization typical of the untreated clusters, AChR assumes a more uniform, diffuse distribution after actin has been removed (Fig. 2 C). This experiment also indicates that the actin of isolated AChR clusters is a peripheral membrane protein. SDS-PAGE of the material retained on the cover slip after extraction of actin revealed that the polypeptide at ∿ 42 kd was the only major macromolecular constituent which was solubilized at low ionic strength. Similarly, treatment of isolated clusters with proteolytic enzymes destroys their ability to be stained with antiactin (Fig. 2 F), and, again, the organization of AChR in the isolated plaques of myotube membrane is disrupted (Fig. 2 E). Actin therefore appears to be involved in maintaining the organization of AChR in isolated clusters.

---

longer present, and AChR have lost their distinctive domain organization. E, F : Clusters treated with chymotrypsin before fixation and staining with antibody. Actin is absent, and AChR are disorga- nized. G, H : Remnants of clusters isolated by saponin extraction of cells treated for 4 hrs. with azide. AChR are depleted, and so is antiactin labeling. The bar in H is 10 microns.

Does the presence of actin have anything to do with the distribution of AChRs in intact cells ? This is considerably more difficult to answer. Antiactin staining of intact, permeabilized myotubes is extremely bright, due to the binding of the antibody to the contractile apparatus and to the cytoskeleton which is not associated with the cluster membrane. To avoid this problem, we have used two alternative methods. In one, we disrupted AChR clusters by treating intact cells with sodium azide (12), then extracted the myotube cultures with saponin, as we would do to obtain AChR clusters. The resulting preparations were depleted of both AChRs (Fig. 2 G) and actin (Fig. 2 H). However, the actin that remained was usually associated with small areas of the membrane which contained AChR (compare panels G and H of Fig. 2). This is consistent with the idea that the intact myotube regulates AChr clustering and associated levels of actin in a similar fashion.

Our second approach to this problem was to prepare "sheared" AChR clusters. Upon incubation with antiactin antibody, these membrane fragments revealed a meshwork of staining, which usually coincided with the R-BT label. Thus, in cell fragments which more closely resemble the intact myotube, actin is also present at AChR domains. In addition, staining in these samples was sometimes seen over contact domains, but this staining was considerably brighter than that over AChR domains and was in dart-like or arrowhead-like structures. This arrangement is reminiscent of the distribution of vinculin at contact domains. As vinculin is known to be present at the intracellular aspect of the plasma membrane where stress fibers insert (46), it seems reasonable to postulate that the structures at contact domains which stain brightly with antiactin are bundled microfilaments. One piece of information which supports this idea is our observation that, in thin sections through intact cells, microfilament bundles are present very close to the membrane, at intervals consistent with their being associated with contact domains (Fig. 3, brackets). Between these bundles is a network of thin filaments, approximately 8 nm in diameter (Fig. 3, small arrows), which may account for the antiactin staining we have observed over AChR domains.

These observations suggest that actin is always present over AChR domains, and that microfilament bundles are associated with at least some contact domains. One question raised by these observations is, "what is the relationship between the actin-containing structures at AChR domains and contact domains ?" One possibility, supported by occasional observations of "sheared" clusters, is that a meshwork containing actin covers the entire face of the cluster membrane in intact cells, but that it is usually removed together with the microfilament bundles during shearing or saponin extraction. The formation of such an actin meshwork may precede the formation of filament bundles, as suggested in other cell-substrate interactions (49). Such a

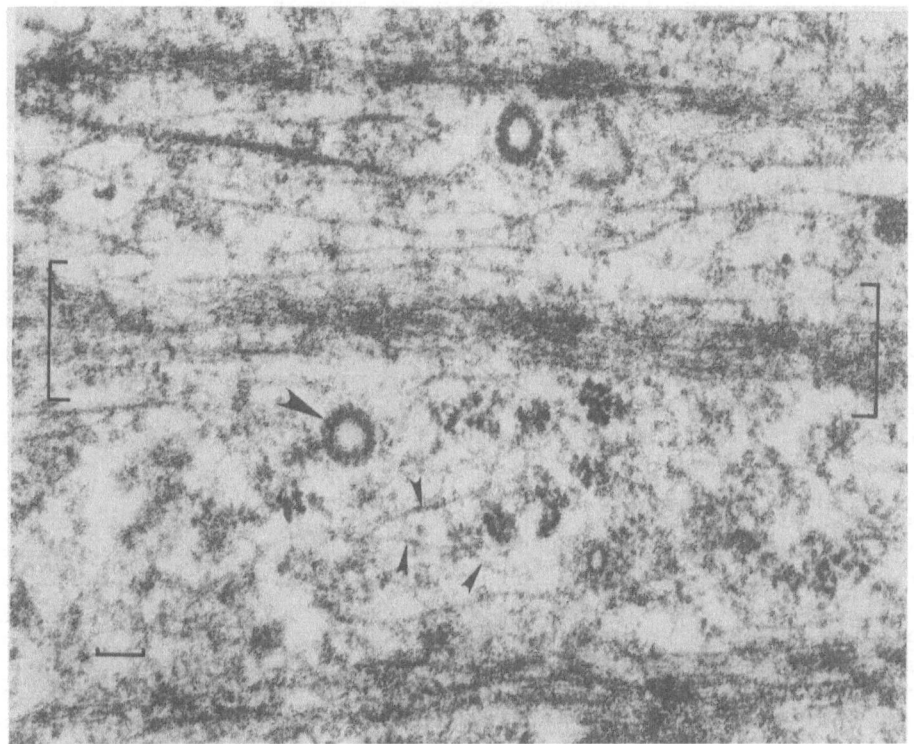

Fig. 3. Thin section view of a portion of an identified
        attachment site. The section was taken parallel to
        the substrate, and is within one thin section
        thickness (60 nm) of the cell membrane. Bundles of
        parallel filaments (one bracketed) flank regions
        containing a looser meshwork of similarly sized
        filaments (small arrows). The filaments are
        approximately 8 nm in diameter. Coated vesicles and
        coated pits (large arrow) are confined to the
        meshwork region. Bar, 100 nm.

meshwork probably includes other actin-binding proteins, such as
those involved in the formation of the cytoskeletal network
underlying the erythrocyte membrane. Further work is necessary to
evaluate these ideas. Nevertheless, our observations strongly
suggest that actin is present at AChR clusters, and is either
directly or indirectly involved in maintaining the organization of
AChR in isolated cluster preparations.

Molecules at sites of myotube-substrate contact

     Despite the fact that cytoskeletal proteins such as vinculin
have been removed during   extraction with saponin, the contact
domains of isolated AChR clusters remain intact and associated with
the tissue culture substrate (48). This suggests that the molecules
involved in myotube-substrate attachment are retained in isolated
AChR clusters. Using selective solubilization techniques, we have
been able to identify some of these molecules as lectin receptors.

     Upon   extraction   of isolated AChR clusters with a low
concentration of a mild neutral detergent, such as Triton X-100 or
Nodinet P40 , most of the AChR and proteins associated with AChR
domains are solubilized and removed from the substrate. After
fixation of the material remaining on the substrate and staining
with fluorescein concanavalin A (Con A), we   have   observed
distinctive structures (Fig. 4). Such structures are reminiscent of
the contact domains, or portions thereof, in intact cells and
isolated clusters ; they are linear, oriented parallel to what was
the long axis of the myotube, and apparently are tightly associated
with the substrate. Similar structures can be visualized after
staining with fluorescein derivatives of lentil and Ricinus
communis lectins, but not with derivatives of lectins from wheat
germ, waxbean, soybean, Limulus polyphemus or Dolichos biflorus.
The ability of Con A and lentil lectin to stain these structures
suggests that molecules containing α-mannoside or α-glucoside
linkages are present. In keeping with this, all staining by
fluorescein concanavalin A is blocked by 10 mM α-methylmannoside.

     We have used selective extraction and lectin binding to
partially   purify   the   Con   A-binding   components in   the
substrate-associated material, or SAM, from myotube cultures. After
extraction of isolated AChR cluster preparations with 0.05% Triton
X-100, SAM was solubilized in 0.25% SDS, diluted into an excess of
Triton X-100 and applied to a Con A-Seraphose column. Unbound and
non-specifically bound material was washed through the column ;
specifically bound material was then eluted with α-methylmannoside
followed by SDS. SDS-PAGE of the specifically bound material shows
the presence of one major band at 36 kd (Fig. 5). We know from
independent experiments that at least the 36 kd band is a Con
A-binding glycoprotein. We do not yet know the nature of the other
bands in the Con A-binding fraction. As SAM contains materials
derived from both myotubes and mononucleate cells in the culture,
we must also learn which of these bands are likely to be derived
from the isolated clusters themselves. Work on this problem and
related questions concerning myotube-substrate attachment are now
in progress in our laboratories.

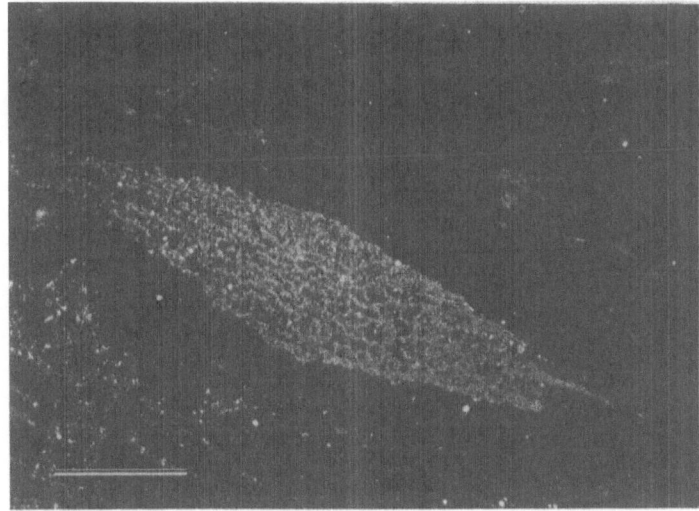

Fig. 4. Fluorescein Con A staining of putative contact
         domains of AChR clusters isolated by saponin ex-
         traction.
         Myotube cultures were extracted with saponin (48),
         followed by 0.05% Triton X-100 in Tris-buffered
         saline. After   fixation   with   paraformaldehyde,
         samples were stained with FITC-conjugated conca-
         navalin A. All AChR have been extracted from this
         preparation   (not   shown), but substrate-attached
         material remains which reacts with the lectin. The
         bar is 10 microns.

AN HYPOTHETICAL MECHANISM FOR AChR CLUSTERING

    We have learned something of how AChR clusters are organized
in cultured rat myotubes. We also know something about the roles of
some of the proteins associated with AChR clusters. It is now
possible to propose a simple model of how AChR clusters form at the
substrate interface of cultured myotubes which incorporates much of
what is known about the properties of the cluster-associated
proteins. This is useful in trying to understand some of the
interactions which may take place during clustering in vitro and in
vivo, and, especially, in designing future experiments.

    We propose that the following sequence of events occurs during
AChR cluster formation in both cultured rat myotubes, and in the
transition from non-clustered to embryonic AChR clusters in the
developing embryo.

    (i) Adhesive interactions dominate the responses of muscle

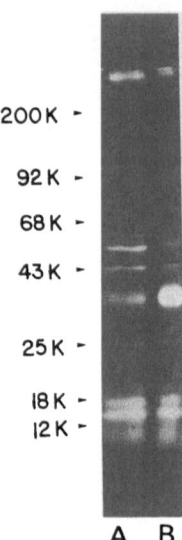

200K -

92K -

68K -

43K -

25K -

18K -
12K -

A   B

Fig. 5.   SDS-PAGE and   autoradiography   of substrate-attached
          material   from   myotube   cultures   extracted   with
          saponin and 0.05% Triton.
          Samples were metabolically radiolabelled with   (35)
          S-methionine,   then   extracted   as   in Fig.   4.
          Substrate-attached material was solubilized in 0.25%
          SDS, and applied to a Con A-Sepharose column. Lane A
          shows the polypeptides in the flow-through ; lane B
          shows   the   polypeptides   bound   to   the   column and
          eluted with 100   mM   α-methylmannoside.   A   band   at
          36,000   is highly enriched in the material bound to
          the column.

which has been contacted by substrate or by nerve.   This   adhesive
interaction eventually triggers AChR clustering.

     This proposition is consistent with the observation that
clustering occurs at sites of muscle attachment to glass, plastic
or   other   materials, that neuromuscular contact is extremely close
(< 10nm membrane-to-membrane distance)   at   early   stages   of
synaptogenesis,   and   that establishment of contact always precedes
the formation of AChR clusters.

     (ii) Molecules on the surface of the muscle which   participate
in   adhesion   accumulate   at   the   region   of   muscle-nerve,   or
muscle-substrate contact.

     This proposition is consistent with observations in other
systems (e.g., 50, 51). It should be noted, however, that molecules

such as N-CAM have not been found to accumulate at sites of nerve-nerve or nerve-muscle adhesion. We do not yet know if the Con A-binding material present in the detergent-resistant residue of contact domains is present at similar densities in other regions of the myotube surface.

(iii) The accumulation of molecules on the extracellular face of the membrane organizes the cytoskeleton by interactions which span the membrane. This cytoskeletal organization involves the localized accumulation of microfilaments.

This proposition is consistent with the observations that some extracellular macromolecules are aligned along intracellular, actin-rich structures (51, 52), and that actin, presumably in filaments, and actin-binding proteins, such as vinculin, α-actinin and filamin, are present at AChR rich-membrane in vitro and in vivo (see Table 2). As in other systems (51), vinculin may be involved in the anchoring of some intracellular microfilaments to the extracellular materials enriched at contact sites.

As we pointed out above, the exact organization of the cytoskeleton in this region is still a matter for conjecture. We currently favor a model in which microfilament bundles capped with vinculin are applied to the membrane at the contact domains. These structures may attach to, or be interspersed within, a cytoskeletal meshwork containing actin, perhaps organized like the actin underlying the erythrocyte membrane.

(iv) The enrichment of actin underlying the domains not involved in contact promotes the accumulation there of AChR.

This is consistent with the frequently observed association of actin with membrane proteins in other systems (53), as well as the importance of actin for maintaining the organization of AChR in isolated clusters, reported here. It is also consistent with the observation (42) that AChR clustering induced by plastic beads is preceded by the appearance of a microfilamentous mass underlying the membrane which contacted by the bead.

(v) The accumulation of AChR in the membrane can act as a template to which other intracellular, and perhaps extracellular, macromolecules can attach.

There is currently no evidence to support this proposition. It does, however, provide an approach to the question of how the embryonic cluster is converted into the neonatal cluster during junctional development. Alternatively, cytoskeletal or extracellular proteins also present at embryonic clusters may fulfill this role.

SIGNIFICANCE

As we have outlined in the earlier sections, the AChR clusters of rat myotubes formed in vivo and in vitro are almost identical. The mechanisms of AChR clustering in the two systems are probably very similar. Likewise, the mechanism of adhesion of the myotube to the glass may resemble the mechanism of adhesion of the developing muscle fiber to the ingrowing motor neuron. One assumption we feel safe in making is that, if these processes are indeed very similar in vivo and in vitro, then the molecules which participate in them will be very similar as well. We have made considerable progress in our studies of the molecules of AChR clusters in vitro. In the future, we should be able to extend these observations to embryonic clusters in vivo, and to learn how the organization of AChR clusters changes during the transition from the embryonic to the neonatal and adult structures. This information should reveal a great deal about the molecular processes involved in synapse formation, and how continued interaction of a nerve with its target modulates synaptic structure.

We are grateful to W.G. Resneck and B. Kidder for their assistance. This work was supported by grants to RJB and DWP from the National Institutes of Health and the Muscular Dystrophy Association, by a grant to RJB from the March of Dimes, and by a fellowship from the Muscular Dystrophy Association to MB.

REFERENCES

1. D.M. Fambrough, Physiol. Rev., 59 : 165-227 (1979).
2. C. Edwards, Neurosci., 4 : 565-584 (1979).
3. J.R. Sanes, Ann. Rev. Physiol., 45 : 581-600 (1983).
4. R.J. Bloch and J.H. Steinbach, In: "Receptors in cellular recognition and developmental processes", ed. R.M. Gorczynski, Academic Press, in the press (1985).
5. M.J. Dennis, L. Ziskind-Conhaim and A.J. Harris, Dev. Biol., 81 : 266-279 (1981).
6. J. Matthews-Bellinger and M.M. Salpeter, J. Neurosci., 3 : 644-657 (1983).
7. J.H. Steinbach, Dev. Biol., 84 : 267-276 (1981).
8. C.R. Slater, Dev. Biol., 94 : 11-22 (1982).
9. D. Axelrod, P. Ravdin, D.E. Koppel, J. Schlessinger, W.W. Webb, E.L. Elson and T.R. Podleski, Proc. Natl. Acad. Sci. USA, 73 : 4594-4598 (1976).
10. C.G. Reiness and C.B. Weinberg, Dev. Biol., 84 : 247-254 (1981).
11. M.M. Salpeter, S. Spanton, K. Holley and T.R. Podleski, J. Cell Biol., 93 : 417-425 (1982).
12. R.J. Bloch, J. Cell Biol., 82 : 626-643 (1979).

13. R.J. Bloch, J. Neurosci., 3 : 2670-2680 (1983).
14. R.J. Bloch and J.H. Steinbach, Dev. Biol., 81 : 386-391
    (1981).
15. G.D. Fischbach and S.M. Schuetze, J. Physiol., 303 : 125-137
    (1980).
16. C.G. Reiness and Z.W. Hall, Dev. Biol., 81 : 324-331 (1981).
17. R.J. Bloch and B. Geiger, Cell, 21 : 25-35 (1980).
18. J.M. Anderson, M.W. Cohen and E. Zorychta, J. Physiol., 268 :
    731-756 (1977).
19. J.M. Anderson and M.W. Cohen, J. Physiol., 268 : 757-773
    (1977).
20. L. Ziskind-Conhaim, I. Geffen and Z.W. Hall, J. Neurosci., 4 :
    2346-2349 (1984).
21. S. Bursztajn, S.A. Berman, J. McManaman and S.H. Appel, Abst.
    Soc. Neurosci., 9 : 1059 (1982).
22. J.A. Connolly, J. Cell Biol., 99 : 148-154 (1984).
23. C. Kalcheim, Z. Vogel and D. Duksin, Proc. Natl. Acad. Sci.
    USA, 79 : 3077-3081 (1982).
24. C. Kalcheim, D. Duksin and Z. Vogel, J. Biol. Chem., 257 :
    12722-12727 (1982).
25. R.M. Nitkin, B.G. Wallace, M.E. Spira, E.W. Godfrey and U.J.
    McMahan, Cold Spring Harbor Symp. Quant. Biol., 48 : 653-665
    (1983).
26. Z.W. Hall, B.W. Lubit and J.H. Schwartz, J. Cell Biol., 90 :
    789-792 (1981).
27. R.J. Bloch and W.G. Resneck, Abst. Soc. Neurosci., 9 : 757
    (1983).
28. R.J. Bloch and Z.W. Hall, J. Cell Biol., 97 : 217-223 (1983).
29. R.J. Bloch, unpublished results.
30. S. Burden, J. Cell Biol., 94 : 521-530 (1982).
31. S. Froehner, J. Cell Biol., 99 : 88-96 (1984).
32. M.J. Anderson and D.M. Fambrough, J. Cell Biol., 97 : 1396-1411
    (1983).
33. F. Moody-Corbett and M.W. Cohen, J. Neurosci., 1 : 596-605
    (1981).
34. M.M. Salpeter, 32 : 379-389 (1967).
35. J.R. Sanes, J. Cell Biol., 93 : 442-451 (1982).
36. Z. Vogel, C.N. Christian, M. Vigny, H.C. Bauer, P. Sonderegger
    and M.P. Daniels, J. Neurosci., 1058-1068 (1983).
37. E.K. Bayne, M.J. Anderson and D.M. Fambrough, J. Cell Biol., 99
    : 1486-1501 (1984).
38. C.N. Christian, M.P. Daniels, H. Sugiyama, L. Jacques and P.G.
    Nelson, Proc. Natl. Acad. Sci. USA, 75 : 4011-4014 (1978).
39. M.H. Buc-Caron, P. Nystrom and G.F. Fischbach, Dev. Biol., 95 :
    378-386 (1983).
40. T.R. Podleski, D. Axelrod, P. Ravdin , I. Greenberg, M.M.
    Johnson and M.M. Salpeter, Proc. Natl. Acad. Sci. USA, 75 :
    2035-2039 (1978).
41. D. Axelrod, J. Cell Biol., 89 : 141-145 (1981).
42. H.B. Peng and K.A. Phelan, J. Cell Biol., 99 : 344-349
    (1984).

43. S.J. Burden, P.B. Sargeant and U.J. McMahan, J. Cell Biol., 82
    : 412-425 (1979).
44. T.G. Burrage and T.L. Lentz, Dev. Biol., 85 : 267-286 (1981).
45. U.J. McMahan and C.R. Slater, J. Cell Biol., 98 : 1453-1473
    (1984).
46. B. Geiger, Cell, 18 : 193-205 (1979).
47. D.W. Pumplin and R.J. Bloch, J. cell Biol., 97 : 1043-1054
    (1983).
48. R.J. Bloch, J. Cell Biol., 99 : 984-993 (1984).
49. J. Boyles and D.F. Bainton, J. Cell Biol., 82 : 347-368
    (1979).
50. C. Ocklind, U. Forsum and B. Obrink, J. Cell Biol., 96 :
    1168-1171 (1983).
51. I.I. Singer, J. Cell Biol., 92 : 398-408 (1982).
52. K. Burridge and J.R. Feramisco, Cell, 19 : 587-595 (1980).
53. J.A. Weatherbee, Int. Rev. Cytol. Suppl., 12 : 113-176 (1981).

THE INTERACTIONS BETWEEN THE CHOLINERGIC RECEPTOR, THE 43K (ν1)

AND CYTOSKELETAL PROTEINS IN THE POSTSYNAPTIC DOMAIN

OF TORPEDO MARMORATA ELECTRIC ORGAN

Jean Cartaud, Hoang-Oanh Nghiêm[1] and Catherine Kordeli

Laboratoire de Microscopie Electronique
Institut Jacques Monod, Université Paris VII
2, Place Jussieu
75251 Paris Cedex 05, France

The postsynaptic membrane of the adult neuromuscular junction, and the closely related electromotor synapse of electric fishes, both contain an exceptional concentration of acetylcholine receptors (Ach-R). These molecules form regular arrays in the plane of the membrane, are strongly immobilized and are metabolically very stable. As a consequence, postsynaptic Ach-R clusters are maintained for a long period following denervation.

It is thought that many aspects of the morphogenesis and stabilization of the synapse can be accounted for by macromolecular interactions between the membrane-bound Ach-R and its immediate environment, such as : i) in-plane protein-protein interactions between the Ach-R molecules, ii) interactions between the externally exposed moiety of the receptor and the basal lamina, and, iii) cross linking of the receptor to cytoplasmic peripheral proteins and to the cytoskeleton. To date, little is known about the molecules which could be involved in such interactions (see also the review by R. Bloch, this volume).

Fish electric organs have provided a uniquely rich source of cholinergic synapses permitting extensive biochemical and structural characterization of the major functional components. For

1 Neurobiologie Moléculaire et Laboratoire Associé au Centre National de la Recherche Scientifique, Interactions moléculaires et cellulaires, Institut Pasteur, 75015 Paris, France.

this reason the electric organ constitutes an excellent model system to study the molecules involved in the immobilization of the Ach-R in the mature postsynaptic domain. Here, we present recent data on the molecular architecture of the postsynaptic domain, particularly on aspects that may be relevant to the immobilization of the Ach-R molecules.

THE TORPEDO Ach-R : MORPHOLOGY AND ORGANIZATION IN THE POSTSYNAPTIC MEMBRANE

In purified fractions of Torpedo receptor, two molecular forms with respective sedimentation coefficient of 9 S and 13 S were observed. Reducing agents, such as β-mercaptoethanol or dithiothreitol convert the 13 S heavy form to the light 9 S form, in vitro. The two molecular species are functionally indistinguishable. The heavy form, thus represents a dimer of two light forms linked by an intramolecular disulfide bridge. Denaturing detergents dissociate the receptor into four different polypeptide subunits that migrate on an SDS gel with apparent molecular weights of 40 K (α) ; 50 K (β) ; 60 K (γ) and 66 K (δ). The stoichiometry of the chains in the 9 S form is α2βγδ (reviewed in 1, 2).

Electron microscopy of negatively stained, purified and membrane-bound receptors discloses ringlike particles, or rosettes, 8 or 9 nm in diameter, with a stain-filled central pit (3, 4). Each rosette represents an axial view of a light form (5). Image analysis reveals, in agreement with the biochemical data, five unequally-sized peaks of electron density distributed around the central pit (6) (Fig. 1). Comparative analysis of two sets of images derived from α-bungarotoxin-labelled and native receptor molecules, discloses the precise location of the two α subunits within the oligomer (Fig. 1) (6, 7). The two α chains are not adjacent and make an angle of 110 - 160°. A similar conclusion was derived from experiments in which the two α-chains were identified using biotinylated α-toxin (8) or monoclonal antibody fragments (9). The probable order α γ α δ β of the subunits in the molecule was derived from cross-linking experiments (10).

When viewed from the side, the light-form molecule appears as a cylinder 11 nm long with an axial well filled with stain, which may represent part of ion channel. The receptor molecule thus appears as a cylindrical bundle with a transverse membrane polarity and featuring a rather uncommon heterologous pentameric shape.

EM and X-ray analyses demonstrated that the receptor molecule extends above the lipid bilayer by 4.5 to 5.5 nm into the synaptic cleft and by about 1.5 to 2 nm on the cytoplasmic side. In addition, selective proteolysis of membrane fragments showed that

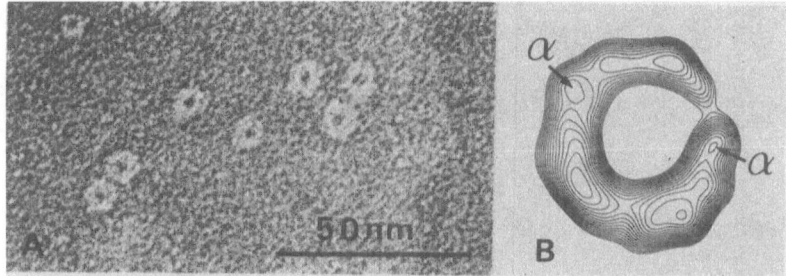

Fig. 1.   Ultrastructure and binding sites of the Acetylcholine
          receptor from Torpedo electric organ.
          A. Electron micrograph of a negatively stained
          preparation of purified Ach-R. The light-form (single
          rosettes) and the dimeric form (doublets)   are ob-
          served. (X 600,000).
          B. Reconstructed image of the light-form pentamer
          obtained by reorientation and computer averaging of
          tens of individual images such as those shown in A.
          The position of the two binding sites   for
          pharmacologically active ligands (α-subunits) is
          indicated (modified from Bon et al., (6)).

the four Ach-R polypeptides have a transmembrane orientation (11).
The recent deciphering of the amino acid sequence of the subunits
(12, 13, 14), further extended our knowledge on the transmembrane
distribution of the protein mass. All four chains show a similar
non-uniform distribution of hydrophobic amino-acids that   are
concentrated into four domains interpreted as transmembrane
α-helices. Another possible transmembrane domain comprising both
hydrophobic and hydrophilic residues was recently identified. This
amphipathic segment presumably participates in the formation of the
ion channel (15). Electron micrographs (Fig. 2) obtained using the
freeze-etching technique, disclose an original arrangement of the
rosettes of receptor in double rows in the plane of the juxtaneural
part of the membrane (16, 17). Within these rows, dimeric
structural units can be recognized. Since the native, heavy form of
the Ach-R represents a dimer formed by the covalent association of
two coplanar Ach-R rosettes, the supramolecular organization of the
receptors in the membrane probably relies upon a linear assembly of
Ach-R doublets. These stable Ach-R clusters do exist without
covalent bonding between individual doublets. They even remain
after reduction of the disulfide bridges between the δ subunits, or
following lipid membrane modifications (18). Extrinsic factors are
thus thought to play a role in the maintenance of the mature
membrane clusters.

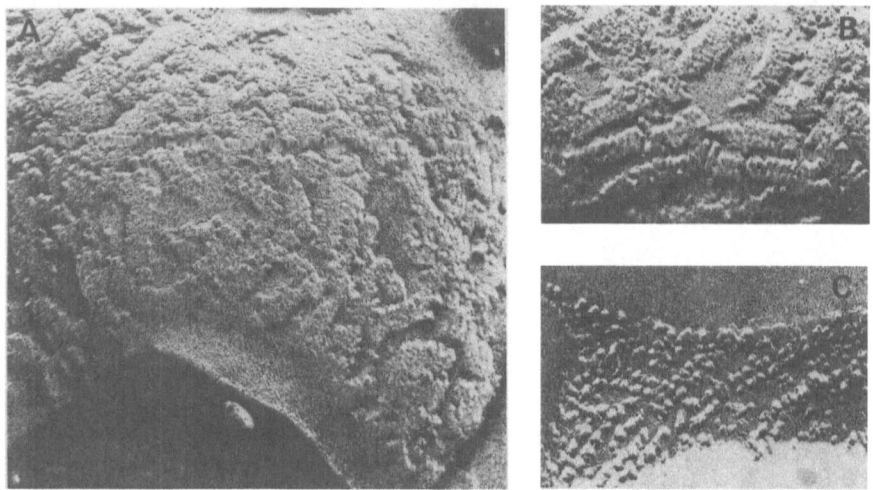

Fig. 2.   The distribution of the Ach-R in the plane of the
          postsynaptic membrane.
          Freeze-etched (A, B) and freeze fractured   (c)
          aspects of the Torpedo postsynaptic membrane. The
          rosettes form alinements at the external surface of
          the   membrane (A). Each track is made by the
          apposition of two adjacent files of paired rosettes
          (B). (X 120,000).

THE MEMBRANE-BOUND 43 K PROTEIN : A MAJOR COMPONENT OF THE
POSTSYNAPTIC DENSITIES THAT BINDS TO THE Ach-R

     The   postsynaptic   membranes   purified from   adult   Torpedo
electric   tissue   retain   most   of   their functional and structural
properties, including the ability to respond to externally   applied
cholinergic agonists by an increase in ionic permeability. The
major structural features associated with   the   membrane   in situ,
such   as   the   Ach-R   clusters   and   the associated postsynaptic
densities, are also conserved in the isolated membrane fragments.
These   membranes   are comprised of the Ach-R subunits (up to 40% of
the total protein mass) and of one major peripheral membrane
protein of apparent molecular mass 43,000 daltons (referred to as
the 43 K or ν1 protein) which is distinct from actin (19).

     Peripheral, non receptor proteins can be extracted from the
membrane by treatment at alkaline pH (20) or by extraction with a
low concentration of lithium diiodosalicylate (21). Extraction of
the peripheral proteins has essentially no effect on the functional
properties (ligand binding and ionic permeability) of the membrane.
However, important "structural" alterations, such as increases in
rotational (22, 23) and translational (24, 25) mobility,

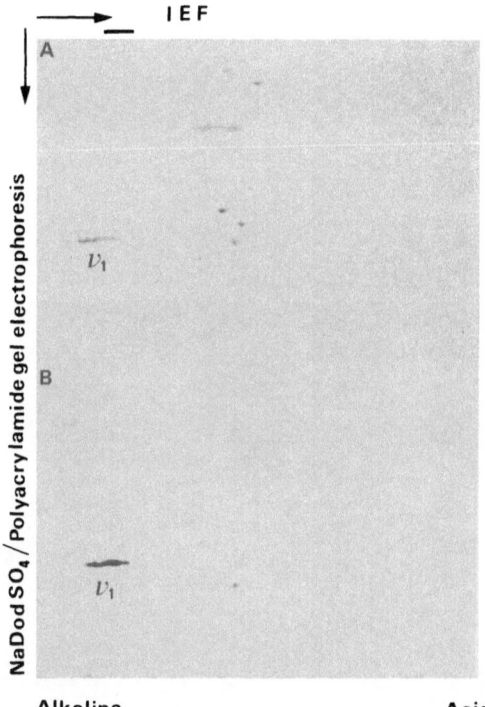

Fig. 3. Demonstration of the anti-43 K νl specificity of monoclonal antibody 418.2.

Ach-R-rich membrane fraction was separated on a two-dimensional isoelectrofocusing/NaDodSO$_4$/polyacrylamide gel electrophoresis according to O'Farrell and Laemmli. Ampholine pH 3.10, servalyte pH 4.9 (1:4 vol/vol)/0.1% NaDodSO$_4$ 10% acrylamide 0.13% bis acrylamide.

A. Coomassie blue staining showing the presence of the 43 K νl polypeptide.

B. Immunostaining of the 43 K νl spot with mAb 418.2 (1:300). The reaction was revealed with a horse-radish peroxydase conjugated sheep anti-mouse (1:250) and rabbit anti-sheep IgG (1:250), from ref. 29.

sensitivity to proteolysis (26) and to heat denaturation (27) of the receptors have been reported. Morphological studies have demonstrated that cytoplasmic condensations, co-extensive with the receptor-rich areas of the membrane, are no longer visible following alkaline extraction (25, 28). These results suggest that the main molecular constituent of the submembraneous condensations, the 43 K protein, occupies the same region of the postsynaptic

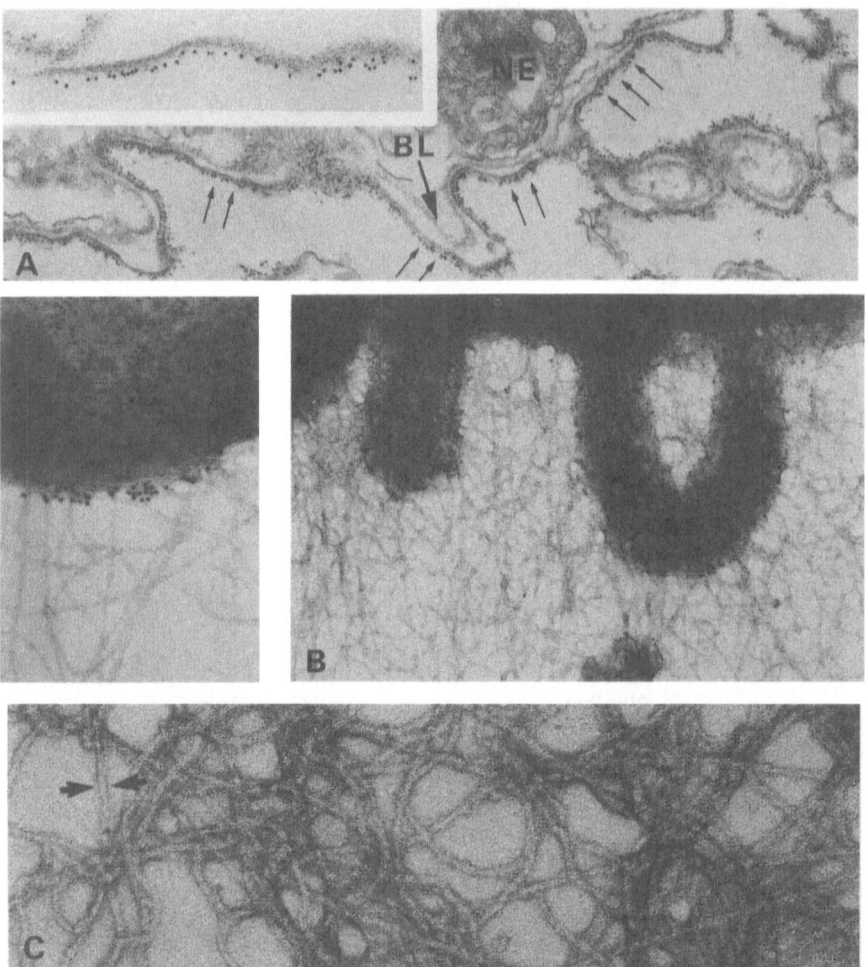

Fig. 4.   Subcellular localization of the 43 K protein.
Membrane   sheets derived from torpedo electric tissue
were incubated with the monoclonal   antibody   against
the   43   K   protein   and immunolabelled with a second
antibody coupled to   electron   dense   colloidal   gold
particles.
A.   Thin   section   through   an epon embedded synaptic
area showing a dense and   regular   labelling   of   the
cytoplasmic face of the postsynaptic membrane.
(X 30,000, inset X 60,000).
B.   After negative staining, the material reveals, in
addition to the heavily   labelled   postsynaptic   membrane,   a dense cytoskeleton. Note that the filaments
are never labelled along their length (inset).
(X 30,000, inset X 60,000).

membrane as the receptor and may contribute to its immobilization via a direct interaction. The demonstration of the codistribution of the Ach-R and 43 K protein constituted an important step in the understanding of the membrane architecture in this system. For this purpose, a direct immunocytochemical localization of the 43 K protein at the ultrastructural level was undertaken. This was achieved by producing a monoclonal antibody against the membrane-bound 43 K protein (29). Since several proteins of the electric tissue do migrate on 1.D SDS gels with an apparent molecular weight of 43 KD (30), it was necessary to test the specificity of the monoclonal antibody. This monoclonal antibody immunostained one band at the 43 K region of 1.D replicas of membrane enriched fraction and did not stain replicas of the cytosolic fraction. Its specificity as anti 43 K (νl) was then tested by immunostaining of 2.D replicas of purified Ach-R-rich membrane fragments (Fig. 3).

The localization of the 43 K protein is limited to the cytoplasmic surface of the Ach-R-rich domains of the postsynaptic membrane (Fig. 4). These experiments further show that immunostaining is totally absent from the bottom of the synaptic folds, areas which are also devoid of receptors (29, 31). High resolution immunocytochemical studies of the tangential distribution of the 43 K protein in isolated sheets derived from the innervated face of the electrocytes provide evidence for a distribution in double rows quite similar to that of the receptors (32). Additional evidence for a topographical proximity of the two proteins in the membrane was recently achieved by crosslinking experiments in which a specific crosslink between the 43 K protein and the β subunit of the receptor was demonstrated (33).

In conclusion, the peripheral 43 K protein that sets in register with respect to the Ach-R, is likely to participate in the structural stabilization of the postsynaptic membrane domain. Its localization at the cytoplasmic face of the membrane raises interesting possibilities for interactions with the cytoskeleton.

CYTOSKELETON - MEMBRANE INTERACTION IN THE ELECTROPLAX

Cytoskeleton - membrane interactions are thought to play a decisive role in the development and maintenance of cell shape and

---

C. High magnification micrograph of the synapse-associated cytoskeleton showing the 10 nm filaments (intermediate - sized filaments). (X 120,000).

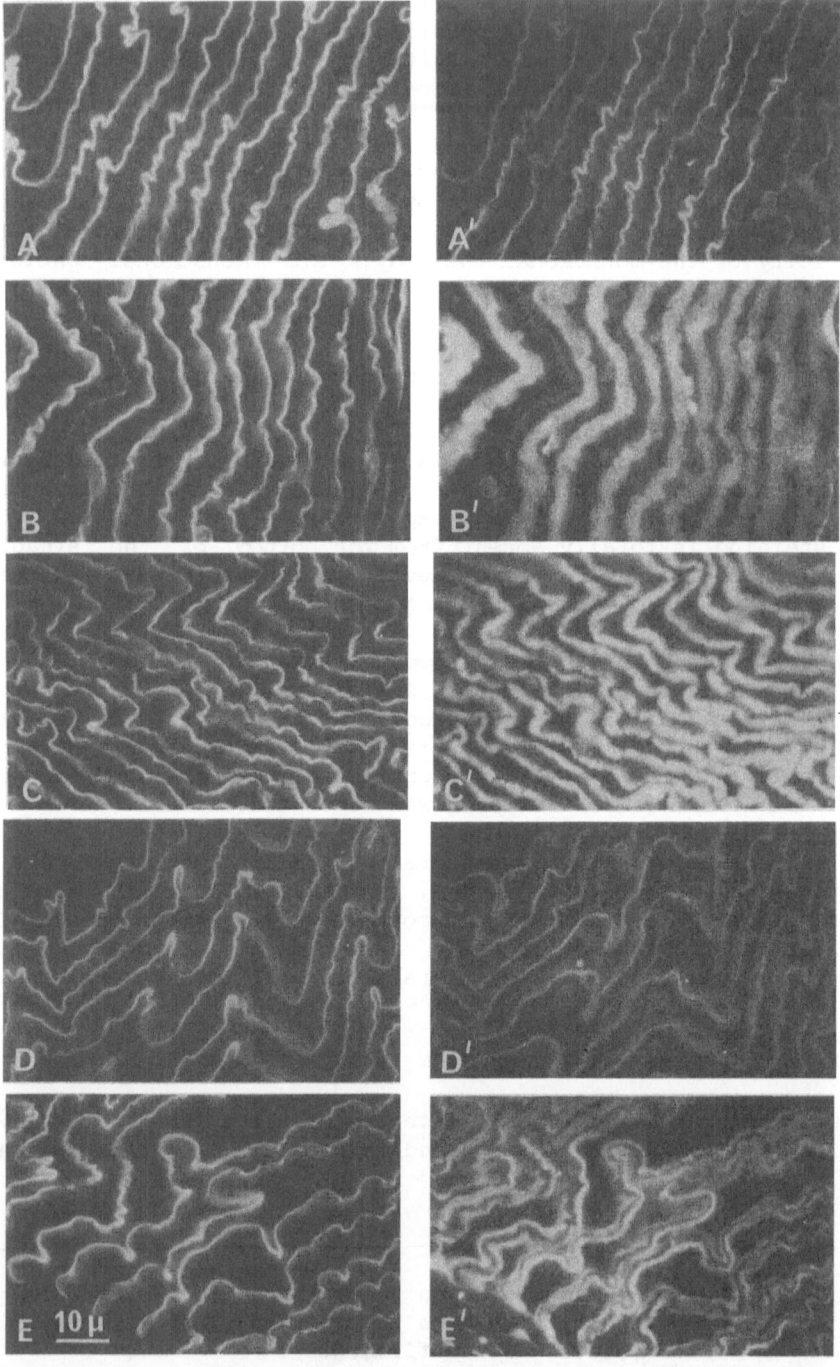

behaviour. Indeed, specialized membrane regions are often characterized by a close association with cytoskeletal components.

The postsynaptic membrane is a specialized membrane domain where Ach-R clusters are associated with differentiations of the underlying cytoplasm : the synaptic densities and attached filaments (34, 35, 36).

The postsynaptic area of the mature neuromuscular junction is associated with cytoskeletal proteins, including a cytoplasmic form of actin (37), several actin-binding proteins (α-actinin, filamin, vinculin (38)), and an intermediate-sized filament-related 51 K protein (39). Torpedo electromotor synapses lack most of the postsynaptic folds. However, sufficient structural similarities exist between the two systems to justify the study of the molecular interactions that take place at the Torpedo postsynaptic membrane, as a simplified model.

Though purified postsynaptic membrane fragments usually retain some cytoskeletal material, such as the postsynaptic densities (25, 28) or actin (40), most of the attached filaments observed in situ are lost. Studies of cytoskeleton - membrane interactions were thus carried out on large sheets derived from the innervated face of electrocytes by gentle homogeneization and low-speed centrifugation. Electron micrographs of deep-etched, rotary-shadowed (16) or negatively-stained (Fig. 4) samples disclosed a dense cytoskeletal network attached to the postsynaptic

Fig. 5.   Immunofluorescence images of Torpedo electric tissue reacted with antibodies directed against cytoskeletal components.
Frozen sections were incubated (A, A') with anti-ν1 monoclonal antibody (m ab.), (B, B') anti-desmin m ab., (C, C') anti-actin m ab., (D, D') anti-fodrin and (E, E') anti-ankyrin. A,B,C,D and E show Rhodamine-α-Bungarotoxin staining, A',B',C', D' and E' show fluorescein - indirect antibody staining. Desmin and actin antibodies stained the whole cytoplasm of the cell (compare to Fig. A, A' in which only the postsynaptic membrane was stained with the anti-ν1 m ab.). Fodrin and ankyrin appeared to be located at the non-innervated face of the electrocytes. The faint labelling observed at the innervated face is due to a reaction with the nerve endings (E.M. data not shown). Using the same immunofluorescence technique, vinculin and filamin were not revealed. (X 800).

membrane. Ultrastructural and immunocytochemical analyses of this cytoskeleton reveal that it consists essentially of inter-mediate-sized filaments, probably desmin filaments, and actin (Fig. 5). As noted previously, the 43 K protein is restricted to the cytoplasmic surface of the postsynaptic Ach-R-rich membrane areas. The filaments observed in contact with the membrane are never labelled on their length with the anti-43 K antibody. Interestingly, the free ends of detached filaments are sometimes observed with a punctual labelling (41). These images suggest that the intermediate-sized filaments attach to the postsynaptic densities, possibly via the 43 K protein. This protein is associated with the Ach-R clusters in the membrane but is also likely to be implicated in the anchorage to the cytoskeleton. It would thus constitute the intermediate link that insures the anchorage of Ach-R clusters to the underlying cytoskeleton.

Antiserum prepared against the Torpedo alkaline extract comprising mainly the 43 K protein, strongly reacts in immu-nofluorescence experiments with the rat neuromuscular junction (42). In addition, several cytoskeletal proteins, including actin and several actin-binding proteins, were also observed (37, 38). It was thus of interest to reinvestigate this point at the Torpedo synapse. Our results, derived from immunofluorescence and electron microscope observations, revealed quite surprisingly that at least actin, vinculin and filamin were not concentrated at the postsynaptic membrane level. However, ankyrin and fodrin were associated only with the non innervated plasma membrane of the electroplax (Fig. 5).

The 43 K protein thus appears as the only extrinsic and specific component that is observed at the postsynaptic membrane in Torpedo. The accumulation of cytoskeletal components at the rat neuromuscular junction suggested their participation in the stabilization of the receptor. Our results are not in agreement with this hypothesis. We believe rather that the actin and actin-binding proteins in the neuromuscular junction participate to the maintenance of the junctional fold complex, almost lacking in Torpedo.

ACKNOWLEDGEMENTS

We thank Drs Daniel Louvard, Denise Paulin and Louise Pradel for gifts of antibodies, Véra Boucharel for typing the manuscript and Louise Labaronne for assistance with the photography.
This work was supported by the Centre National de la Recherche Scientifique, the Institut National de la Recherche Médicale, The Collège de France and the Université Paris VII. Catherine Kordeli was supported by a grant from the Association des Myopathes de France.

REFERENCES

1. J.P. Changeux, A. Devillers-Thiery and P. Chemouilli, Science,
    225 : 1335-1345 (1984).
2. J.L. Popot and J.P. Changeux, Physiological Rev., (in press)
    (1984).
3. J. Cartaud, E.L. Benedetti, A. Sobel and J.P. Changeux, J. Cell
    Sci., 29 : 313-337 (1978).
4. M.W. Klymkovsky and R.M. Stroud, J. Mol. Biol., 128 : 319-334
    (1979).
5. J. Cartaud, J.L. Popot and J.P. Changeux, FEBS Letters, 121 :
    327-332 (1981).
6. F. Bon, E. Lebrun, J. Gomel, R. Van Rapenbusch, J. Cartaud,
    J.L. Popot and J.P. Changeux, J. Mol. Biol., 176 : 205-237
    (1984).
7. H.P. Zingsheim, F.J. Barrantes, J. Frank, W. Hänicke and D.C.
    Neugebauer, Nature (London), 299 : 81-84 (1982).
8. E. Holtzman, D. Wise, J. Wall and A. Karlin, Proc. Natl. Acad.
    Sci. USA, 79 : 310-314 (1982).
9. R.H. Fairclough, J. Finner-Moore, R.A. Love, D. Kristofferson,
    P.J. Desmeules and R.M. Stroud, Cold Spring Harbor Symp.
    Quant. Biol., 48 : 9-16 (1983).
10. A. Karlin, R. Cox, R.R. Kaldany, P. Lobel and E. Holtzman,
    ibid., 1-8 (1983).
11. L.P. Wennogle and J.P. Changeux, Eur. J. Biochem., 106 :
    381-393 (1980).
12. A. Devillers-Thiery, J. Giraudat, M. Bentaboulet and J.P.
    Changeux, Proc. Natl. Acad. Sci. USA, 80 : 2067-2071
    (1983).
13. T. Claudio, M. Ballivet, J. Patrick and S. Heineman, Proc.
    Natl. Acad. Sci. USA, 80 : 1111-1115 (1983).
14. M. Noda, H. Takahashi, T. Tanabe, M. Toyosato, S. Kikyotani, Y.
    Furutani, T. Hirose, H. Takashima, S. Inayama and S. Numa,
    Nature (London), 302 : 528-532 (1983).
15. J. Finner-Moore and R.M. Stroud, Proc. Natl. Acad. Sci. USA, 81
    : 155-160 (1984).
16. J.E. Heuser and S.R. Salpeter, J. Cell Biol., 82 : 150-173
    (1979).
17. J. Cartaud, In: "Ontogenesis and Functional Mechanisms of
    Peripheral Synapses", J. Taxi editor, Elsevier, North
    Holland-Amsterdam-New York, 199-210 (1980).
18. A. Rousselet, J. Cartaud and P.F. Devaux, Biochim. Biophys.
    Acta, 648 : 169-185 (1981).
19. A. Sobel, M. Weber and J.P. Changeux, Eur. J. Biochem., 80 :
    215-224 (1977).
20. R.R. Neubig, E.K. Krodel, N.D. Boyd and J.B. Cohen, Proc. Natl.
    Acad. Sci. USA, 76 : 690-694 (1979).
21. S. Porter and S.C. Froehner, J. Biol. Chem., 258 : 10034-10040
    (1983).
22. M.M.S. Lo, P.B. Garland, J. Lamprecht and E.A. Barnard, FEBS

(Fed. Eur. Biochem. Soc.) Lett., III : 407–412 (1980).

23. A. Rousselet, J. Cartaud, P.F. Devaux and J.P.Changeux, EMBO J., 1 : 439–445 (1982).

24. F.J. Barrantes, W.Ch. Neugebauer and H.P. Zingsheim, FEBS (Fed. Eur. Biochem. Soc.) Lett., 112 : 73–78 (1980).

25. J. Cartaud, A. Sobel, A. Rousselet, P.F. Devaux and J.P. Changeux, J. Cell Biol., 90 : 418–426 (1981).

26. M. Klymbowsky, J.E. Heuser and R.M. Stroud, J. Cell Biol., 85 : 823–838 (1980).

27. T. Saitoh, L.P. Wennogle and J.P. Changeux, FEBS (Fed. Eur. Biochem. Soc.) Lett., 108 : 489–494 (1979).

28. R. Sealock, J. Cell Biol., 92 : 514–522 (1982).

29. H.O. Nghiêm, J. Cartaud, C. Dubreuil, C. Kordeli, G. Buttin and J.P. Changeux, Proc. Natl. Acad. Sci. USA, 80 : 6403–6407 (1983).

30. R. Gysin, W. Wirth and S.D. Flanagan, J. Biol. Chem., 256 : 11373–11376 (1981).

31. R. Sealock, B. Wray and S.C. Froehner, J. Cell Biol., 98 : 2239–2244 (1984).

32. C. Kordeli, J. Cartaud, H.O. Nghiêm, D. Paulin, L. Pradel and J.P. Changeux (in preparation).

33. S.J. Burden, R.L. DePalma and G.S. Gottesman, Cell, 35 : 687–692 (1983).

34. M.R. Couteaux, J. Neurocytol., 10 : 947–962 (1981).

35. R. Birks, H.E. Huxley and B. Katz, J. Physiol. (London), 150 : 134–144 (1960).

36. M.H. Ellisman, J.E. Rash, L.A. Staehelin and K.R. Porter, J. Cell Biol., 68 : 752–774 (1976).

37. Z.W. Hall, B.W. Lubit and J.H. Schwartz, J. Cell Biol., 90 : 789–792 (1981).

38. R.J. Bloch and Z.W. Hall, J. Cell Biol., 97 : 217–223 (1983).

39. S.J. Burden, J. Cell Biol., 94 : 521–530 (1982).

40. C.O. Strader, E. Lazarides and M.A. Raftery, Biochem. Biophys. Res. Commun., 92 : 365–373 (1980).

41. J. Cartaud, C. Kordeli, H.O. Nghiêm and J.P. Changeux, C.R. Acad. Sci. Paris, 297 : 285–289 (1983).

42. S.C. Froehner, V. Gulbrandsen, C. Hyman, A.Y. Jeng, R.R. Neubig and C. Cohen, Proc. Natl. Acad. Sci. USA, 78 : 5230–5234 (1981).

CELL SURFACE COMPONENTS

AND DIFFERENTIATION IN NEUROBLASTOMA CULTURE

Gabriella Augusti-Tocco

Dipartimento di Biologia Cellulare e dello Sviluppo
Universita di Roma "La Sapienza"
Piazzale A. Moro, I-00185 Roma, Italy

INTRODUCTION

In the development of the nervous system, cells are committed to become neurons and glial cells well before any sign of cellular differentiation is apparent. Cytodifferentiation is a complex process which occurs gradually and progressively restricts the developmental potentialities of the stem cells. Successive stages of neuron cytodifferentiation have been indicated as germinal cells, neuroblasts and neurons. As cells progress through these stages many cellular and molecular events occur, such as :
- restriction of mitotic activity,
- increase of cell size,
- acquisition of typical cell shape, characterized by fiber formation,
- acquisition of action potential mechanism,
- appearance of neurotransmitter synthesizing enzymes and mechanism of neurotransmitter storage and release,
- appearance of specific receptors,
- formation of synaptic contacts with target cells.

The emergence of several classes of neurons with different specific properties in the course of development, increases the degree of complexity of the system. This makes it very difficult to study the regulatory mechanisms for the selection of the developmental program by a certain group of cells and for the sequence of events leading from an undifferentiated cell to a mature neuron.

These considerations have brought in neurobiology to the development of cell culture as simple experimental system. In

culture, neurons can still exhibit their specific properties ; at
the same time experimental conditions can be manipulated in order
to interfere with the expression of a specific developmental
program and thus gain an insight in the mechanisms underlying the
cellular events in progress.

ORIGIN AND GENERAL PROPERTIES OF MOUSE NEUROBLASTOMA CELL LINES

     Mouse neuroblastoma cell lines have been isolated from a
transplantable tumor of spontaneous origin, the mouse neuroblastoma
C-1300. Cells growing in the tumor do not show any morphological
characteristic of neurons and appear as round undifferentiated
cells ; they however synthesize some enzymes, such as tyrosine
hydroxylase, cholinacetyl transferase and acetylcholinesterase,
which are specific neuronal markers (1, 2). An interesting feature
of neuroblastoma cells is that in culture they can maintain the
undifferentiated morphology typical of the tumor growth or extend
long fibers, thus assuming the characteristic shape of mature
neurons (3-5). Many clonal lines have been isolated and their
ability to express several neuronal properties has been extensively
investigated. The general conclusions emerging from these studies
can be summarized in three main points :

     a) Several neuronal properties can be expressed independently
of the expression of others. Thus, clones have been described which
are able to grow fibers, but do not contain enzymes for transmitter
synthesis (6). Also the presence of an action potential mechanism
can be found in absence of fibers (7).

     b) Fiber outgrowth can be observed in the large majority of
clones which have been isolated with different techniques. Its
expression is dependent on the culture conditions and can be
experimentally controlled. The factors which appear to be critical
for the formation of fibers are the interaction of cells with an
appropriate substrate and the cellular level of cyclic AMP (see for
review 3-5).

     c) Neuroblastoma cells are unable, under the experimental
conditions sofar tested, to establish synaptic contacts with cells
of the same type or with putative target cells. Functional synapses
have however been observed in hybrid lines obtained by fusion of
neuroblastoma with rat glioma, hamster brain and rat liver cells
(8). These findings suggest that neuroblastoma cells are defective
in some components, which can be successfully supplied by the other
parental lines, and which are necessary for synaptogenesis.

     In conclusion, in neuroblastoma culture, by selecting the
appropriate cell lines and culture conditions one can obtain
homogeneous cell populations, which in a controlled way undergo

independently two critical steps of neuron maturation, namely growth of fibers and synaptogenesis. They can therefore be considered as representative of successive stages of neuronal maturation.

CHARACTERIZATION OF MORPHOLOGICAL DIFFERENTIATION IN NEUROBLASTOMA CULTURES

Neurite extension in neuroblastoma cultures can be elicited by a number of different treatments, such as transfer from suspension to monolayer cultures, addition to the culture medium of cyclic AMP, bromodeoxyuridine, dimethylsulfoxide, hexamethylene-bis-acetamide, or serum removal from medium. For discussion of the various factors capable of inducing this phenomenon and their possible mechanism of action the reader is referred to previous reviews on the subject (3-5, 9).

A relevant question to the use of neuroblastoma cultures as a model of neuronal differentiation is whether neurite extension in the system represents only a modification of cell shape or is accompanied by other cellular and molecular events, occurring in the course of development. In fact, when neuroblastoma cells are induced to extend neurites several other events occur, which in vivo characterize the transition from undifferentiated cells to neurons. The very diverse nature of the agents used as inducers of differentiation in neuroblastoma culture and the various markers chosen to evaluate this effect, makes it rather difficult to have a general view of the phenomenon. However, as it has been reviewed (3), there are some effects which, whenever tested, have always been observed as accompanying fiber outgrowth. This is the case for the increase of cell size, protein and RNA content. They also occur in the course of neuron differentiation in vivo after cell division has come to a stop and neurons begin to acquire their specific traits.

Neurite formation is accompanied by the appearance in the elongating processes of neurotubules and neurofilaments (7, 10). The appearance of these structures typical of nerve fibers has been reported to be dependent on the appearance of a high molecular weight microtubule-associated-protein responsible for the polymerization of preexisting tubulin subunits (11).

Other markers of neuron function have been considered in this respect. An increase in the activity of enzymes related to the synthesis (tyrosine hydroxylase and cholinacetyl transferase) and inactivation (acetylcholinesterase) of neurotransmitter molecules can be induced in some neuroblastoma clones by cyclic AMP (12-14). However, an increase in the level of synthesizing enzymes is not consistently found in association with fiber outgrowth. On the

contrary   the   AChE   level,   whenever   examined,   has   been   found
increased in response to the various treatments capable of inducing
fiber outgrowth.

An interesting study reports that induction of fiber formation
in N18 clone by removal of serum from medium, also brings   about   a
modification   in   the   electrical   properties of the cell membrane,
represented from from Ca^{++}-dependent to Na^{+}-dependent   spikes
of electrical activity (15). Unfortunately the lack of this type of
study on other clonal lines, does not allow to establish how strict
is   the   association   between   fiber   formation   and   appearance of
functional Na^{+} channels in the cell membrane.

In conclusion, from a survey of a   large   number   of   studies,
using   several   clones   independently   isolated   and with different
properties, fiber outgrowth in these cells appears to be associated
to   a   reduction   of   cell proliferation, an increase in cell size,
protein and RNA   content   and   modifications   of   several   specific
neuronal components.

LEVEL OF CONTROL MECHANISM : TRANSLATION AND TRANSCRIPTION

As   it   has   been   described   in   the   previous   section,
neuroblastoma cells under the influence of some   agents   undergo   a
number   of   modifications   which   also   occur   in   neuronal
differentiation. It is thus   interesting   to   examine   the   control
mechanism   responsible for these changes in neuroblastoma cultures.
Do   they   require   synthesis   of   new   RNA   transcripts   or
post-translational   events   occur,   which   bring   about a different
utilization of preexisting mRNA ? Most likely both types of control
operate   in this system. Evidence for post-translational control of
ribosomal RNA content (17) and   protein   synthesis   (18)   has   been
reported.   These   observations however do not exclude the existence
of processes leading to transcription of new sets   of   genes,   when
neuroblastoma   cells   are induced to differentiate. Several authors
have addressed this problem (3, 4), but some difficulties have been
encountered   to   obtain a clear answer. In fact, one has to keep in
mind that the underlined undifferentiated neuroblastoma cells already   express
several neuronal specific properties. Thus the new proteins and new
mRNAs related to the acquisition of the more   differentiated   state
represent   a   small   fraction   of   the   total   mRNA   population.
Experiments of hybridization of poly-A   mRNA   to   cDNA   have   shown
that,   out of a common background, there are mRNA species unique to
differentiated and undifferentiated cells (19).

A different approach has recently been used to investigate the
mechanism   of   gene   regulation   operating   in   neuroblastoma
differentiation. It is known that sensitivity to virus is not   only
species-specific,   but   is   also dependent on the state of cellular

differentiation. Thus, viruses can be used as probes for studying cell differentiation (20). Following this line teratocarcinoma cells have been shown to be susceptible to polyoma and other DNA virus infection only in differentiated state (21). The response of neuroblastoma cells to polyoma virus has been tested under culture conditions allowing or no differentiation (that is suspension and monolayer culture). No major difference has been detected in the sensitivity to virus infection in the two cases. However virus mutants isolated from a neuroblastoma clone carry different rearrangements of their regulatory region depending on the differentiation state of the cells during virus mutant isolation (22). This finding suggests that factors regulating the expression of the genome somehow differ in the two culture conditions, which resemble different stages of differentiation.

ANALYSIS OF CELL SURFACE COMPONENTS

     Fiber elongation and formation of synaptic contacts require the ability of neurons to respond to some external signals. Both processes thus involve specific interactions which are mediated by the neuron surface. It is therefore conceivable to expect surface modifications occurring when these processes take place. As has been discussed in previous sections neuroblastoma cultures provide an experimental system in which, by changes in the culture conditions and adequate selection of cell lines, one can obtain cell populations corresponding to three successive steps of neuronal maturation :
     - cells without fibers ;
     - cells actively engaged in fiber outgrowth ;
     - cells capable of synaptogenesis.
This system appears particularly suitable to investigate whether changes in membrane components can be specifically related to successive developmental events.

Membrane chemical composition

     Modifications of the cell surface has been studied especially in relation to the process of fiber growth. Different experimental approaches have been used, as already extensively reviewed (3-5, 9). The study of cell surface antigens using immunological methods has shown that brain specific antigens are expressed in neuroblastoma cells ; they are shared by cells with and without fibers (23-24). This is not surprising since, as already mentioned, "undifferentiated" cells have already acquired many specific properties of neurons. However, Akeson and Herschman, comparing undifferentiated cells and cells induced to grow fibers by serum withdrawal (25) and clones unable to grow fibers or with poor fiber outgrowth (26), were able to demonstrate a class of antigens specific to differentiated cells. These antigens are also detected in mouse brain.

A great number of studies have analysed the protein components of cell surfaces by PAGE. Proteins were metabolically labelled using aminoacids or sugars, as fucose and glucosamine. Alternatively cells were externally labelled with 125-I. The mono- or bidimensional electrophoretic pattern of protein bands was compared in cells induced or not to form fibers. The induction treatments included the passage from suspension to monolayer cultures and the addition to the culture medium of bromodeoxyuridine or cyclic AMP. Also clones differing for their ability to grow fibers were compared. In spite of the diversity of cell lines and treatments used in the experiments, it is possible to draw the following very general conclusions. Very little differences are seen in the protein pattern when aminoacids are used as precursors (27-29). On the other hand labelling with fucose or glucosamine demonstrates significant changes, mainly in the degree of glycosylation of defined protein bands (28-29). These data suggest that the polypeptide chains remain largely constant and that modifications of surface components in the maturational process characterized by fiber outgrowth, is mainly confined to the carbohydrate moiety of glycoproteins. The discrepancies among experiments using aminoacids or sugars as precursors has also been tentatively ascribed to different rates of turnover of the polypeptide and carbohydrate part of glycoproteins (4). This hypothesis may find some grounds in the heterogeneous rates of protein turnover described in neuroblastoma cells (30). However, the results of these studies aiming to a chemical characterization of surface components specific to the two maturational states remain rather elusive.

Using a hybrid line of glioma x neuroblastoma, bidimensional electrophoresis of glycoprotein fractions purified by affinity chromatography on lectins, has led to the identification of 12 new proteins appearing when cells are induced to differentiate by prostaglandins. At the same time, a large number of proteins disappear, while many others show quantitative variations (8). It is tempting to ascribe these more extensive changes in the expression of protein species, as compared to parental neuroblastoma, with the ability of hybrid cells to proceed further in the developmental program of neuron differentiation, as indicated by the formation of synaptic contacts. However the identification of a larger number of proteins involved in significant changes may also be dependent on the use of a purification step (affinity chromatography on lectins).

A different approach has been to search for molecular species characteristic of different cellular compartments. Neurites have been isolated from the cell bodies of neuroblastoma cells. Electrophoretic analysis of plasma membrane preparations showed that a glycoprotein of an approximate molecular weight 200000 is considerably enriched in the neurite fraction as compared to cell bodies (31). The presence of a glycoprotein of similar characteristics in the culture media of undifferentiated cells suggests that a mechanism responsible for differential retention in the cell membrane, rather than biosynthesis of the protein, may be acting in this case.

Evidence of a different composition among cell bodies and neurites in surface glycoprotein derives also from experiments in which the distribution of Con A receptors has been studied in neuroblastoma cells (32). Using FITC-Con A, binding is evident on the cell body and absent on the fibers. Contrary to the results discussed above of increased glycosylation accompanying fiber formation, quantitative measurements of 3H-Con A binding shows that cells grown in suspension bind more Con A than cells grown in monolayer (3, 33). The observation that in the hybrid cells fibers are stained by FITC-Con A, has suggested that the appearance of Con A receptors on the fibers may be related to their ability to form synapses (3). This hypothesis is supported by a parallel study on developing chick dorsal root ganglia (DRG), which also display changes in Con A binding pattern at successive stages of development (32). Also in neurons, staining of the fibers with FITC-Con A is found only after the neurons have established functional contacts, while staining of the cell body can be observed at an earlier developmental stage.

These experiments do not provide informations on the functional role related to the different glycoconjugates, which become available for binding at the cell surface. It is pertinent however to mention that evidence of differential distribution of specific receptors on cell body and fibers has been reported in neuroblastoma cultures for acetylcholine receptors (41) and in DRG neurons for Ca^{++} and Na^{+} channels (16).

Glycosaminoglycans (GAGs) are another class of cell glycoconjugates which play an important role as regulatory factors in cell interactions with their extracellular environment. The ratio between heparan sulphate and hyaluronic acid is higher in cells growing in monolayer than in suspension cultures. Analysis of culture medium shows that in the latter, heparan sulphate is preferentially released by the cells in the medium, while in the former it remains associated to the cell surface (34). As in the case of the 200000 d proteins of neurites, also for heparan sulphate it seems that undifferentiated cells, as compared to differentiated ones, lack some factor(s) necessary for its association to the cell surface. This different ability to maintain heparan sulphate cell-associated, depending on the culture conditions, may be critical for the process of fiber growth. This hypothesis finds some support in the more recent observation that, in culture of sympathetic neurons, neurite extension can be supported in the absence of nerve growth factor, by a factor containing heparan sulphate (56).

Membrane dynamic properties

The dynamic behaviour of lipid and protein components of the plasma membrane has been studied in relation to neuroblastoma differentiation. Measurements of the microviscosity of membrane lipids has shown that differentiation of neuroblastoma cells is accompanied by an increased membrane fluidity (35). On the other hand, a modification of the membrane lipid composition which reduces membrane fluidity also prevents neurite outgrowth (35). Lateral mobility of proteins in the membrane was also studied by fluorescence photobleaching recovery. Most interestingly, neurites were found to behave differently than the cell body, showing higher lateral diffusion coefficients than the latter (36). This is not surprising in view of the fact that neurite outgrowth requires insertion of new membrane components, which is believed to occur in the growth cone region (55). In other studies no significant variations of the membrane fluidity was reported (37) or a decrease of membrane component mobility in differentiated cells was observed (38, 39). Interestingly, receptors for Con A and WGA behave differently in DMSO-induced differentiation. In fact, mobility of Con A receptors decreases, while that of WGA receptors remains unchanged (40). The heterogeneous response of the two receptors may reflect the existence of independent domains in the membrane organization.

Although the functional significance of these variations has to be demonstrated, these studies show that considerable changes in the membrane architecture occur at the time of fiber outgrowth in neuroblastoma cultures.

Specific receptors and ion channels

Functional characterization of neuroblastoma cells has included the study of specific surface receptors. Early electrophysiological characterization of neuroblastoma cells showed the presence of acetylcholine receptors and their specific localization on the cell body and terminal region of the fibers (41). Later studies have shown that both muscarinic and nicotinic receptors can be found in clonal lines (42). Dopamine (43) and histamine receptors (44) have also been demonstrated. Opiate receptors have been found mostly in glioma x neuroblastoma hybrid lines (45). Ca^{++} channels, predominantly responsible for the production of action potentials in undifferentiated cells, are gradually replaced by Na^{+} channels in the course of differentiation in culture (15). Similar changes of the action potential mechanism have been described in the development of neurons in vivo and in vitro (46, 47). Furthermore, in hybrid glioma x neuroblastoma lines, activation of voltage-sensitive Ca^{++} channels occurs when cells are induced to differentiate by prostaglandins (8).

Acetylcholinesterase

Acetylcholinesterase (AChE) is present in several lines of neuroblastoma (1, 6) and its level increases in differentiated cells (48). AChE is known to exist in multiple forms, which are preferentially located in different cell compartments ; only tetrameric forms have been reported as associated to the cell surface, which is the site for the physiological action of the enzyme (49). Thus, regulation of AChE monomer polymerization and of its cellular localization is at least as important as regulation of monomer biosynthesis. An increase of the 11S form at the expense of the 4S form has been reported in various clones, when cells are induced to differentiate (50, 51).

In a previous section the changes in Con A receptors in neuroblastoma and hybrid glioma x neuroblastoma cells and their possible relation to synaptogenesis have been discussed. However, the functional counterparts of Con A receptors are not known. AChE is a glycoprotein, known to bind Con A (52) ; it then appeared of interest to investigate its cellular localization in parental neuroblastoma and hybrid lines. The visualization of enzymatic activity at the E.M. showed that in hybrid cells, AChE is predominantly associated to the cell surface, while in the parental cells it remains in the cytoplasmic compartment (53). This finding is of particular interest in view of the diversity in functional properties of the two cell lines. The difference in localization could be related to a different distribution of AChE molecular forms. However, sucrose gradient sedimentation of AChE showed in both cases the presence of G1 and G4 forms, without significant variations in their relative distribution. Thus the lack of association of AChE to the plasma membrane in parental cells is not dependent on their inability to assemble the monomeric polypeptide chains in a higher molecular weight complex. Several alternative hypotheses are possible to explain this observation. AChE molecules synthesized by parental neuroblastoma cells may carry some minor modifications, which do not alter the aggregation of monomers into the tetrameric structure ; they may however affect the interaction of the enzyme with membrane components. In their turn, modifications of these components, shown to occur during differentiation (see previous sections), may be responsible for the altered interaction. Also, the mechanism of insertion of AChE into the plasma membrane may be somehow deficient in parental cells. The observation that these cells release more AChE in the culture medium than the hybrid cells, may however be in favor of a defect in the AChE association to the plasma membrane.

It is pertinent to recall that at early stages of DRG maturation a progressive shift from a perinuclear to a cell surface localization has been described for AChE (54). These observations indicate that the cellular localization of AChE is developmentally regulated. They also show, as is the case for Con A receptor distribution, that in neuroblastoma cells developmental events occurring in DRG can be observed. Therefore further studies to elucidate the mechanism responsible for AChE cellular localization in parental and hybrid cells, may throw some light on processes relevant to neuronal maturation in DRG.

CONCLUDING REMARKS

In the preceding sections, data have been reviewed showing that neuroblastoma cells can mimick in culture several events occurring in neuronal maturation in vivo. Attention has been focused on modifications of the cell surface, which appear to have a major role in the regulation of the expression of several differentiated properties. In particular the hypothesis appears worthy of attention that the presence of some specific molecules in the differentiated cells can be regulated, not only by activation of their biosynthesis, but also by changes in membrane structure, which in turn may or may not result in a functional association of cellular components. In the neuroblastoma system this mechanism seems to be operating in more than one case. The tumor origin of these cells and the high number of cell generation in an artificial environment may be responsible for this. Therefore, generalization of this hypothesis to developmental events in vivo deserves caution. However, studies on Con A receptor distribution and AChE localization, have shown a parallel behaviour in neuroblastoma cultures and developing DRG : these observations appear to support the relevance of studies on neuroblastoma cultures for development in vivo.

REFERENCES

1. G. Augusti-Tocco and G. Sato, Proc. Natl. Acad. Sci. USA, 64 : 311-315 (1969).
2. D. Schubert, S. Humphreys, C. Baroni and M. Cohn, Proc. Natl. Acad. Sci. USA, 64 : 316-323 (1969).
3. S. Denis-Donini and G. Augusti-Tocco, Curr. Top. Devel. Biol., 16 : 323-348 (1980).
4. Y. Kimhi, In: "Excitable Cells in Tissue Cultures", ed. Nelson P.J., Plenum Press, New York, pp. 173-243 (1982).
5. S.W. De Laat and P.T. van der Saag, Int. Rev. Cyt., 74 : 1-54 (1982).
6. T. Amano, E. Richelson and M. Nirenberg, Proc. Natl. Acad. Sci. USA, 69 : 258-263 (1972).

7.   G. Augusti-Tocco, G. Sato, P. Claude and D.D. Potter, Symp.
     Int. Soc. Cell Biol., 9 : 109-120 (1970).
8.   M. Nirenberg, S.P. Wilson, H. Higashida, A. Rotter, K. Kreuger,
     N. Busis, R. Ray, G. Kenimer, M. Adler and H. Fukni, Cold
     Spring Harbor Quant. Symp., 48 : 707-715 (1983).
9.   R. Akeson, Curr. Top. Devel. Biol., 13 : 215-236 (1979).
10. J. Ross, J.B. Olmsted and G.L. Rosenbaum, Tissues and Cells, 7
     : 107-136 (1975).
11. N.W. Seeds and R.B. Maccioni, J. Cell Biol., 76 : 547-555
     (1978).
12. J.C. Waymire, N. Weiner and K.N. Prasad, Proc. Natl. Acad. Sci.
     USA, 69 : 2241-2246 (1972).
13. E. Richelson, Nature New Biol., 242 : 175-177 (1973).
14. K.N. Prasad, S. Kumar, K. Gilmer and A. Vernadakis, Biochem.
     Biophys. Res. Comm., 50 : 973-977 (1973).
15. M. Miyake, Brain Res., 143 : 349-354 (1978).
16. M.A. Dichter and G.D. Fischbach, J. Physiol., 267 : 281-298
     (1977).
17. L. Casola, M. Romano, G. Di Matteo and G. Augusti-Tocco, Dev.
     Biol., 41 : 371-379 (1974).
18. F. Zucco, M. Persico, A. Felsani, S. Metafora and G.
     Augusti-Tocco, Proc. Natl. Acad. Sci. USA, 72 : 2289-2293
     (1975).
19. L.D. Grouse, B.K. Schrier, C.H. Letendre, M.Y. Zubairi and P.G.
     Nelson, J. Biol. Chem., 255 : 3871-3877 (1980).
20. W. Maltzman and A.J. Levine, Adv. Virus Res., 26 : 65-116
     (1981).
21. P.E. Swartzendruber and M. Lehman, J. Cell Physiol., 85 :
     179-187 (1975).
22. R. Maione, C. Passananti, G. Augusti-Tocco and P. Amati, Atti
     Convegno ABCD - AGI - SIBBM, p. 354 (1984).
23. M. Schachner, Nature New Biol., 243 : 117-119 (1973).
24. S.E. Martin, Nature, 249 : 71-73 (1974).
25. R. Akeson and H. Herschman, Proc. Natl. Acad. Sci. USA, 71 :
     187-191 (1974).
26. R. Akeson and H. Herschman, Exp. Cell Res., 93 : 492-495
     (1975).
27. G.Augusti-Tocco, E. Parisi, F. Zucco, L. Casola and M. Romano,
     In: "Tissue Culture of the Nervous System", ed. Sato G.,
     Plenum Press, New York, pp. 87-106 (1973).
28. R. Truding, M.L. Shelanski, M.P. Daniels and P. Morell, J.
     Biol. Chem., 249 : 3973-3982 (1974).
29. J.H. Garvican and G.L. Brown, Eur. J. Biochem., 76 : 251-261
     (1977).
30. R.H. Mathews, T.C. Johnson and J.E. Hudson, Biochem J., 154 :
     57-64 (1976).
31. U.Z. Littauer, M.Y. Giovanni and M.C. Glick, J. Biol. Chem.,
     255 : 5448-5453 (1980).
32. S. Denis-Donini, M. Estenoz and G. Augusti-Tocco, Cell Diff., 7
     : 193-201 (1978).

33. S.B. Rosenberg and F.C. Charalampous, Arch. Biochem. Biophys.,
    181 : 117-127 (1977).
34. G. Augusti-Tocco and V. Chiarugi, Cell Diff., 5 : 161-170
    (1976).
35. S.W. de Laat, P.T. Van der Saag, S.A. Nelemans and M.
    Shinitzky, Biochem. Biophys. Acta, 509 : 188-193 (1978).
36. S.W. de Laat, P.T. Van der Saag, E.L. Elson and J. Shlessinger,
    Biochem. Biophys. Acta, 558 : 247-250 (1979).
37. T. Koike, Biochem. Biophys. Acta, 509 : 429-439 (1978).
38. L.J. Erkell, FEBS Letter, 77 : 187-190 (1977).
39. Y. Kawasaki, N. Wakayama, T. Koike, M. Kawai and T. Amano,
    Biochem. Biophys. Acta, 509 : 440-449 (1978).
40. M.C. Fishman, P.R. Dragsten and I. Spector, Nature, 290 :
    781-783 (1981).
41. H.J. Harris and M.J. Dennis, Science, 167 : 1253-1255 (1970).
42. R. Siman-Tov and L. Sachs, Proc. Natl. Acad. Sci. USA, 70 :
    2902-2905 (1973).
43. K.N. Prasad and K.N. Gilmer, Proc. Natl. Acad. Sci. USA, 71 :
    2525-2529 (1974).
44. E. Richelson, Science, 201 : 69-71 (1978).
45. W.A. Klee and M. Nirenberg, Proc. Natl. Acad. Sci. USA, 71 :
    3474-3477 (1974).
46. N.C. Spitzer and P.I. Baccaglini, Brain Res., 107 : 610-616
    (1976).
47. N.C. Spitzer and J.E. Lamborghini, Proc. Natl. Acad. Sci. USA,
    73 : 1641-1645 (1976).
48. A. Blume, F. Gilbert, S. Wilson, J. Farber, R. Rosenberg and M.
    Nirenberg, Proc. Natl. Acad. Sci. USA, 67 : 786-792 (1970).
49. J. Massoulié and S. Bon, Ann. Rev. Neurosci., 5 : 57-106
    1982).
50. C. Vimard, C. Jeantet, Y. Netter and F. Gros, Biochemie, 58 :
    473-478 (1976).
51. U.Z. Littauer, C. Palfrey, Y. Kimhi and I. Spector, Nat. Cancer
    Inst. Monograph., 48 : 333-337 (1976).
52. J. Prives, L. Hoffman, R. Tarrab-Hazdai and S. Fuchs, Life
    Sci., 24 : 1713-1718 (1979).
53. G. Augusti-Tocco, M.A.R. Melone, A. Longo and C. Taddei, Abstr.
    X Int. Meeting Int. Soc. Neurochem. (1985).
54. E. Pannese, L. Luciano, S. Iurato and E. Reale, J. Ultrast.
    Res., 36 : 46-67 (1971).
55. K.H. Pfenninger and M.F. Maylié-Pfenninger, J. Cell Biol., 89 :
    547-559 (1981).
56. A.D. Lander, D.K. Fuji, D. Gospodarowicz and L.F. Reichardt, J.
    Cell Biol., 94 : 574-585 (1982).

SECTION IV

THE BASICS OF ELECTRICAL COUPLING

Luca Turin

UA 671 du CNRS, Station Zoologique La Darse
06230 Villefranche-sur-mer, France

HOW COUPLING IS MEASURED

Cells are said to be electrically coupled when ionic current injected into one cell passes to the cell's neighbour. Fig. 1 a illustrates the electrical circuit in the simplest case, when only two cells are present. In a), the cells are simply apposed, and the membranes are assumed to be homogeneous, that is to say no junctions are present. Two microelectrodes are inserted in cell 1, the first injects current into the cell, the second measures the voltage across the cell membrane, by comparison with a reference electrode in the bath. Current pulses give rise to voltage deflections proportional to the membrane resistance (Ohm's law). These deflections are not instantaneous, because the membrane acts as a capacitor, which is charged and discharged at each current pulse. Cell 2 is impaled with a second voltage-recording microelectrode. What does this electrode see ? In the absence of true electrical coupling, the membranes are still close enough to one another to be capacitively coupled, but no steady current can flow from one cell to another. If one looks more closely at the space separating the two membranes, it becomes clear that ions, having crossed the first membrane under the influence of the current pulse, have so to speak a choice between entering the second cell or going to earth. Even if the cells are pushed against one another, the resistance to earth is much lower than that of the other cell's membrane, and all the current goes that way.

Things would be different if a circular electrical seal (for example a tight junction) existed around the periphery of the membranes in contact, as in Fig. 1 b. Under these conditions, despite the fact that the apposed membranes have no special properties, the current now finds it easier to enter the second

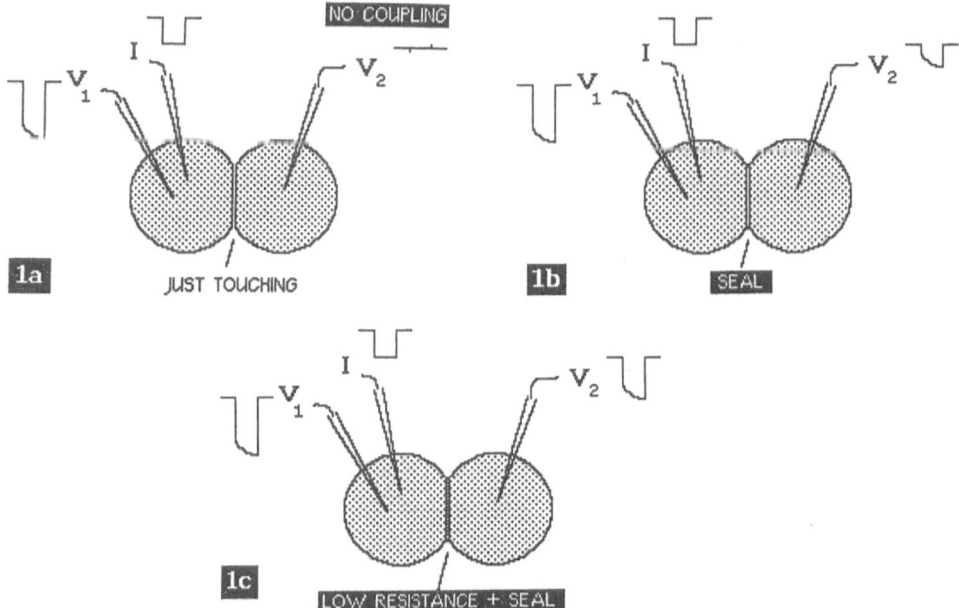

Fig. 1. a) Two spherical cells are pushed together without
any form of specialized contact between them.
Current pulses are injected into cell 1 through the
I microelectrode, and V1 records deflections in the
membrane potential of cell 1. In the neighbouring
cell electrode V2 sees capacitive "glitches" at the
beginning and end of each current pulse, but no
steady current flows from cell to cell.
b) A seal between the intercellular space and earth
is added, for example a tight junction. Some current
now goes into cell 2 and V2 records a small
deflection.
c) The seal is still present but in addition the
membranes facing each other contain many more ion
channels than the rest of the cell. Coupling is now
good.

cell than to escape to the bath, and some coupling is present. The
second microelectrode then records a much attenuated version of the
voltage deflection seen in cell 1. Suppose now that in addition to
the seal, the membranes which face each other have a say, tenfold
lower electrical resistance than the rest of the cell membrane (in
other words, there are ten times more ion channels in the apposed
membranes). Under these conditions (Fig. 1 c), the majority of the
current injected into cell 1 ends up in cell 2, and electrical
coupling is good. The coupling ratio V2/V1 is then high, say 0.9.

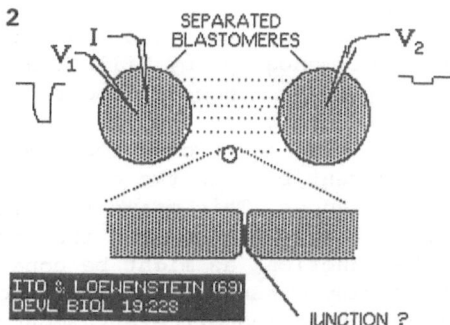

Fig. 2.   Coupled early embryo cells (blastomeres) remain so
          when separated, owing to the presence of thin
          fingers of membrane presumably bearing gap junctions
          at their ends.

Finally, suppose that one clumsily or deliberately damages cell 2,
say by poking a hole into its membrane. Electric current continues
to come from cell 1, but membrane resistance of cell 2 is now so
low that no voltage deflections are seen in V2 (Ohm's law again).
Coupling has now disappeared, although the intercellular pathway is
still present.

HOW THE CURRENT GETS ACROSS

     This hypothetical situation illustrates two important facts
about electrical coupling. Firstly, that it requires both a current
conducting pathway and a seal to earth between the cells, and
secondly that its detection depends on the electrical properties of
the entire cell membrane, not just of the part mediating
intercellular current flow. In real life, coupling is thought to be
mediated by specialized membrane structures, gap junctions, which
combine in a compact way current-passing channels and an electrical
seal to earth. How do we know that gap junctions are small, and
that junctional channels are very densely packed ?   Much
circumstancial evidence points in this direction, but perhaps the
clearest demonstration comes from the work of Ito and Loewenstein
(1969). As is shown in Fig. 2, separated cells of an early
amphibian embryo remain coupled even if they are joined only by
thin threads of membrane. They uncouple completely when a needle is
passed between the cells to break the threads. This illustrates
both the smallness and the solidity of gap junctions, which
seemingly refuse to split down the middle even when pulled hard.

     For the electrophysiologist, the gap junction is only one of a
great variety of membrane channels and carriers capable of

transporting current across cell membranes. For a more general view of membrane channels, placing the gap junction into a broader context, see the excellent monograph by Hille (1984). It has unique properties which make it at once interesting and difficult to study: it is made up of two membrane proteins in series, i.e., ions have to cross two channels to get from cytoplasm to cytoplasm, and it carries electrical current between two compartments not normally accessible to solution changes. This means that it is not possible to test junctional permeability by changing the ionic composition on either side of the junction, as might be done for example when characterising a potassium channel by looking at membrane properties in solutions containing different amounts of potassium.

The gap junction seen in the electron microscope, first in thin sections, (Mc Nutt and Weinstein, 1973) later in freeze-fracture (Peracchia, 1980) has the morphology which one might expect from its behaviour. Densely-packed, sometimes paracrystalline arrays of membrane proteins face each other "in register" across a narrow gap which gives the junction its name. Despite the great jump forward in resolution afforded by recent advances in electron-microscope techniques (Unwin and Zampighi, 1980), it is worth emphasizing that the "pore" at the centre of the gap junction has never been filled from end to end with the stain used to obtain contrast in the electron micrographs, and that its existence is inferred from the fact that junctions seem to allow the passage of many small molecules from cell to cell, rather than from the structure data.

To be exact, the distinction between a pore and a carrier cannot be made on the basis of size of permeant ion alone, and requires rather more sophisticated measurements to be made (saturation, unit conductance), none of which have been satisfactorily performed on the gap junction. The comprehensive review by Lefevre (1975) will help dispel any notion that carriers are limited in their scope and efficiency of transport.

WHAT JUNCTIONS MAY BE DOING

The basic question about gap junctions to which we would like an answer is "why are they there in electrically inactive tissues?". Good functional reasons for the presence of ion channels, (particularly in excitable tissues such as nerve and muscle) are usually easy to find. In heart, smooth muscle, electrical synapses and excitable epithelia, gap junctions allow the action potential to propagate from cell to cell. In embryos, we know that the cells are communicating with one another during orderly development. To give but one example among hundreds, if the first two blastomeres of the sea urchin embryo are separated, each gives a complete, though small larva. Clearly, each cell "knows"

whether or not it has a neighbour. For a comprehensive review of
the theory and practice of experimental embryology, see Slack
(1984). Gap junctions, electrical coupling and dye coupling are
common in embryos, as needed in most tissue sheets with epithelial
structures. During development, cells are thus both communicating
and communicant, and one is naturally led to suspect that gap
junctions mediate intercellular communication. It is worth noting
here that embryo membranes sometimes contain conductances which
they do not seem to need, such as neurotransmitter receptors
(Kusano et al., 1982) and channels typical of nerve tissues
(Miyazaki et al., 1974). The mere presence of gap junctions is thus
no proof of their importance.

It is also fair to say that other candidate mechanisms for
intercellular communication have received less attention for
reasons which may have more to do with fashion than with science.
For example, transfer of macromolecules from cell to cell can
occur, though of course not through gap junctions (Kolodny, 1973).
Similarly, a recently proposed structure for the tight junction
(Kachar and Reese, 1982) raises the interesting possibility of
continuity of the membrane outer leaflet in epithelia connected by
tight junctions. Much of pattern formation occurs in quasi
two-dimensional tissues (a few cells thick at most) with an
epithelial arrangement. With a little poetic licence, it may be
said that if this idea is correct, then from the standpoint of a
substance diffusing within the outer membrane leaflet there are no
cell boundaries and a metazoan early embryo looks just like a large
protozoan. See Frankel (1984) for a review of intracellular
communication in protozoa.

HOW TO MAKE COUPLING GO AWAY

How then will we find out whether gap junctions are indeed
important in development or whether they serve some essential but
more mundane function, e.g., making sure pH is equal in all cells ?
This calls for some pharmacology, and the situation was until
recently rather hopeless on this front, the only reasonably
specific means known to close gap junctions being calcium ions,
hydrogen ions and transjunctional voltage (see Spray et al.,
1984).

Intracellular free calcium ion is usually kept at very low
levels by the cell (around .1 micromolar). This level can be
raised, either by poisoning mitochondria where Ca is stored, by
helping it enter from the outside with ionophores, or by
microinjecting it directly into a cell. The clearest demonstration
of the effect of calcium on junctions comes from the work of Rose
and Loewenstein (1976). They microinjected calcium into coupled
cells of the insect salivary gland and visualised the calcium

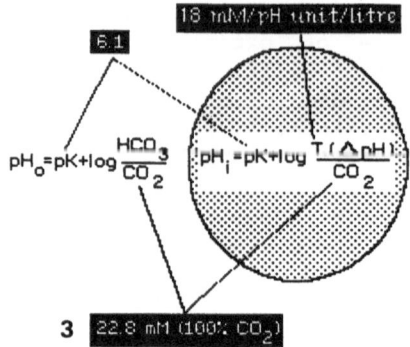

Fig. 3.   Changing intracellular pH with a weak acid, in this
          case   carbon   dioxide.   The   Henderson-Hasselbalch
          applies   both outside and inside the cell (circle at
          right).   The   undissociated   form   of   the   acid   is
          assumed   to equilibrate rapidly across the membrane,
          and all charged species are supposed   impermeant   in
          the short term. The value given for T, the buffering
          power, is that appropriate for Xenopus blastomeres.
          The   curve   relating   carbon   dioxide   tension   and
          intracellular pH obtained by numerical   solution   of
          the equation is given in Turin and Warner (1980).

"cloud" in   the   cell   using   aequorin, a protein each molecule of
which emits light (once) when it binds calcium. Calcium, because of
its   coordination   chemistry,   easily   achieves   both   potency   and
specificity, but is alas used in a wide variety of   other   cellular
control   mechanisms,   and   is (perhaps as a result) highly toxic to
cells at concentrations which close junctions (1-10 micromolar).

     It is easy to raise   cytoplasmic   hydrogen   ion   concentration
(lower   intracellular   pH),   because   weak   acids   of low molecular
weight such as acetic acid, carbonic acid, etc... will   all   cross
membranes in their uncharged form, and dissociate inside the cell,
thus releasing hydrogen ions (Sharp and   Thomas,   1981)   (Fig.   3).
Intracellular pH can be measured with pH-sensitive microelectrodes.
Rapid and reversible excursions of pH are possible in this way, and
they   are   very   effective   in   closing   junctions   reversibly.
Unfortunately, junctions   are   not   alone   in   being   pH-sensitive.
Indeed,   it   is   difficult to even think of something less specific
than the hydrogen ion in its interactions with   cell   constituents.
No unambiguous embryological experiments have been done using Ca or
pH as tools to close junctions.

     Voltage across junctions can be changed very rapidly by   using
a   double   voltage   clamp technique (Harris et al., 1981) (Fig. 4).

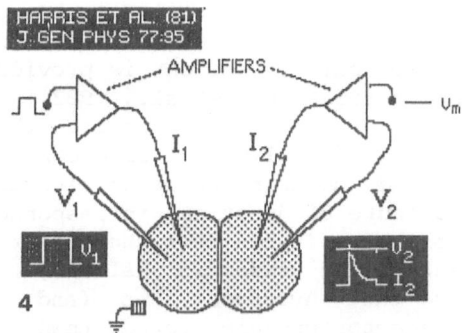

Fig. 4. Voltage clamp of a pair of cells. V1 is varied under
feedback control while V2 is held steady. I2 shows a
deflection   as V1 is varied due to the current which
came through the   junction. The   current   decreases
exponentially during the voltage step, showing that
the junction is voltage dependent. The   junction is
symmetric :   A   change in V1 in the other direction
would have given the   same   effect. The   uncoupling
reverses   as   rapidly   as it appeared if V1 is again
made equal to V2.

The principle is very simple. Each amplifier compares membrane
voltage to   a   reference   level,   and   injects current in either
direction to maintain it. If the cells are   coupled,   and   the   set
levels   are identical, no current will flow across the junction. If
the set level of one of the two cells is raised   or   lowered by a
certain   amount,   the   current   flowing   across the junction to the
neighbouring cell will be detected by the second   amplifier   trying
to   keep its cell's voltage steady. This arrangement eliminates all
the complications due to non-junctional membranes, provided one can
put   four   electrodes   into   a   cell   pair. Voltage uncoupling is
perfectly reversible, rapid and harmless, but   cannot   be   used   to
close   many junctions in a tissue at once, and is therefore of more
biophysical than developmental interest. The   question   of   whether
any or all   of these three mechanisms occur during development to
control the degree of opening of intercellular junctions remains
unanswered,   though   these mechanisms may be used in the adult (see
for example Giaume and Korn, 1983).

RECENT ADVANCES IN PHARMACOLOGY

A ray of hope is provided by the recent raising of antibodies
to gap junctions able,   when injected into embryo cells to block
intercellular   current   flow,   abolish   dye   coupling and disrupt
development   (Warner et al., 1984). Antibodies do not readily enter

cells and must be microinjected, so the technique is necessarily laborious, but its results definitive. A first step in a more classically pharmacological direction is provided by the work of Peracchia and coworkers (Peracchia et al., 1983 ; Peracchia, 1984). They showed that drugs known to inhibit the response of the calcium-binding protein calmodulin (but also capable of acting as local anaesthetics and dopamine receptor blockers (Seeman, 1972) also abolished the closure of junctions in response to a lowered pH. In their preparation, it had been shown that lowering pH does not lead to a generalized rise in calcium, so their result presumably suggests that hydrogen ions (and calcium ions) act through a specific receptor protein, rather than by protonating a site on the ion channel itself.

If the results obtained with antibodies cited above prove to be generally applicable, the role of gap junctions in development will at last have been established. The question will then shift to the identification of the intercellular messenger molecule. In the absence of an a priori guess as to what this "morphogen" might be, one is reduced to asking what junctions are permeable to. Unfortunately, they seem to be far from specific. (See Loewenstein, 1981, for review). It has been repeatedly stated that junctional channels behave like sieves selecting molecules on the basis of size only, much like other water-filled pores known to biophysicists (Hille, loc. cit) but larger, allowing molecules smaller than 700-1500 daltons in molecular weight to go through. This seems reasonable in view of the wide variety of transported molecules (see article by Pitts in this volume), but the methods used for detection (e.g. microinjected tracers detected by fluorescence) are inherently difficult to quantitate. It has only recently become possible to show that current flow and dye coupling were related, again by antibody injection, which suppresses both (Warner et al., loc. cit), and by voltage uncoupling of a junction between two excitable cells (Biaume and Korn, 1984). Both these studies used Lucifer yellow, (see article by W.W. Stewart in this volume). It is therefore still far from clear whether or not the many different molecules which are found to diffuse from cell to cell go through the same pathway as the current-carrying ions or through some other as yet undiscovered structure associated with cell contacts.

MAKING JUNCTIONS MORE ACCESSIBLE

Quantitative measurements of membrane permeability require access to at least one side, preferably both sides of the membrane under study. Cytoplasmic perfusion methods have been developed for use with both tubular and spherical cells, but the replacement of the cell's cytoplasm with an artificial medium, as may be expected, is not without unwanted side-effects, and perfusion media are

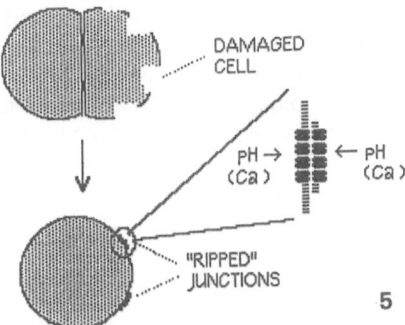

Fig. 5. "Ripped" junctions on the surface of a mechanically
        isolated blastomere. Membrane channels with the
        properties expected of gap junctions can be detected
        in single blastomeres by using pH changes. They are
        selective for potassium and hydrogen ions. The
        external pH effect is abolished by 200 nM
        concentration of the calmodulin blocker calmi-
        dazolium in the presence of 10 mM extracellular
        calcium, suggesting that sensitivity to pH has
        little to do with Ca (Turin and Bernardini, MS in
        preparation).

seldom physiological even in their gross composition if membrane
stability is to be ensured (see Tasaki et al. (1965) for a
systematic study of this problem). For example junctions of the
crayfish septate axon longer respond to changes of intracellular pH
or calcium (Johnston and Ramon, 1981).

    A perfusion technique similar in principle has been used by
Spray et al. (1982) to study the pH and calcium ion dependence of
junctions. Turin and Bernardini (Ms in preparation) have shown that
when cells from a blastula are mechanically dissociated (Fig. 5), a
conductance remains on the cell surface which has the pH dependence
expected of a whole gap junction with its previously cytoplasmic
face exposed : it closes if either extracellular pH or
intracellular pH is made acid. The selectivity of these membrane
channels is easily measured by conventional electrophysiological
methods. Interestingly the sensitivity to extracellular pH is
suppressed by low concentrations (200 nanomolar) of calmidazolium,
a calmodulin antagonist. This fact suggests that the effect of this
class of drugs on junctions may be unrelated to their effect on
calmodulins, since in this case they abolish the pH dependence in
the presence of several millimoles/litre of calcium. To a
calmodulin, this would be a saturating concentration. What is
puzzling in that these "ripped" junctions are highly selective for
hydrogen and potassium ions, and do not therefore behave like the

large, nonselective pores one might have expected. Naturally, the cell may have other means of knowing that another cell is no longer present across the junction and may alter junctional permeability to prevent leakage of larger cellular components. "Ripped" junctions could be very different in permeability from proper ones. It remains to be seen whether they allow dyes to enter the cell, and whether they respond to antibodies.

CONCLUSION

It has been said that the only biological function of gap junctions in inexcitable tissues which was established beyond reasonable doubt was to provide grants for the scientists who studied them. This quip may now be losing some of its sting, but a lot remains to be done with the new tools at our disposal to establish whether they are indeed a communication channel over which cells talk to each other, and maybe even say something interesting.

I wish to thank Walter Stewart for several stimulating discussions in Banyuls-sur-mer and the CNRS for support.

REFERENCES

Frankel J., 1984, Pattern formation in ciliated protozoa, In: "Pattern formation", Malacinski and Bryant Eds., Macmillan, New York, pp. 163-196.
Giaume C. and Korn H., 1984, Voltage-dependent dye coupling at a rectifying electrotonic synapse of the crayfish, J. Physiol., 356 : 151-157.
Giaume C. and Korn H., 1983, Bidirectional transmission at the rectifying electrotonic synapse : a voltage-dependent process, Science, 220 : 84-87.
Harris A.L., Spray D.C. and Bennett M.V.L., 1981, Kinetic properties of a voltage-dependent conductance, J. Gen. Physiol., 77 : 95-117.
Hille B., 1984, Ionic channels of excitable membranes, Sinauer Associates, Sunderland Mass USA.
Ito S. and Loewenstein W.R., 1969, Ionic communication between early embryonic cells, Devl. Biol., 19 : 228-243.
Johnston M.F. and Ramon F., 1981, Electrotonic coupling in internally perfused crayfish segmented axons, J. Physiol., 317 : 509-518.
Kachar B. and Reese T.S., 1982, Evidence for the lipidic nature of tight junction strands, Nature, 296 : 464.
Kolodny G.M., 1973, Transfer of nuclear and cytoplasmic proteins between cells in culture, J. Mol. Biol., 78 : 197-210.

Kusano K., Miledi R. and Stinnakre J., 1982, Cholinergic and catecholaminergic receptors in the Xenopus oocyte membrane, J. Physiol., 328 : 143-170.

LeFevre P.G., 1975, The current state of the carrier hypothesis, Curr Top Membr. Transp. 7 :

Loewenstein W.R., 1981, Junctional intercellular communication : the cell-to-cell membrane channel, Physiol. Rev., 61 : 829-913.

MC Nutt N.S. and Weinstein R.S., 1973, Membrane ultrastructure at mammalian intercellular junctions, Prog. Biophys. Mol. Biol., 26 : 45-173.

Miyazaki S., Takahashi K. and Tsuda K., 1974, Electrical excitability in the egg cell membrane of the tunicate, J. Physiol., 238 : 37-54.

Peracchia C., 1980, Structural correlates of gap junction permeability, Int. Rev. Cytol., 66 : 81-146.

Peracchia C., Bernardini G. and Peracchia L.L., 1983, Is calmodulin involved in the regulation of gap junction permeability ?, Pflugers Archiv., 399 : 152-154.

Peracchia C., 1984, Communicating junctions and calmodulin : inhibition of electrical coupling in Xenopus embryos by calmidazolium, J. Membrane Biol., 81 : 49-50.

Rose B. and Loewenstein W.R., 1976, Permeability of a cell junction and the local cytoplasmic free ionized concentration : a study with aequorin, J. Membr. Biol., 28 : 87-119.

Seeman P., 1972, The membrane actions of anaesthetics and tranquilizers, Pharm. Rev., 24 : 583-651.

Sharp A. and Thomas R.C., 1981, The effects of chloride substitution on intracellular pH in crab muscle, J. Physiol., 312 : 71-80.

Slack J.M.W., 1984, From egg to embryo, Cambridge University Press, Cambridge.

Spray D.C., Stern J.H., Harris A.L. and Bennett M.V.L., 1982, Comparison of sensitivities of gap junctional conductance to H and Ca ions, Proc. Natl. Acad. Sci. USA, 79 : 441-445.

Spray D.C., White R.L., Campos de Carvalho A., Harris A.L. and Bennett M.V.L., 1984, Gating of gap junctional channels, Biophys. J., 45 : 219-230. (see also other articles in the same issue).

Tasaki I., Singer I. and Takenaka T., 1965, Effects of internal and external ionic environment on excitability of squid giant axon : a macromolecular approach, J. Gen. Physiol., 48 : 1095-1123.

Turin L. and Warner A.E., 1980, Intracellualr pH in early Xenopus embryos : its effect on current flow between blastomeres, J. Physiol., 300 : 489-504.

Warner A.E., Guthrie S.C. and Gilula N.B., 1984, Antibodies to gap junctional protein selectively disrupt junctional communication in the early amphibian embryo, Nature, 311 : 127-131.

Zampighi G. and Unwin P.N.T., 1979, Two forms of isolated gap
        junctions, <u>J. Mol. Biol.</u>, 135 : 451-464.

LUCIFER DYES AS BIOLOGICAL TRACERS : A REVIEW

Walter W. Stewart and Ned Feder

National Institute of Arthritis, Diabetes, and Digestive
and Kidney Diseases, National Institutes of Health
Bethesda, Maryland 20205, USA

Lucifer dyes are sulfonated 4-aminonaphthalimides. Most have a strong yellow fluorescence in aqueous solution. This and other properties make them useful as biological tracers. We briefly review here the synthesis of some Lucifer dyes, as well as their chemical and biological properties. Some points we present here have been discussed in detail elsewhere (1-3).

CHEMISTRY

Lucifer dyes are prepared in a synthesis that starts with the commercial dye brilliant sulfoflavine (Fig. 1 A). One must be careful in the choice of starting material , since some commercial dye sold under the name brilliant sulfoflavine may not have the structure shown in Fig. 1 A. Some may even be a mixture of dyes. Also, commercial brilliant sulfoflavine is contaminated by large amounts of inorganic salts. The analytically pure potassium salt of brilliant sulfoflavine can be prepared by procedures that have been described (2).

Two good methods for characterizing brilliant sulfoflavine and the Lucifer dyes are thin-layer chromatography and proton nuclear magnetic resonance spectroscopy (^{1}H NMR). Using the latter method it is possible to get information about both the purity of the sample and its chemical structure. Since ^{1}H NMR gives signals from carbon-bound protons, most organic impurities give signals that can be identified. Inorganic impurities, on the other hand, can not be identified with ^{1}H NMR, but can be determined by elemental analysis. D_2O is usually the most suitable solvent for ^{1}H NMR of Lucifer dyes. If signals are too weak in a saturated solution at

297

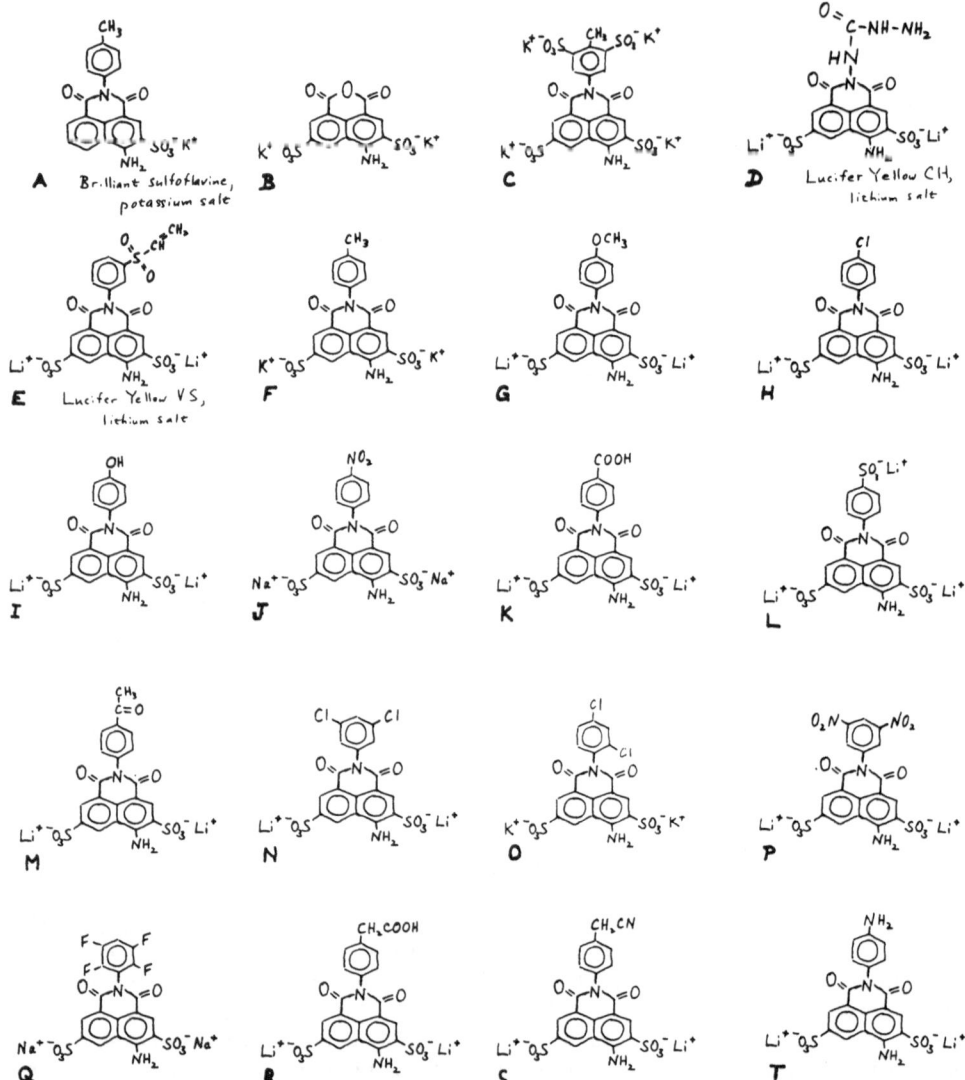

Fig. 1.   Lucifer Dyes (A, C-T) and 4-amino-
3,6-disulfonaphthalic anhydride (B).

room temperature, a heated saturated solution will usually contain
enough dye to give a good spectrum.

Brilliant sulfoflavine (Fig. 1 A) was converted to dipotassium
4-amino-3,6-disulfonaphthalic anhydride (Fig. 1 B) - a convenient
intermediate for the preparation of disulfonated Lucifer dyes.
Despite repeated trials, the imide ring of brilliant sulfoflavine

could not be hydrolyzed under either acidic or basic conditions. It proved necessary to sulfonate the dye to tetrapotassium 4-amino-N-(4-methyl-3,5-disulfophenyl)-naphthalimide-3,6-disulfonate (Fig. 1 C), which could be hydrolyzed to the desired anhydride. Even then, it was surprisingly difficult to find conditions for the production of pure anhydride in reasonable yield. Basic hydrolysis of the tetrasulfonate at 100° C followed by acidification gave a 30% yield of anhydride contaminated by a compound with an aromatic methyl group. It was discovered that simply by repeating the cycle of basic hydrolysis followed by acidification, both the yield and the purity of the final product could be increased. The conditions finally chosen involved cycling the tetrasulfonate four times (without isolation) from base to acid at 50° C ; these rather unusual conditions gave a 98% yield of pure anhydride.

This anhydride (Fig. 1 B) can also be synthesized by reduction of 4-nitronaphthalic anhydride to 4-aminonaphthalic anhydride (4, 5), and subsequent sulfonation (6).

It is interesting that 4-amino-3,6-disulfonaphthalic anhydride (Fig. 1 B) is sufficiently stable to hydrolysis to be recrystallized from boiling water. The dipotassium salt of the anhydride is sparingly soluble in cold water and about 1.2% soluble in boiling water. It dissolves readily in cold aqueous base, presumably by formation of the dicarboxylate, and crystallizes out again as the anhydride when the solution is acidified.

The anhydride reacts readily with a wide variety of aromatic amines to give imides, but fails to react with its own amino group (which has an exceptionally low pK - less than 1). The stability of the anhydride in boiling water was imporatnt in designing simple conditions for the synthesis of naphthalimides. As previously reported (2), most primary amines, both aliphatic and aromatic, react on heating with the anhydride to give the expected N-substituted imide. Aliphatic amines were reacted in a pH 5.0 buffer, aromatic amines in 5% aqueous acetic acid. The products were isolated as the crystalline potassium, sodium, or lithium salts, with the choice depending on solubility of the particular compound ; the potassium salts were generally the least soluble, the sodium salts were intermediate in solubility, and the lithium salts were the most soluble. The products were characterized by ^{1}H NMR. Some of the compounds synthesized in this way are shown in Fig. 1 (D, F-T) and Fig. 2 (A-X). Two compounds in particular, Lucifer Yellow CH (Fig. 1 D) and Lucifer Yellow VS (Fig. 1 E, synthesized from the compound in Fig. 2 D), have proved useful in biological tracing, but others may eventually prove useful as well.

Small differences in spectral properties and a considerable difference in solubility are associated with different degrees of

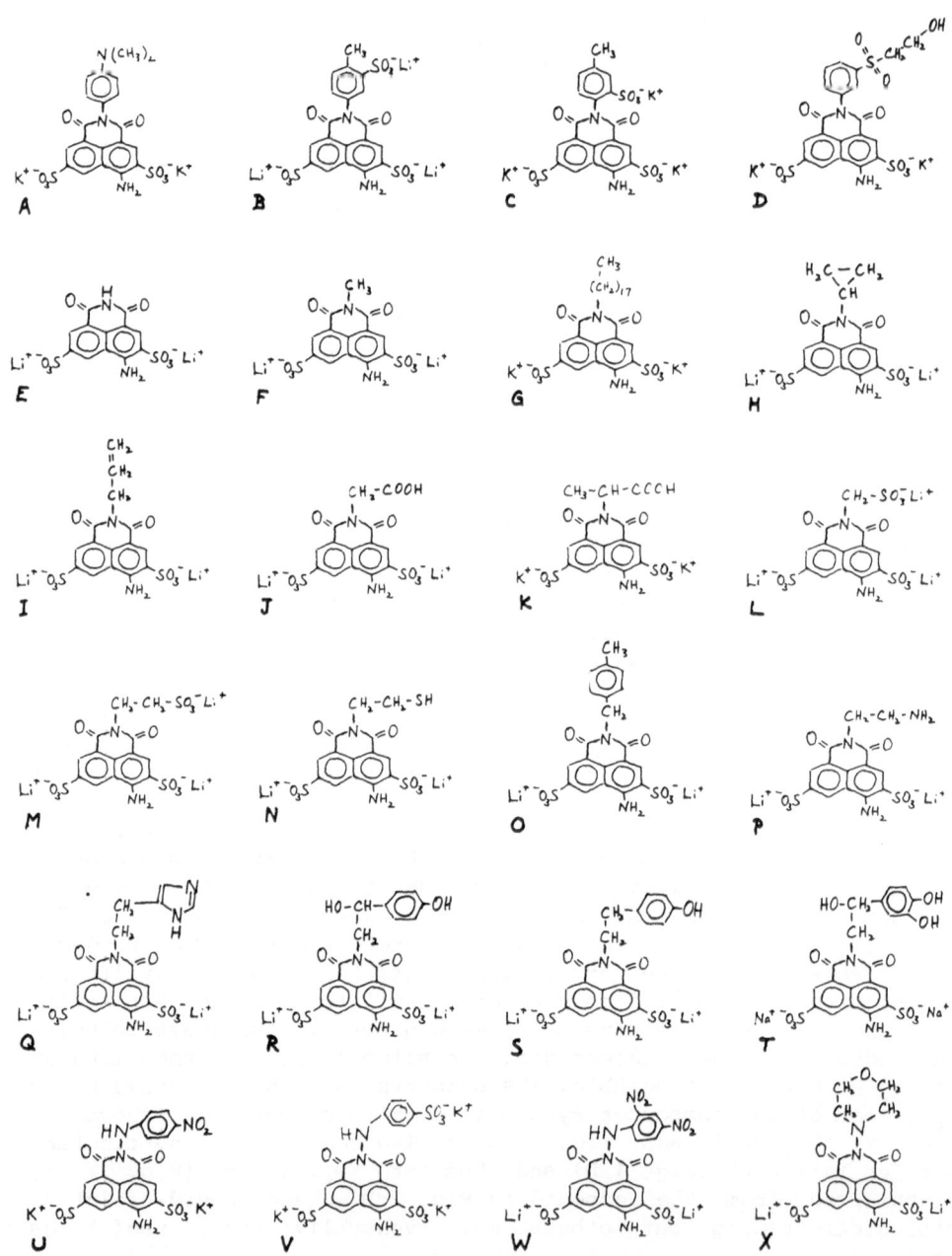

Fig. 2. Lucifer dyes.

sulfonation. Typically, on going from a 3-sulfo compound to the corresponding 3,6-disulfo compound, there is a slight red shift in the peak of the absorption spectrum (from $\sim$ 420 nm to $\sim$ 430 nm),a decrease in the quantum yield (from $\sim$ 0.5 to $\sim$ 0.25), and an increase in solubility.

Since their introduction in 1978, Lucifer Yellow CH and Lucifer Yellow VS have been used with considerable success as tracers in a variety of biological systems. The dyes differ in their chemical properties, but both can be bound to tissue, though by different chemical mechanisms. Lucifer Yellow CH contains a hydrazido group by which the dye is presumably bound covalently to tissue by means of aldehyde fixatives. Lucifer Yellow VS binds rapidly and directly to tissue, presumably by addition of the vinyl sulfone of the dye to the sulfhydryl and amino groups of protein ; this dye has been found to be highly reactive with amino groups, but is extremely stable in water, with a half-life in water of about one week at room temperature.

Several published papers, unfortunately, report on "Lucifer Yellow" without indicating which of the two dyes was used.

The dyes have similar spectral properties : absorption maxima at 280 nm and 430 nm, corrected emission maxima near 540 nm (1, 7), and quantum yields of about 0.25 (1). The high quantum yields make possible the detection of the dyes at low concentration, and the wide separation between the absorption and emission maxima facilitates excitation at one wavelength and observation of fluorescence at another.

Lucifer dyes are stable in the dark but fade under intense illumination. However, even after continuous irradiation for 7 days, a cell that had been strongly marked with Lucifer Yellow CH was still clearly detectable (1).

BIOLOGICAL PROPERTIES

The intense fluorescence and low toxicity of the Lucifer Yellow CH make it possible to carry out tracer experiments which otherwise are difficult or impossible - for example, experiments that involve examination of live tissue preparations. Futhermore, even after the tissue is fixed, the possibility of examining marked cells in wholemount has several advantages. (a) It saves time : material marked too faintly can be discarded. (b) In fixed, embedded material that will be sectioned, the optimum plane of section can be determined. (c) The complex shape of a neuron is usually more apparent in wholemount than in the sections prepared from it. The technical details of cell marking with Lucifer Yellow CH are described elsewhere (1, 3) and have been reviewed at length (8).

Neurons and other cells can be marked with Lucifer Yellow CH by injecting the dye through a micropipette using pressure or iontophoresis. The dye spreads rapidly throughout an injected cell. Furthermore, it is retained within a cell for many hours and perhaps as long as the cell remains healthy. The intense fluorescence of the dye permits several kinds of studies. (a) A cell can be stained in its entirety with a relatively small amount of dye. In preparations which are sufficiently transparent, injected cells can be viewed while alive ; illumination sufficient for photomicrography of live cells does not noticeably after their physiology. (b) One can determine whether the distribution of dye found after fixation is the same as that seen in the living state. (c) The dye frequently can be seen to spread in vivo from the injected cell to certain nearby cells. The movement of dye from cell to cell has been termed dye-coupling (1). It is perhaps associated with the dye's relatively low molecular weight (443 for the dye anion).

In some cases, cells found by Lucifer injection to be dye-coupled have also been shown to be electrically coupled to each other. There is a wide range of cell types which are both dye-coupled and electrically coupled, for example : adjacent neurons at the crayfish giant motor synapse (9) ; Leydig cells of the leech (3, 10) ; a pair of identified neurons in the buccal ganglion of a pulmonate snail (11) ; epithelial cells of Necturus gallbladder (12) ; islet cells (13-15) and acinar cells (16) of the pancreas ; myocardial cells (17, 18) ; horizontal cells in the retina of turtle (1, 19), salamander (20), and carp (21, 22) ; frog spinal motoneurons (23) ; hippocampal neurons (24) and nigral dopaminergic neurons (25) of rat brain ; and embryonic cells in afrog (26) and a teleost fish (27), as well as other vertebrate and invertebrate embryos. Many other examples are now known of cells that are both Lucifer dye-coupled and electrically coupled. Presumably in most but not necessarily all instances, dye-coupling and electrical coupling (sometimes called "ionic coupling") have a common basis. Nevertheless, significant electrical coupling is occasionally found without detectable dye-coupling (11, 28-30), and similarly perhaps for dye-coupling without electrical coupling (31). Electrical coupling and dye-coupling are clearly two different experimental findings, and thus one should not refer to dye-coupled cells as electrically coupled or vice versa unless there is experimental evidence for that type of coupling in the particular cell system under consideration. The procedure for showing dye coupling (which is a single-electrode experiment) is generally easier to carry out than the two-electrode procedures needed to show electrical coupling ; in most tissues where dye-coupling has been observed, the experiment has not been done showing whether or not the types of cells that are dye-coupled also show electrical coupling. In any case, dye-coupling provides a convenient method of recognizing certain functional connections between cells by morphological means. Lucifer Yellow injection often unveils a spectacular view of one kind of tissue unit - a unit within which one may presume that the cells are constantly exchanging a multitude of physiologically active substances, and doing so rapidly and far less selectively than cells not so coupled.

The anatomical basis of dye-coupling is also not known with certainty, though there is now support for the idea that channels within gap junctions are generally the path by which Lucifer Yellow (and other ionic substances of similar size) move directly from cell to cell without entering the extracellular space. In some cases (for example, refs. 32 and 33 ; see also reviews, refs. 34 and 35), this conclusion is based on such evidence as a correlation between numbers of gap junctions (or the sizes or substructure of gap junctions) and rate of cell-to-cell dye movement, and also on the absence of other structures that are good candidates as channels for dye movement from cell to cell - evidence that is persuasive though not conclusive. In other cases reported in the literature, however, no such evidence is given (e. g., in refs. 18, 19 and 36) ; it is simply assumed that gap junctions must be the pathways by which dye has passed from cell to cell. It would seem preferable in these cases to avoid making the assumption that Lucifer Yellow passed through gap junctions. Instead one could use a term such as "dye-channels", defined as the spaces or channels through which dye moves when dye put into one cell is later observed in another. Explicit arguments could then be given favoring the hypothesis that the dye-channel (a physiological entity) is mediated by the gap junction (an ultrastructural entity) in the particular tissue under consideration. For a similar viewpoint, see ref. 34, p. 844-845. Certain reported observations are worth noting : Lucifer dye-coupling could not be demonstrated between some cells where gap junctions were present in seemingly sufficient numbers (29, 30, 36-38) ; for related observations and discussion of "uncoupled" gap junctions, see refs. 35, 39 and other references cited therein. Futhermore, in a few cases Lucifer Yellow moves from cell to cell by pathways probably not involving gap junctions (40, 41).

Some of the studies in which Lucifer Yellow CH and VS have been used are summarized briefly below. References are given for the less common uses.

INTRACELLULAR INJECTION

Lucifer Yellow CH, injected intracellularly, has been used to study neuronal structures in many species, mainly among vertebrates, mollusks, arthropods, and in the leech. Non-neuronal cell types have also been examined, including neuroglia, muscle, and various secretory and nonsecretory epithelia. Among the other invertebrates studied by intracellular injection of Lucifer Yellow are various jellyfish (42, 43), a hydra (37), a ctenophore (44), a polyclad flatworm (45), and the sea urchin embryo (46).

There are now numerous studies in which injection of cells in early embryos has permitted (a) construction of fate maps and (b) identification (by dye-coupling) of groups of functionally related cells. Lucifer Yellow CH has been used to monitor dye-coupling in regenerating neurons (47). Intracellular injection has also been used to study neurons and other cells in culture.

Lucifer Yellow CH has been injected into and found to be retained by tobacco cell protoplasts (48).

UPTAKE OF DYE BY CELLS

1. Neurons can be filled with dye through their cut processes rather than by injection of dye into their somata. Backfilling, introduced by Iles and Mulloney (49), is carried out simply by immersing the cut end of a nerve in a droplet of tracer solution. The tracer moves up the processes of the nerve, either by diffusion or in response to an externally applied electric field ; the cell bodies of the neurons with processes in that nerve become filled with tracer. The original report described backfilling with Procion Yellow ; for backfilling with Lucifer Yellow CH, see refs. 3, 50 and 51. Since neither Lucifer Yellow CH nor the procedure of backfilling appears to alter physiological properties noticeably, and since the dye can usually be seen in vivo, one can first identify a backfilled cell by its fluorescence, then impale it with a micropipette and investigate its electrophysiology.

2. Retrograde axonal transport has been demonstrated with Lucifer Yellow VS (3, 52) : dye injected into the chick eye was later found in the brain within cell bodies of neurons having axon terminals within the retina. Retrograde transport of Lucifer Yellow has also been reported in the rat brain (53).

3. Lucifer Yellow CH was taken up fairly selectively from an extracellular medium by certain retinal cells in the turtle (54), chicken (55), and rat (56).

4. Wilcox (57) has made the interesting observation that when housefly retina was bathed in Lucifer Yellow CH solution, dye was taken up by photoreceptor cells - but only if the retina was illuminated.

5. Cells in culture took up Lucifer Yellow CH dissolved in the culture medium into endocytotic or pinocytotic vesicles ; uptake

has been observed in several types of cells : cultured fibroblasts
(1), kidney cells (58), and amebae (59). Furthermore, when certain
fatty amines were added to the culture medium bathing the kidney
cells, dye was released from the vesicles into the cytoplasm (58).
These amines are known to be cytotoxic, and the above observation
was viewed as supporting the hypothesis that their toxicity results
from the release of lysosomal enzymes into the cytoplasm. In
another study, amebae phagocytosed yeast cells labelled with
Lucifer Yellow VS about as fast as they phagocytosed unlabelled
yeast cells (60) ; when the two populations of yeast cells were
presented separately to the amebae and phagocytosed in sequence,
the fate of each population could be followed.

LUCIFER YELLOW IN CONJUNCTION WITH OTHER TECHNIQUES

    1. Lucifer Yellow CH has been used as an intraneuronal marker
along with horseradish peroxidase, which was also used as an
intraneuronal marker in the same tissue specimen (61). It was found
necessary to modify the standard histological procedure for
peroxidase so that both tracers could be viewed in the same
specimen. When the two tracers were injected separately into
different neurons, the contacts - presumably synaptic contacts -
between the injected neurons were apparent on inspection. When the
two tracers were injected into the same neuron, it was possible to
identify the site of a presumed electrical synapse between that
neuron (containing both tracers) and the post-synaptic neuron
(containing only Lucifer Yellow, which had traversed the synapse).

    2. Lucifer Yellow dyes do not contain a heavy atom, and so
they have no useful electron density. However, their intense
fluorescence allows them to be used as tracers for electron
microscopy in an indirect procedure. An ultrathin section was
examined in a light microscope with fluorescence optics, and the
resulting light micrograph was compared with the subsequently
obtained electron micrograph of the same section (3, 51). An
advantage of an ultrastructural tracer that does not rely on
electron density for its detection is that the normal
ultrastructure of the marked cell is relatively unaffected.

    3. In a different approach to electron microscopy, Maranto
(62) has reported that when a neuron filled with Lucifer Yellow CH
was irradiated with intense light in the presence of
3,3'-diaminobenzidine, an osmiophilic polymer was deposited within
the injected cell and could be visualized with the electron
microscope. Two other groups have reported using this procedure
(63, 64).

    4. Taghert and coworkers (65) have reported a method of
converting Lucifer Yellow staining, by means of an anti-Lucifer

antibody, into an osmiophilic polymer. Tissue containing neurons marked with Lucifer Yellow was fixed and rinsed, incubated with rabbit antiserum to Lucifer Yellow, rinsed, incubated with goat anti-rabbit IgG labelled with horseradish peroxidase, rinsed, and stained for peroxidase activity with a procedure using 3,3'-diaminobenzidine and hydrogen peroxide. The final step generated an osmiophilic polymer. The tissue was then osmicated, and the osmiophilic polymer, presumably superimposed on the original sites of Lucifer Yellow staining, was visualized with a light microscope having interference contrast optics. In a neuronal growth cone, fine filopodia that were studied only with difficulty by their Lucifer fluorescence could be routinely viewed and studied after the procedure noted above.

5. Tissue containing neurons marked intracellularly with Lucifer Yellow has also been stained immunocytochemically with rhodamine-labelled antibodies (66-71). The usual goal of this approach was to characterize the immune reactivity as well as the electrophysiology of the same cell.

6. Lucifer Yellow VS labels proteins rapidly and covalently under mild conditions. The conjugates generally have quantum yields between 0.1 and 0.2 (R. Chen and W. Stewart, unpublished observations), so Lucifer-conjugated antibodies would be expected to be suitable for immunofluorescence staining. Indeed this technique has been used to visualize cytoplasmic actin filaments (3); the results were similar to those obtained with fluorescein-labelled anti-actin antibody. Lucifer Yellow VS has also been used as a label for serum albumin (72) and for peptides (73). Unlike most fluorescent dyes, Lucifer Yellow dyes have a fluorescence intensity that is relatively unchanged from pH 2 to pH 10, which makes these dyes suitable for monitoring pH-induced conformational changes in macromolecules.

MISCELLANEOUS

1. An individual Lucifer-filled neuron or part of a neuron can be selectively damaged by means of an intense, focused spot of light having a wavelength strongly absorbed by the dye (74). This procedure is sometimes referred to as cell killing and sometimes as photoinactivation. As the originators of the technique have pointed out (75), it is not known whether the affected cell is killed or simply inactivated temporarily ; see also ref. 76. (A closely related procedure employs a different dye, carboxyfluorescein, injected intracellularly into neurons (77). Several days after irradiation of a small part of an axon filled with carboxyfluorescein, there was a profuse sprouting of neurites proximal to the region of irradiation - sprouting similar to that seen proximal to an ordinary surgical axotomy.)

When one cell of a two-cell <u>Arbacia</u> embryo was injected with Lucifer Yellow CH, development was normal, unless the embryo was irradiated with intense blue light. After irradiation, the uninjected cell continued to divide normally, while the injected cell failed to divide (46).

2. Lucifer has been used to follow changes in junctional permeability caused by moderate transjunctional polarization (78).

3. The surfaces of live thymocytes oxidized with periodate or galactose oxidase were labelled with Lucifer Yellow CH (79). The binding of dye was presumably covalent, through formation of a carbazone. The cell surface appeared evenly stained, and no intracellular staining was observed. The labelled cells appeared to be fully viable.

4. Lucifer Yellow CH has been injected into and found to be retained by giant mitochondria (80) and by protein-lipid vesicles prepared <u>in vitro</u> (81).

5. A marked improvment in intensity of staining after fixation and paraffin embedding has been described for neurons injected with an equimolar mixture of Lucifer Yellow CH and formaldehyde (82). The authors report that, unexpectedly, there was no apparent toxic effect of this mixture on the electrical activity of the injected cells. In a study done by other investigators, dye-coupling was demonstrated with the Lucifer-formaldehyde mixture (83), but no parallel experiment was carried out with Lucifer Yellow alone to see if the rate of dye transfer from cell to cell was affected by the presence of formaldehyde.

6. In plastic sections of marked cells, Lucifer Yellow has been reported to fade much more quickly when the sections are counterstained with toluidine blue (16).

7. Many of the studies employing Lucifer Yellow CH have been carried out exclusively on living cells or tissues ; at no point in these studies were the tissues fixed in solutions of formaldehyde or in other fixatives. Thus the reactive moeity of Lucifer Yellow CH - the hydrazido group - would appear to be superfluous in these studies ; it is possible that the hydrazido group has an unexpected (and as yet unrecognized) effect of some sort on the injected cells. For such studies a more inert Lucifer dye might be useful, for example, the compounds in Fig. 1 A, C, K, L ; Fig. 2 F, J, M ; and others. In only a very few biological studies have any such compounds been used (e. g., the adduct of 1,4-diaminobutane with the anhydride (Fig. 1 B) in ref. 1 ; the compound in our Fig. 2 B, which was used in ref. 84). More widespread use of such Lucifer dyes would seem worthwhile.

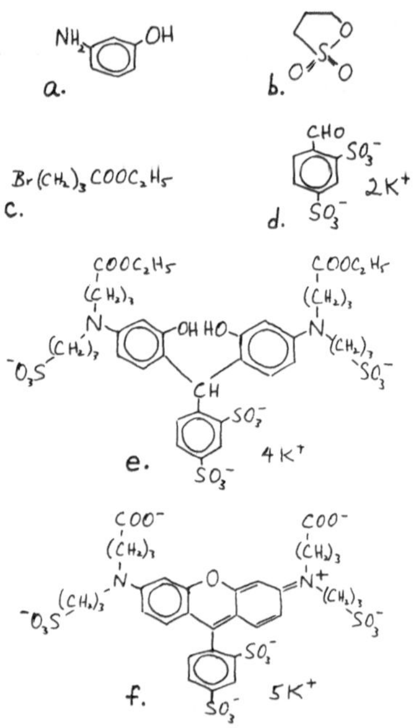

Fig. 3. A rhodamine (f), and the interme-
diates (a-e) used in its synthesis.

8. Lucifer Yellow CH is ordinarily used for intracellular
injection as its dilithium salt, sometimes with added lithium
acetate or lithium chloride. Leakage of lithium ion into a neuron
from a microelectrode containing lithium acetate and the lithium
salt of Lucifer Yellow CH was reported to prolong markedly the
action potential of that neuron, thus seriously compromising the
characterization of its normal electrophysiological properties
(85).

A SECOND FLUORESCENT TRACER

There are some circumstances in which it would be useful to
study a pair of neighbouring neurons by means of two fluorescent
tracers with different colors. See, for example, refs. 86-89. We
are attempting to prepare a red-fluorescing tracer to be used in
conjunction with Lucifer Yellow.

Though many fluorescent dyes are known, few have a red
fluorescence. Among the latter are the N, N, N', N'-tetra-alkylated
rhodamines (emission maximum around 580 nm) ; these dyes have a
high quantum yield and are said to be resistant to photobleaching.

For reasons given elsewhere (1) we also wanted a dye with high solubility in water, with a "fixed" net negative charge, and with a functional group enabling the dye to be bound to tissue by aldehyde fixatives. A dye derived from the rhodamine shown in Fig. 3 f, could have all these properties. The synthesis in brief is to N-alkylate m-aminophenol (Fig. 3 a) successively with propane sultone (Fig. 3 b) and ethyl bromobutyrate (Fig. 3 c), then to condense 2 moles of the dialkylated aminophenol with 1 mole of benzaldehyde disulfonic acid (Fig. 3 d) to give a triphenylmethane (Fig. 3 e). The latter cyclodehydrated to a rhodamine leucodye, which is then oxidized to the rhodamine in Fig. 3 f. From this dye we hope to synthesize a suitable tracer.

REFERENCES

1. W.W. Stewart, Cell, 14 : 741-759 (1978).
2. W.W. Stewart, J. Am. Chem. Soc., 103 : 7615-7620 (1981).
3. W.W. Stewart, Nature, 292 : 17-21 (1981).
4. British Intelligence Objectives Sub-commitee, B.I.O.S., final report number 959, London, H.M. Stationery Office (about 1946).
5. K. Venkataraman, "The chemistry of Synthetic Dyes", vol. 2, Academic Press, New York, pp. 1188-1189 (1952).
6. M. Scalera and W.S. Forster, U.S. Patent 2 455 095. See Chem. Abstr., 43 : 7710 f (1949).
7. J.N. Miller, Analyst., 109 : 191-198 (1984).
8. N.J. Strausfeld, H.S. Seyan, D. Wohlers and J.P. Bacon, "Functional Neuroanatomy", ed. N.J. Strausfeld, Springer-Verlag, Berlin, 132-155, 386-419 (1983).
9. J.F. Margiotta and B. Walcott, Nature, 305 : 52-55 (1983).
10. K.T. Keyser, B.M. Frazer and C.M. Lent, J. Comp. Physiol., A, 146 : 379-392 (1982).
11. A.D. Murphy, R.D. Hadley and S.B. Kater, J. Neurosci., 3 : 1422-1429 (1983).
12. J.A. Jarrell, Am. J. Physiol., 244 : C419-C421 (1983).
13. G.T. Eddlestone and E. Rojas, J. physiol., 303 : 76P-77P 1980).
14. H.P. Meissner, Nature, 262 : 502-504 (1976).
15. R.L. Michaels and J.D. Sheridan, Science, 214 : 801-803 (1981).
16. I. Findlay and O.H. Petersen, Cell tissue Res., 232 : 121-127 (1983).
17. J.M. Burt, J.S. Frank and M.W. Berns, J. Membrane Biol., 68 : 227-238 (1982).
18. W.C. De Mello , M.G. Castillo and P. van Loon, J. Mol. Cell Cardiol., 15 : 637-643 (1983).
19. H.M. Gerschenfeld, J. Neyton, M. Piccolino and P. Witkovsky, Biomed. Res. Suppl., 3 : 21-34 (1982).
20. J. Skrzypek, Vision Res., 24 : 701-711 (1984).

21. A. Kaneko and A.E. Stuart, Neurosci. Lett., 47 : 1-7 (1984).
22. T. Teranishi, K. Negishi and S. Kato, J. Neurosci., 4
    :1271-1280 (1984).
23. G.L. Brenowitz, W.F. Collins III and S.D. Erulkar, Brain Res.,
    274 : 371-375 (1983).
24. B.A. MacVicar, N. Ropert and K. Krnjevio, Brain Res., 238 :
    239-244 (1982).
25. A.A. Grace and B.S. Bunney, Neurosci., 10 : 333-348 (1983).
26. S.C. Guthrie, Nature, 311 : 149-151 (1984).
27. C.B. Kimmel, D.C. Spray and M.V.L. Bennett, Dev. Biol., 102 :
    483-487 (1984).
28. G. Audesirk, T. Audesirk and P. Bowsher, J. Neurobiol., 13 :
    369-375 (1982).
29. C.W. Lo and N.B. Gilula, Cell, 18 : 411-422 (1979).
30. A.E. Warner and P.A. Lawrence, Cell, 28 : 243-252 (1982).
31. W.D. Knowles, P.G. Funch and P.A. Schwartzkroin, Neurosci., 7 :
    1713-1722 (1982).
32. D.J. Meyer, S.B. Yancey and J.P. Revel, (with appendix by
    Peskoff A.), J. Cell Biol., 91 : 505-523 (1981).
33. K. Willecke, D. Müller, P.M. Drüge, U. Frixen, R. Schäfer, R.
    Dermietzel and D. Hülser, Exp. Cell Res., 144 : 95-113
    (1983).
34. W.R. Loewenstein, Physiol. Rev., 61 : 829-913 (1981).
35. C. Peracchia, Int. Rev. Cytol., 66 : 81-146 (1980).
36. D.F. Mülser and F. Brümmer, Biophys. Struct. Mech., 9 : 83-88
    (1982).
37. S.W. de Laat, L.G.J. Tertoolen and C.J.P. Grimmelikhuijzen,
    Nature, 288 : 711-713 (1980).
38. A.W.C. Dorresteijn, H.A. Wagemaker, S.W. de Laat and J.A.M. van
    den Biggelaar, Roux's Arch. Dev. Biol., 192 : 262-269
    (1983).
39. C.R. Green and N.J. Severs, J. Cell Biol., 99 : 453-463
    (1984).
40. E.S. Schulze and S.H. Blose, Exp. Cell Res., 151 : 367-373
    (1984)
41. T.A. Viancour, G.D. Bittner and M.L. Ballinger, Nature, 293 :
    65-67 (1981).
42. R.A. Satterlie and A.N. Spencer, J. Comp. Physiol. A, 150 :
    195-206 (1983).
43. C. Weber, C.L. Singla and P.A.H. Kerfoot, Cell Tissue Res., 223
    : 305-312 (1982).
44. P.A.V. Anderson, J. Comp. Physiol. B, 154 : 257-268 (1984).
45. L. Keenan and H. Koopowitz, J. Exp. Zool., 215 : 209-213
    (1981).
46. M.B. Pochapin, J.M. Sanger and J.W. Sanger, Cell tissue Res.,
    234 : 309-318 (1983).
47. S.A. Scott and K.J. Muller, Dev. Biol., 80 : 345-363 (1980).
48. H.H. Steinbiss and P. Stabel, Protoplasma, 116 : 223-227
    (1983).
49. J.F. Iles and B. Mulloney, Brain Res., 30 : 397-400 (1971).

50. V.O. Adanina, A.I. Shapovalov, B.I. Shiriaev and Z.A. Tamarova, Neuroscience, 9 : 453-461 (1983).
51. S. Heinrichs, Mikroskopie, 40 : 79-86 (1983).
52. J.H. LaVail, "Neuroanatomical Research Techniques" ed. R.T. Robertson, Academic Press, New York, pp. 355-384 (1978).
53. P. Sloniewski and C. Pilgrim, Neurosci. Lett., 49 : 29-32 (1984).
54. P.V. Sarthy, S.M. Johnson and P.B. Detwiler, J. Comp. Neurol., 206 : 371-378 (1982).
55. P.G. Layer, G. Vollmer and S. Kotz, Neurosci. Lett., 35 : 239-245 (1983).
56. P.V. Sarthy and H.S. Hilbush, Dev. Brain Res., 11 : 275-280 (1983).
57. M. Wilcox and N. Franceschini, Science, 225 : 851-854 (1984).
58. D.K. Miller, E. Griffiths, J. Lenard and R.A. Firestone, J. Cell Biol., 97 : 1841-1851 (1983).
59. T.C. Hohman and B. Bowers, J. cell Biol., 98 : 246-252 (1984).
60. B. Bowers and T.E. Olszewski, J. cell Biol., 97 : 317-322 (1983).
61. E.R. Macagno, K.J. Muller, W.B. Kristan , S.A. DeRiemer, R. Stewart and B. Granzow, Brain Res., 217 : 143-149 (1981).
62. A.R. Maranto, Science, 217 : 953-955 (1982).
63. T.A. Reaves Jr., R. Cumming, M.T. Libber and J.N. Hayward, Neurosci. Lett., 29 : 195-199 (1982).
64. S.M. Schuetze and S. Vicini, J. Neurosci., 4 : 2297-2302 (1984).
65. P.H. Taghert, M.J. Bastiani, R.K. Ho and C.S. Goodman, Dev. Biol., 94 : 391-399 (1982).
66. J.C. Bornstein, M. Costa, J.B. Furness and G.M. Lees, J. physiol., 351 : 313-325 (1984).
67. J.N. Hayward, T.A. Reaves Jr., R.S. Greenwood and R.B. Meeker, Methods in Enzymology, 103 : 132-147 (1983).
68. M. Kawata, Y. Sano, K. Inenaga and H. Yamashita, Histochem., 78 : 21-26 (1983).
69. H. Kettenmann, R.K. Orkand and M. Schachner, J. Neurosci., 3 : 506-516 (1983).
70. H. Kettenmann, M. Wienrich and M. Schachner, Neurosci. Lett., 41 : 85-90 (1983).
71. K.G. Smithson, P. Cobbett, B.A. MacVicar and G.I. Hatton, J. Neurosci. Meth., 10 : 59-69 (1984).
72. M.P. Bailey, B.F. Rocks and C. Riley, Ann. Clin. Biochem., 20 : 213-216 (1983).
73. J.L. Meek, J. Chromatog., 266 : 401-408 (1983).
74. J.P. Miller and A.I. Selverston, Science, 206 : 702-704 (1979).
75. A.I. Selverston and J.P. Miller, J. Neurophysiol., 44 : 1102-1121 (1980).
76. J.S. Eisen and E. Marder, J. Neurophysiol., 48 : 1392-1415 (1982).
77. C.S. Cohan, R.D. Hadley and S.B. Kater, Brain Res., 270 : 93-101 (1983).

78. D.C. Spray, A.L. Harris and M.V.L. Bennett, Science, 204 :
    432-434 (1979).
79. S. Spiegel, M. Wilchek and P.H. Fishman, Biochem. Biophys. res.
    Commun., 112 : 872-877 (1983).
80. C.I. Bowman and H. Tedeschi, Biochem Biophys. Acta, 731 :
    261-266 (1983).
81. A. Darszon, C.A. Vandenberg, M. Schönfeld, M.H. Ellisman, N.C.
    Spitzer and M. Montal, Proc. Natl. Acad. Sci. USA, 77 :
    239-243 (1980).
82. M. Kanou and T. Shimozawa, Stain Technol., 58 : 189-192
    (1983).
83. Y. Kondoh and M. Hisada, J. Exp. Biol., 107 : 515-519 (1983).
84. P.R. Brink, V. Verselis and L. Barr, Biophys. J., 45 : 121-124
    (1984).
85. M.L. Mayer, V. Crunelli and J.A. Kemp, Brain Res., 293 :
    173-177 (1984).
86. P. Gilbert, H. Kettenmann, R.K. Orkand and M. Schachner,
    Neurosci. Lett., 34 : 123-128 (1982).
87. A. Kaneko, J. Physiol., 213 : 95-105 (1971).
88. D.I. Vaney, Proc. Roy. Soc. Lond., B, 220 : 501-508 (1984).
89. G.L. Westbrook and P.G. Nelson, Methods in Enzymology, 103 :
    111-132 (1983).

SECTION V

THE DESMOSOMAL DOMAIN, AN EXAMPLE OF CELL-CELL AS WELL AS MEMBRANE-CYTOSKELETON INTERACTION

Werner W. Franke, Hans-Peter Kapprell and Pamela Cowin

Institute of Cell and Tumor Biology
German Cancer Research Center
D-6900 Heidelberg, F.R.G.

Desmosomes are intercellular junctions characterized by (1) a midline structure containing carbohydrate residues of glycoproteins, (2) the membrane proper, and (3) a cytoplasmic plaque which is insoluble in low and high salt buffers, non-denaturing detergents and thiol agents and is remarkably resistant to treatment with chaotropic and denaturing agents. Major polypeptides of this plaque structure have been characterized by biochemical and both conventional and monoclonal immunological methods. They contain one or two predominant non-glycosylated polypeptides, desmoplakins I (M_r 250K) and II (M_r 215K) of almost neutral isoelectric pH, which are closely related to each other and have been fairly well conserved in diverse epithelia and during evolution. In addition, at least in epidermis, the plaques contain a non-glycosylated polypeptide of M_r 83K ("band 5") isoelectric with serum albumin and non-glycosylated basic polypeptide of M_r 75K ("band 6"). The latter two polypeptides have been identified as translational products of epidermal mRNA. Guinea pig antibodies and monoclonal murine antibodies to such plaque proteins have detected similar antigens in epithelial, arachnoidal and myocardial cells, including Purkinje fibre cells of the heart, as well as in tumors derived from epithelial cells, notably carcinomas. These proteins provide biochemical markers for a category of membrane domains which include diverse morphological appearances such as typical desmosomes, hemidesmosomes, endocytotically internalized desmosomal material and "nascent desmosomes". Other plasma membrane-associated plaque structures, including those associated with junctions of the adhaerens type, do not contain these proteins. In epithelial tissues, desmosomal plaques represent specific membrane-attachment domains anchoring intermediate-sized filaments (IF) of the cytokeratin type. However, desmosomal plaques are able to bind

315

other types of IF. For example, in cardiac cells and Purkinje
fibers desmin IF are found attached to such plaques.
Correspondingly, arachnoidal cells of brain and meningiomas and
certain cultured epithelia-derived cells with very low amounts of
cytokeratin and a relative abundance of vimentin, show attachment
of vimentin IF to these plaque domains. This indicates that the
specificity of attachment to this membrane domain is controlled by
a homologous portion of IF structure. The location of desmoplakins
in the desmosomal plaque has not only been shown by
immunolocalization on fixed or frozen cells but also by
microinjection of desmoplakin antibodies into living cultured
cells.

Epidermal desmosomes also contain two major types of
glycosylated proteins. The "band 3" protein is a group of
glycopolypeptides of apparent M_r 165-175K which is relatively rich
in glucose and galactose. On the basis of amino acid composition,
peptide map glycosylation profile and reaction with both monoclonal
and polyclonal antibodies the band 3 glycopolypeptides are clearly
distinct from those of "band 4". These consist of a number of very
acidic isoelectric variants which migrate in SDS as two distinct
groups ("4a" of $\sim M_r$ 130K and "4b" of $\sim M_r$ 115K). Both monoclonal
and polyclonal antibodies have indicated a close immunological
relationship between these groups and have located them to the
mid-line structure and membranes of the desmosomes. This, together
with their loss during desmosomal splitting by 9.5 M urea,
indicates an important role in desmosomal adhesion for these
glycoproteins. Components 4a and 4b show very similar amino acid
profiles but differ considerably in their glycosylation. Both
polyclonal and monoclonal antibodies indicate that polypeptides 4a
and 4b are major components of desmosomes of stratified tissues,
however, cell type-specificities exist in the expression of these
polypeptides or at least certain epitopes.

As shown by microinjection of mRNA coding for cytokeratins
into cells devoid of desmosomes and containing only vimentin IF,
cytokeratin filaments can be produced and correctly assembled in
the absence of desmosomal structures and proteins, indicating that
desmosomes are not obligatory nucleation sites for IF assembly.
However, in all desmosome-containing cells the association of IF
with the desmosomal plaque is maintained even after splitting of
the desmosome with EGTA or by trypsin treatment, resulting in the
endocytotic internalization of these membrane domains. This
emphasized the high stability of this type of filament-membrane
interaction.

The amino acid and carbohydrate composition of the diverse
desmosomal proteins has been determined and will be discussed in
relation to compositional data of other cytoskeletal and
membrane-bound proteins.

REFERENCES

Cohen, S.M., Gorbsky, G., and Steinberg, M.S., 1983, Immunochemical
    characterization of related families of glycoproteins in
    desmosomes, J. Biol. Chem., 258 : 2621-2627.
Cowin, P., and Garrod, D.R., 1983, Antibodies to epithelial
    desmosomes show wide tissue and species cross-reactivity,
    Nature, 302 : 148-150.
Drochmans, P., Freudenstein, C., Wanson, J.C., Laurent, L., Keenan,
    T.W., Stadler, J., Leloup, R., and Franke, W.W., 1978,
    Structure and biochemical composition of desmosomes and
    tonofilaments isolated from calf muzzle epidermis, J. Cell
    Biol., 79 : 427-443.
Franke, W.W., Schmid, E., Grund, C., Mueller, H., Engelbrecht, I.,
    Moll, R., Stadler, J., and Jarasch, E.D., 1981, Antibodies
    to high molecular weight polypeptides of desmosomes :
    Specific localization of a class of junctional proteins in
    cells and tissues, Differentiation, 20 : 217-241.
Franke, W.W., Moll, R., Müller, H., Schmid, E., Kuhn, C., Krepler,
    R., Artlieb, U., and Denk, H., 1983, Immunocytochemical
    identification of epithelium-derived human tumors with
    antibodies to desmosomal plaque proteins, Proc. Natl. Acad.
    Sci. USA, 80 : 543-547.
Franke, W.W., Moll, R., Schiller, D.L., Schmid, E., Kartenbeck, J.,
    and Müller, H., 1982, Desmoplakins of epithelial and
    myocardial desmosomes are immunologically and biochemically
    related, Differentiation, 23 : 115-127.
Franke, W.W., Müller, H., Mittnacht, S., Kapprell, H.P., and
    Jorcano, J.L., 1983, Significance of two desmosome
    plaque-associated polypeptides of molecular weights 75 000
    and 83 000, EMBO J., 2 : 2211-2215.
Franke, W.W., Kapprell, H.P., and Mueller, H., 1983, Isolation and
    symmetrical splitting of desmosomal structures in 9 M urea,
    Eur. J. Cell Biol., 32 : 117-130.
Geiger, B., Schmid, E., and Franke, W.W., 1983, Spatial
    distribution of proteins specific for desmosomes and
    adhaerens junctions in epithelial cells demonstrated by
    double immunofluorescence microscopy, Differentiation, 23 :
    189-205.
Kartenbeck, J., Schmid, E., Franke, W.W., and Geiger, B., 1982,
    Different modes of internalization of proteins associated
    with adhaerens junctions and desmosomes : experimental
    separation of lateral contacts induces endocytosis of
    desmosomal plaque material, EMBO J., 1 : 727-732.
Kartenbeck, J., Franke, W.W., Moser, J.G., and Stoffels, U., 1983,
    Specific attachment of desmin filaments to desmosomal plaque
    in cardiac myocytes, EMBO J., 2 : 735-742.
Kreis, T.E., Geiger, B., Schmid, E., Jorcano, J.L., and
    Franke,W.W., 1983, De novo synthesis and specific assembly
    of keratin filaments in nonepithelial cells after

microinjection of mRNA for epidermal keratin, <u>Cell</u>, 32 : 1125-1137.

Müller, H., and Franke, W.W., 1983, Biochemical and immunological characterization of desmoplakin I and II, the major polypeptides of the desmosomal plaque, <u>J. Mol. Biol.</u>, 163 : 647-671.

MEMBRANE CYTOSKELETON INTERACTIONS, A MODEL SYSTEM :

THE INTESTINAL MICROVILLI

Hubert Reggio, Daniel Louvard[1] and Evelyne Coudrier[1]

Université d'Aix-Marseille
Centre de Marseille-Luminy
70 Route Léon Lachamp
13288 Marseille Cedex 9, France

Epithelial cells display a striking structural and functional polarity. Inside the cell, the arrangement of different organelles is highly ordered and their function is coordinated. For example, secretory proteins, plasma membrane proteins and at least some of the lysosomal enzymes are assembled in the rough endoplasmic reticulum and then pass through the Golgi complex before reaching their final destination inside or outside the cell. In addition, the plasma membrane contains at least two specialized domains : the apical face which is involved in absorption or secretory processes and is covered with microvilli, whereas the basolateral faces are involved in exchanges with the blood front and are generally free of microvilli. Each domain possesses a characteristic set of proteins specific for their particular function. These two surfaces are separated from one another by junctional complexes that limit the penetration of macromolecules into the intercellular spaces. The tight junction is also throught to prevent lateral diffusion of proteins from one surface domain to the other.

Superimposed on the asymterical organization of the plasma membrane one can observe that specific sets of cytoskeletal proteins are associated with each cell surface specialization. It is reasonable to postulate that this specialisation is the consequence of specific interactions between the two types of structures.

---
1) Unité de Biologie des Membranes, Département de Biologie Moléculaire, Institut Pasteur, 25 Rue du Docteur Roux, 75724 Paris Cedex 15.

We shall review here one of this special type of interaction, between membrane and cytoskeleton in the system of intestinal microvilli. The apical membrane of the absorptive cell of the intestinal epithelium contains numerous well-organized microvilli forming the so called brush border. In these microvilli bundles of microfilaments are connected to their surrounding membranes by means of end-on association and by lateral bridges oriented perpendicularly to the axis of the core filaments. These filaments contain actin (43 kd) associated with a specific set of cytoskeletal proteins : calmodulin (17 kd), fimbrin (69 kd), villin (95 kd) and a 110 kd protein. Villin and fimbrin are actin binding proteins which are involved in the control of bundle formation while the 110 kd protein connects laterally the core filaments to the lateral membrane of each microvillus. An amphipathic membrane glycoprotein (M_r 140 kd) is consistently found in the Triton X-100 insoluble pellet associated with the core structure of microvilli. This protein can be solubilized by dialysis against a low ionic strength buffer in the presence of chelating agents or by controlled proteolysis with papain using right side out sealed vesicles. In this way, the 140 kd protein can be purified to homogeneity. Antibodies were prepared against the 110 kd and the 140 kd proteins in order to study their localization by immunocytochemical techniques and their topological organization with respect to the membrane bilayer.

Triton X-114 extraction of the papain of low ionic strength solubilized forms led to the partition of the 140 kd protein in two different phases. The material extracted at low ionic strength bound to the detergent, whereas the protease-solubilized form did not. Both forms comigrate on SDS polyacrylamide gels. These observations are consistent with the presence of a large glycosylated hydrophilic domain accessible from the outside to papain and of a small hydrophobic domain presumably involved in anchoring this protein to the microvillar membrane. These structural features strikingly resemble those described for aminopeptidase, a major amphipathic transmembrane glycoprotein of microvilli. However, this newly described protein has a distinct behaviour because of its strong association with microfilaments. We have recently demonstrated a specific binding of the amphipathic form of this protein with the 110 kd protein (a major polypeptide of the lateral bridges) in vivo. These observations suggest that this 140 kd polypeptide is a transmembrane protein and might provide in vivo attachment sites for the cytoskeletal lateral bridges.

RELATED ANTIGENS IN SKELETAL MUSCLE

On the basis of structural observations it has been proposed that the cytoskeleton of intestinal microvilli could be related to striated muscle in terms of its organization (Tilney and Mooseker, 1976). This hypothesis was tested by immunocytochemical (frozen sections) and immunochemical techniques (Western blotting).

Thin frozen sections of striated muscle were prepared according to the technique of Tokuyasu and visualized by indirect immunofluorescence using antibodies against the 100 kd and 140 kd proteins. These antibodies gave a specific staining pattern which was similar in both cases. In longitudinal sections the labelling was concentrated mainly in the area of the I band, in cross sections a honeycomb pattern was observed suggesting that the recognized antigens were most probably associated with the periphery of the myofibrils. Thin frozen sections prepared for electron microscopy revealed that both antigens were closely associated with the membrane of the sarcoplasmic reticulum.

In muscle extracts the anti-140 kd antibodies recognized a protein of 100 kd which comigrates with the Ca^{++}ATPase in the muscle extract, they also bind to a purified preparation of sarcoplasmic reticulum Ca^{++}ATPase. Particularly interesting is the fact that this cross reactivity lies in the domains of the two molecules which contain the membrane spanning hydrophobic segment.

Although these proteins, expressed in cells of totally different function, have a different molecular weight, they share a common structural domain responsible for their cross reactivity. This domain may be responsible for a common function. Such a function could be the lateral bridging of actin filaments to membranes. This would be accomplished by the interaction of a protein of the 110 kd type (an actin binding protein) with an anchoring membrane protein (140 kd or Ca^{++}ATPase).

REFERENCES

Bretscher A. and Weber K., 1980 a, Villin is a major protein of the microvillus cytoskeleton, which binds both G and F actin in a calcium-dependent manner, Cell, 20 : 839-847.
Bretscher A. and Weber K., 1980 b, Fimbrin, a new microfilament-associated protein present in microvilli and other cell surface structures, J. Cell Biol., 86 : 335-340.
Coudrier E., Reggio H. and Louvard D., 1981, Immunolocalization of the 110,000 molecular weight cytoskeletal protein of intestinal microvilli, J. Mol. Biol., 152 : 49-66.
Howe C.L. and Mooseker M.S., 1983, Characterization of the 110-kdalton Actin calmodulin and membrane-binding protein from microvilli of intestinal epithelial cells, J. Cell Biol., 92 : 974-985.
Matsudaira P.T. and Burgess D.R., 1979, Identification and organization of the components in the isolated microvillus

cytoskeleton, J. Cell Biol., 83 : 667-673.

Matsudaira P.T., Mandelkow E., Renner W., Hesterberg L.K. and Weber
      K, 1984, Role of fimbrin and villin indetermining the
      interfilament distance of actin bundles, Nature, 301 :
      209-214.

Mooseker M.S, Pollard T.D. and Fujiwara K., 1978, Characterization
      and localization of myosin in the brush border of intestinal
      epithelial cells, J. Cell Biol., 79 : 444-453.

Mooseker M.S., Bonder E.M., Conzelman K.A., Fishkind D.J., Howe
      C.L. and Keller T.C.S., 1984, III. Brush border cytoskeleton
      and integration of cellular functions, J. Cell Biol., 99 :
      104S-112S.

Reggio H., Coudrier E., Tokuyasu K. and Louvard D., 1984,
      Ca^{2+}-ATPase of the sarcoplasmic reticulum shows a common
      domain with a membrane glycoprotein associated with the
      cytoskeleton of microvilli, Proc. Natl. Acad. Sci. USA, 81 :
      1130-1134.

Tilney L.G. and Mooseker M.S., 1971, Actin in the brush border of
      epithelial cells of the chicken intestine, Proc. Natl. Acad.
      Sci. USA, 68 : 2611-2615.

Tokuyasu T.K., 1973, A technique for ultracryotomy of cell
      suspension and tissue, J. Cell Biol., 57 : 551-565.

JUNCTIONAL COMMUNICATION AND

COMMUNICATION COMPARTMENTS IN DEVELOPMENT

John D. Pitts, Malcolm E. Finbow, T. Eldridge,
J. Buultjens, Ephraim Kam and John Shuttleworth

The Beatson Institute for Cancer Research
Garscube Estate, Switchback Road
Bearsden, Glasgow G61 1BD, Scotland

INTRODUCTION

Two quite different lines of research led to independent discoveries of junctional communication between animal cells. Neurophysiologists interested in the interactions between nerve cells showed the presence at some sites of low resistance pathways or junctions which allow the direct movement of current carrying ions between the cytoplasms of adjacent cells (1). This work was extended in two important ways to show first that these low resistance junctions are also present (and much more frequently) between cells in non-excitable tissues (2, 3), and second that certain small molecular weight fluorescent dyes could also pass through the intercellular junctions (2, 3). Since these early observations, electrical coupling and dye coupling have been extensively used to define the nature and roles of the junctional pathway and these studies are described in other chapters of this book.

The other discovery of junctional communication is yet another example of the role of serendipity in scientific research. John Subak-Sharpe, Bobby Burk and John Pitts (4-6) were setting up reconstruction experiments to test the feasibility of detecting hypoxanthine-guanine phosphoribosyl transferase positive ($HGPRT^+$) transformants of $HGPRT^-$-BHK cells by autoradiography. A few (1 in 10^6 to 1 in 10^2) wild-type $HGPRT^+$ cells were added to cultures of $HGPRT^-$ cells in the presence of ^{3}H-hypoxanthine. Only those cells containing active HGPTR were expected to incorporate the radioactive precursor into nucleic acid (see the metabolic pathways shown in Fig. 1) so after acid fixation and autoradiography only

323

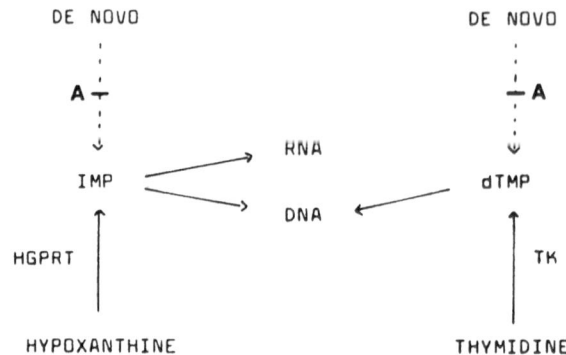

Fig. 1   Metabolic pathways leading to purine
          nucleotide and dTMP synthesis. A –
          pathway blocked by aminopterin.

the HGPRT+ cells should be covered with black grains.

    When the autoradiographs were examined, it was easy, as
anticipated, to locate even 1 in 10^6 black cells, but surprisingly,
each black cell was surrounded by a ring of cells showing an
intermediate level of black grains. Either the HGPRT+ cells were
present in groups and in each group only one cell was incorporating
the expected high level of ^{3}H-hypoxanthine, or else HGPRT− mutant
cells around each wild-type had incorporated the precursor even
though they had no enzyme. The effect was shown to be due to the
latter, that is to the phenotypic modification of the mutant cells
by contact, either directly or through other mutant cells, with the
HGPRT+ cells. This effect was termed metabolic cooperation, a
fortunate choice of term as it turned out.

METABOLIC COOPERATION

    The phenomenon of metabolic cooperation could in theory be
explained by the transfer from the HGPRT+ to the HGPRT− cells of
any of the following : ^{3}H-purine nucleotides, ^{3}H-RNA, the enzyme
HGPRT, the mRNA or gene to make the enzyme or of some control
molecule (the molecular basis of the HGPRT− lesion was not known).
Further work showed, however, that nucleotide transfer was the
correct explanation (7-9).

    Direct experiments showed that neither the enzyme HGPRT nor
any protein labelled with ^{3}H-leucine, is transferred in amounts
which can be detected between these or any other cells tested
(7-9).

The simplest way to show nucleotide transfer between cultured cells is to prelabel cells with ^{3}H-uridine, wash the cells to remove unused uridine and then co-culture the labelled, washed cells (called donor cells) with unlabelled cells (called recipient cells). ^{3}H-uridine nucleotides are transferred to recipient cells in contact with donors, either directly or through other recipient cells, and these ^{3}H-nucleotides are then incorporated into recipient cell nucleic acid which can be detected by auto-radiography (see ref. 9).

The washed donor cells contain both ^{3}H-nucleotides and ^{3}H-RNA, but if these are kept in unlabelled medium for a period (say 24 h), the radioactive nucleotide pool (which is replenished by the breakdown of short-lived ^{3}H-nuclear RNA species) is eventually chased into stable forms of RNA (mostly ribosomal RNA). Such chased donors, when co-cultured with unlabelled cells show little transfer of radioactivity. In every case the extent of transfer (measured by grain counting) from the donor cells to the recipients is directly proportional to the activity of the donor uridine nucleotide pool and is quite unrelated to the activity of the donor RNA (see ref. 9).

The tansfer of ^{3}H-uridine nucleotides between cells in culture functionally defines the presence of intercellular junctions. The cytoplasmic membranes of cells are impermeable to nucleotides (as they are to all metabolites, control molecules, macromolecules, etc., which have to be retained within the cell), so the movement of nucleotides between cells means that the membranes at the points of contact must contain some structure or mechanism to permit the transfer. Subsequent work (10, 11) has shown that this structure is the gap junction.

It has also been shown these junctions are permeable to a wide range of different metabolites, small ions, small molecular weight control molecules, and a vitamin derived co-factor tetrahydrofolate (12, 13) but not to RNA, DNA or protein (see Table 1).

The junctions are also permeable to small molecular weight dyes injected into the cells but not the large dyes. All molecules with a molecular weight less than about 900 which have been tested pass through while all larger molecules do not. It seems reasonable to conclude that the junctions contain channels which are permeable to all small molecules below are certain cut-off size (M_r 900) but not to larger molecules. This is consistent with the known structure of the gap junction (12, 13) which is described by Gilula in this volume. Gap junctions are areas of close cell-cell contact containing closely packed particles (connexons) each of which is made of protein, in the form of a cylinder with a central hole or channel, which traverses both cytoplasmic membranes and the small intercellular space to join the two cytoplasmic compartments.

Table 1. Permeability   of vertebrate gap
          junctions. For further information,
          see Refs. 9, 12 and 24.

Pass through the junctional channels	Do not pass through the junctional channels

nucleotides
sugar phosphates
choline phosphate
amino acids
tetrahydrofolate
small inorganic
   ions
	DNA
	RNA
	protein
	tetrahydro-
	   folate-glutamate$_4$

dyes M_r < 900	dyes M_r > 900

CONSEQUENCES OF JUNCTIONAL COMMUNICATION BETWEEN CELLS

1. Electrotonic synaptic transmission. Gap junctions between
heart cells and between some nerve cells allow, by the movement of
ions through the junctional channels, the transmission of action
potentials from one cell to the next. Transmission via these
electrotonic synapses is significantly faster than transmission via
chemical synapses as the former avoids the transmitter release,
diffusion, receptor binding and activation steps.   Electrical
transmission, on the other hand, is probably less susceptible to
modulation but there are several instances where the junctional
mechanism has been preferred during evolution. It is the pathway
for transmission in heart and smooth muscle. The contractions in
these tissues require less control than in skeletal muscle where
gap junctions between the muscle fibres would result in all or
nothing contractions (so they disappear during the final stages of
myoblast differentiation). Junctional communication is also used
for nervous transmission in a limited number of specialised
circumstances. The electrical conductance of gap   junctional
channels is discussed in more detail elsewhere in this volume.

2. Phenotypic modification. In any cell population or tissue
where cells express different enzymic activities, and where the
products of such enzymes are small enough to pass through the gap
junctional channels, the cells can share each other's activities.

An example of this, which in principle is the same as the
metabolic cooperation seen in culture, was first reported by Moor

and his colleagues (14). The mammalian oocyte, during its maturation, is surrounded in the follicle by cumulus cells. Processes from these cells penetrate the intervening barrier (zona pellucida) and form gap junctions with the oocyte. Isolated oocyte cumulus complexes incorporate various radioactive precursors (including uridine and inositol) into macromolecules of both the cumulus cells and the oocyte. The two cell types can be physically separated after which the cumulus cells still incorporate these precursors while the oocyte does not. Different precursors (e.g., cycloleucine) are incorporated into the oocyte whether or not it is associated with cumulus cells. It appears that the developing oocyte does not express the activities required for the incorporation of some precursors and that it overcomes these deficiencies by using the intermediate metabolites produced in the cumulus cells and then transferred to the oocyte by way of gap junctions.

     3. Tissue phenotypes. The exchange of metabolites or other small cellular ions or molecules between cells joined by gap junctions means that no one cell has to produce all that is required for growth or survival. The population can have some characteristic property (tissue phenotype), such as survival, which is not seen in any of the separated cells. For example, neither HGPRT$^-$ cells nor cells lacking thymidine kinase (TK$^-$ cells ; see Fig. 1) can survive in medium containing aminopterin, a drug which blocks the de novo synthesis of purine nucleotides and dTMP.

     However, mixed cultures of the two cell types do grow, by mutual nucleotide exchange (Fig. 2). The HGPRT$^-$ cells provide the dTMP for both cell types while the TK$^-$ cells provide all the purine nucleotides. Excess of either cell type is starved of a required metabolite (as there are too few of the other cell type to provide the necessary amount), so the population tends to a self-stabilizing 1 : 1 mixture.

     At the moment, there is no example of such tissue phenotypes in vivo, although because there is no evolutionary constraint on the production of new phenotypes (by changes in gene expression in differentiated cells) against loss of activities such as those used in the model system, it seems almost inevitable that such situations will have arisen. Examples may be hard to find as it is not unusual for cells to change their properties when isolated, or indeed for cells to fail to grow under such conditions. It is normally assumed that these changes are an artefact of cell culture techniques but they may, as suggested, be reflecting important tissue features.

     The model system, using mixed cultures of HGPRT$^-$ and TK$^-$ cells also allows the calculation of the rate of metabolite exchange between cells coupled by gap junctions. All the purine nucleotides

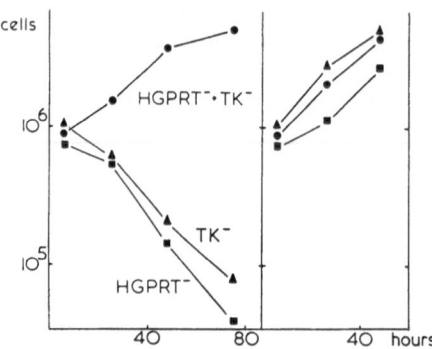

Fig. 2   A tissue phenotype. The panel on the left is cell
          growth in the presence of aminopterin and the panel
          on the right is cell growth without the drug. In the
          presence of aminopterin both mutant cell cultures
          degenerate and detach from the culture dish and are
          lost. The mixed culture survives and grows at a rate
          similar to those of all the cultures without aminop-
          terin. For experimental conditions see Ref. 8 except
          that aminopterin was added at twice the concen-
          tration.

required for the growth of the HGPRT⁻ cells are derived by
junctional transfer from the other cell type. It is simple to
calculate the total number of purine nucleotides present in the DNA
of these cells and, knowing that the division time under these
conditions is about 15 h to estimate the number of purine
nucleotide molecules which must be transferred per second per cell
pair. The number is about 10^6 but this is likely to be a serious
underestimate for two reasons. First, much of the RNA made in
animal cells is unstable so the amount synthesised during one cell
cycle will be more than the figure used in the calculation and,
second, the traffic through the junctions is two-way. It is
sufficiently fast in the required direction to allow cell growth
but it may be much faster in both directions. Furthermore, all
other metabolites, ions and small molecular weight cell components
will be moving back and forth at similar rates, so it is clear that
when thinking of junctional communication, a huge traffic must be
considered and not just trivial 'leakage' between adjacent cells.

    4. Stabilization by junctional communication. Again it seems
obvious that in a population of cells where each is able separately
to exert imperfect control on say the intracellular level of a
metabolite or ion, the joint control by all the cells, if the
metabolite or ion is shared between them, will be much better. An
example of this is given by the Ca^{++}-dependent potassium channels
of pancreatic acinar cells. These channels are voltage and

Ca^{++}-dependent but because there are only 50 per cell and under normal calcium conditions only some are open, activity in any one cell is unstable. When many cells are coupled, however, fluctuations in activity are evened out to produce stable levels of response (15). The evolution of such finely modulated membrane channels may not have been possible in the absence of the stabilizing property afforded by gap junctions.

5. Intercellular control of activity. An enzyme activity which is influenced by the concentration of a small intracellular ion or molecule will be under the control of the tissue level of that ion or molecule in coupled cell populations. An increase or decrease in activity by any one cell type in the tissue can therefore be compensated for by changes in activity in another cell type (16).

6. The tissue as a polyplast. It is often a mistake to attempt to coin new words, but the recent appreciation of the consequences of junctional communication may be best understood by the introduction of the term polyplast. A syncytium results from cell fusion and all the molecules (small and large) and organelles share a common cytoplasm (as in skeletal muscle cells). In the polyplast, a very much more common form of tissue organization than the syncytium, all the cell are joined by gap junction and therefore share common pools of all small molecular weight cytoplasmic components but the macromolecules and organelles remain segregated within the cells where they were synthesised (or in their daughter cells after cell division).

The polyplast is now a concept which must always be borne in mind when considering tissue function and intercellular signalling.

ROLE OF JUNCTIONAL COMMUNICATION IN DEVELOPMENT

1. Communication compartments. Various mammalian cell types (e.g., liver epithelial cells and fibroblasts) when grown in mixed cultures, segregate into areas containing either one cell type or the other. Each group of homologous cells is a coupled 'compartment' (i.e., all the cells are joined by gap junctions) which is separate from adjacent coupled compartments of the other cell type. Boundaries between such compartments are characterized by much reduced communication between the cells on opposite sides. This reduced heterologous communication appears to be due to a lower rate of junction formation between heterologous cells (17).

Intercellular signalling by molecules which use the junctional pathway (i.e., small molecules which cannot cross the cytoplasmic membrane) is thus restricted by the compartmentation. The boundaries, even though there is a low level of heterologous

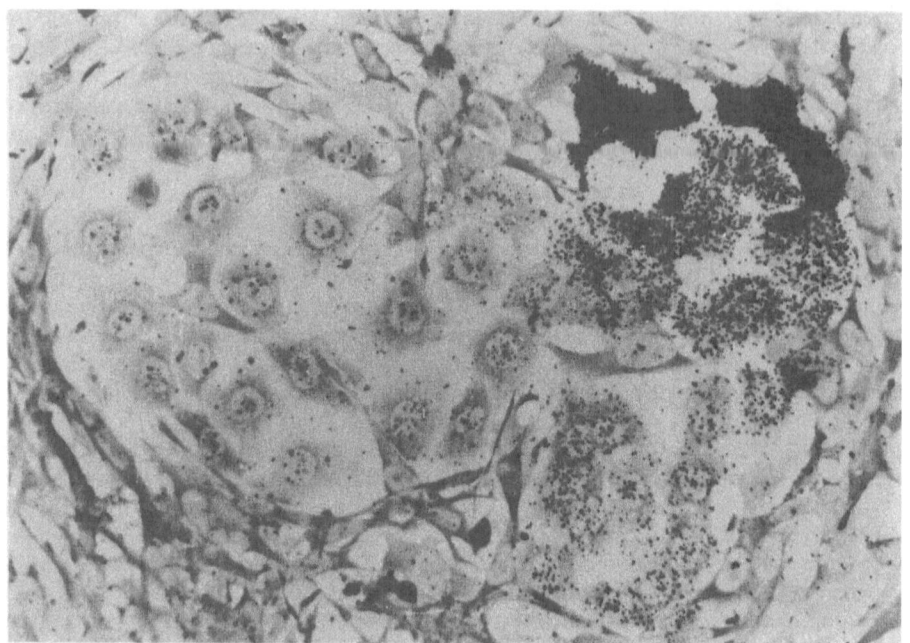

Fig. 3   Communication compartments in tissue culture.
          ^{3}H-uridine nucleotides have passed from the added,
          prelabelled donor cells (black cells) to other epi-
          thelial cells but not detectably to the fibro-
          blasts. For experimental conditions see Ref. 17.

communication, are accentuated because relatively slow leakage
across a boundary is accompanied by faster intercellular movement
on the far side in the next highly coupled population. Small
molecules (analogous to the uridine nucleotides diffusing through
cells shown in Fig. 3) can set up gradients and influence cell
activity sufficiently quickly to account for many of the
developmental phenomena thought to be due to signalling by
positional information (18). Several hundred cells can receive
detectable levels within just a few minutes.

     There is as yet no direct evidence that junctional
communication does provide a mechanism for developmental
signalling, but it is of interest that in insect systems, groups of
cells known to share common developmental fates and thus forming
separate developmental compartments (19), also in the few instances
which have been analysed, form communication compartments with
similar properties to the model described above (20-22).

     2. Proliferative control. Proliferative control must play an

important role in the processes which mold the development of the
tissues of an embryo and in this context, it is interesting to note
that the developmental compartments described above are, at least
in one type of instance, also elements of proliferative control. It
is possible to induce mutations in cells in Drosophila larvae, so
that the mutant cells carry a characteristic marker (e.g., bristle
morphology) and also divide at a faster rate than the other cells.
In such cases (23) the faster growing cells fill the compartment
but do not transgress across the boundaries into adjacent
compartments.

The idea of proliferative control at the level of the
communication compartment is very attractive, as individual cells
are able to sense, in a variety of ways (see Ref. 24) the total
volume and hence the cell number in the coupled population. A
signal molecule (small and which cannot cross the membranes) if
made for a short period and asynchronously in the different cells
in a small coupled population, will diffuse into the surrounding
cells and reach a final concentration which will be less the more
cells it has to diffuse into (i.e., the greater the compartment).
If it must reach a certain concentration to activate the processes
leading to the cell cycle, the cells will only divide if the
population is small enough for that critical concentration to be
reached. The eventual size of the population will depend on the K_m
of the signal molecule for its receptor, a property which will be
sensitive to mutations in the receptor protein. Compartments can
therefore evolve which grow to different terminal sizes, an
important requirement of any proliferative control theory.

3. Control of differentiation. It has frequently been
suggested that embryonic induction may operate by signal molecules
which pass through gap junctions. From our present knowledge of
compartmentation, such junctional communication between cells in
different tissue types may seem unlikely, but it must be emphasised
that the reduced level of communication across boundaries may still
be important and sufficient to allow tissue-tissue signalling. As
yet, however, there is no evidence to support the suggestion.

There is better information in a somewhat different system of
differentiation control. Thymic cells appear to be instrumental in
directing the differentiation of immature, uncommitted
T-lymphocytes along different pathways (towards helper, suppressor
or cytotoxic T-cells). Peripheral blood lymphocytes, like certain
other terminally differentiated cells (skeletal muscle cells and
some nerve cells) have lost the ability to form gap junctions, but
it has recently been shown (25) that immature thymocytes (which
will form the T-cells) can couple to other cells. This fits
particularly well with the idea put forward (26) that the lethal
accumulation of deoxynucleotides found in developing thymocytes
(due to losses of specific enzymic activities) may be relieved by

metabolic cooperation with the thymic stromal and accessory cells
(which have the necessary activities). In this way, the thymic
tissue cells could selectively rescue the T-cells and thus
influence   the   rates of access to different differentiation
pathways.

SIGNAL MOLECULES WHICH USE THE JUNCTIONAL PATHWAY

       It is relatively easy to purify and identify extracellular
signal molecules such as hormones. Even though they occur in low
concentrations, they may fill significant parts of the body volume.
The specificity of this signalling system, which is not contained
within specific cell populations, is governed by the distribution
of the receptors. When the hormones are isolated, even in crude
form and at low concentration, they can be detected by adding them
back to whole animals, tissues or cells and measuring the response
they produce.

       Signal molecules which operate via the junctional pathway will
be much more difficult to identify. It is a requirement that they
cannot cross the cytoplasmic membranes and they are likely to be
produced in amounts which provide low concentrations in just a few
hundred cells or so. If such small amounts can be isolated, they
will have no effect when added back to tissues or intact cells
because they cannot enter across the membranes ; to see the effect,
needed to identify their presence, they will have to be injected
into sensitive cells. This introduces a new range of experimental
difficulties which will have to be overcome before these molecules
can be found.

INHIBITORS OF JUNCTIONAL COMMUNICATION

       In the absence of any chemical identification of morphogens or
signal molecules which use the junctional pathway, it is necessary
to consider alternative ways of studying the role of gap junctions
in development.

       If specific inhibitors of gap junction formation, or gap
junction permeability were available, these could be used to block
the pathway at specific times during embryogenesis and the
consequences used to define the role of junctional communication.
Unfortunately, to date no such specific inhibitors have been found
although there are several ways in which the junctional channels
can be rapidly and reversibly closed (or gated). The channel is
sensitive to pH (closing at more acid pH, usually brought about by
exposing the cells or tissue to carbon dioxide (27)), to voltage
(28), to octanol (29) and retinoic acid (30). None of these

treatments or reagents can reasonably be beleived to only affect gap junctions and therefore cannot be used in the manner described.

The most optimistic approach in the search for a specific inhibitor is to raise antibodies to the structural protein of the gap junction.

The major protein associated with gap junctions extracted from a wide variety of tissues from many different vertebrate sources has a molecular weight of 16,000 (31, 32). There are various other reports in the literature which suggest that the junctional protein has a different molecular weight but the only other popular contender is a protein of molecular weight 27,000 (for a description of this protein see the article by Gilula). At the moment there is no definitive evidence to decide between these two. However, antisera raised against chicken liver gap junctions and against lobster hepatopancreas gap junctions react with the mouse and chicken 16K gap junctional proteins and with the lobster 18K junctional protein (33) in Western blot analyses (but not with the 27K protein). Antibodies affinity purified against the purified proteins bind to isolated junctions and the binding can be visualised by protein A or anti-rabbit IgG coated gold particles in EM cytochemical analysis (32).

The antibodies do not block junction formation when added to the medium of cultured cells but this is probably to be expected as vertebrate cells from apparently any species can form junctions with each other as well as they can among themselves. It is likely therefore that at least a part of the extracellular face of the junctional protein, that part which is involved with the interaction with a similar component on the adjacent cell to form a new cell-cell junctional channel, is highly conserved and as a consequence will not be immunogenic (because the rabbits will be tolerant).

When the antibodies are injected into cells, on the other hand, they rapidly (whithin a minute) block cell-cell communication (34). This is an interesting result of great potential value, but it will still take further work to make certain that the antibodies which block communication when injected into cells (only a few million molecules should be sufficient) are the same as the major component in the sera or affinity purified sera which is detected by Western blotting. But when this has been done, the technology will be available to begin to unravel the roles of junctional communication in developmental signalling. This approach is described in greater detail in the next chapter.

ACKNOWLEDGEMENT

This work was supported by the Cancer Research Campaign.

REFERENCES

1. E.J. Furshpan and D.D. Potter, J. Physiol., 145:289-325
    (1959).
2. W.R. Loewenstein and Y. Kanno, J. Cell. Biol., 22:565-586
    (1964).
3. E.J. Furshpan and D.D. Potter, Curr. Topics Dev. Biol.,
    3:95-127 (1968).
4. J.H. Subak-Sharpe, R.R. Burk and J.D. Pitts, Heredity, 21:342
    (1966).
5. J.H. Subak-Sharpe, R.R. Burk and J.D. Pitts, J. Cell Sci.,
    4:353-367 (1969).
6. R.R Burk, J.D. Pitts and J.H. Subak-Sharpe, Exp. Cell Res.,
    53:297-301 (1968).
7. R.P. Cox et al., Proc. Natl. Acad. Sci. USA, 67:1573-1579
    (1970).
8. J.D. Pitts, Ciba Found. Symp. Growth Control in Cell Cultures,
    pp. 89-105 (1971).
9. J.D. Pitts and J.W. Simms, Exp. Cell Res., 104:153-163 (1979).
10. N.B. Gilula, O.R. Reeves and A. Steinbach, Nature, 235:262-265
    (1972).
11. M.E. Finbow, Brit. Soc. Cell Biol. Symp., 5:1-37 (1982).
12. M.E. Finbow and J.D. Pitts, Exp. Cell Res., 131:1-13 (1980).
13 J.D. Sheridan, M.E. Finbow and J.D. Pitts, Exp. Cell Res.,
    123:111-119 (1979).
14. L.D. Makowski et al., J. Cell Biol., 74:605-628 (1977).
15. P.N.T. Unwin and G. Zampighi, Nature, 283:545-549 1980).
16. O.L. Petersen, Cold Spring Harbor Banbury Proc., (in press).
17. J.D. Pitts and E. Kam, Exp. Cell Res., (in press) (1984).
18. L. Wolpert, in: Intercellular Junctions and Synapses, eds. J.
    Feldman, N.B. Gilula and J.D. Pitts (Chapman and Hall, London),
    pp. 81-86 (1978).
19. F.H.C. Crick and P.A Lawrence, Science, 189:340-347 (1975).
20. A.E. Warner and P.A. Lawrence, Cell, 28:243-252 (1982).
21. M.P. Weir and C.W. Lo, Proc. Natl. Acad. Sci. USA, 79:3232-3235
    (1982).
22. S. Caveney and R. Berdan, in: Insect Ultrastructure, eds. R.C.
    King and H. Akai (Plenum, New-York), pp. 434-470 (1982).
23. G. Morata and P. Ripoll, Dev. Biol., 42:211-222 (1975).
24. W.R. Loewenstein, Biochim. Biophys. Acta, 560:1-65 (1979).
25. E. Carolan and J.D. Pitts, (in preparation).
26 D.D.F. Ma et al., Immunol. Today, 3:65-68 (1983).
27 L. Turin and A.E. Warner, Nature, 270:56-57 (1977).
28. A.L. Harris, D.C. Spray and M.V.L. Bennett, J. Neurosci.,
    3:79-100 (1983).
29. D.C. Spray et al., Proc. VII Int. Biophys. Cong., pp. 155
    (1981).
30. J.D. Pitts, R.R. Burk and J.P. Murphy, Cell Biol. Int. Rep., 5
    (Supp. A):45 (1981).

31. M.E. Finbow, J. Shuttleworth, A.E. Hamilton and J.D. Pitts, EMBO J., 2:1479-1486 (1983).
32. M.E. Finbow, T.E.J. Buultjens, E. Kam, J. Shuttleworth and J.D. Pitts, Cold Spring Harbor Banbury Proc. (in press) (1984).
33. M.E. Finbow, T.E.J. Buultjens, N.J. Lane, J. Shuttleworth and J.D. Pitts, EMBO J., 3:2271-2278 (1984).
34. T.E.J. Buultjens, E. Kam, M.E. Finbow and J.D. Pitts, (in preparation).

GAP JUNCTIONAL COMMUNICATION BETWEEN CELLS DURING DEVELOPMENT

C.R. Green and N.B. Gilula

Department of Cell Biology, Baylor College of Medicine
One Baylor Plaza
Houston, TX 77030, U.S.A.

INTRODUCTION

In the quarter century since low resistance intercellular pathways were first discovered (1), considerable knowledge has been gained about gap junctions, the membrane specializations generally considered to provide the means for ionic and metabolic coupling between cells. Since most of this work has been extensively reviewed in recent years (2-12), we only intend to provide a general background to gap junction communication, with selected examples, concentrating more on the potential function of this specific cell-cell interaction event in developing systems.

GAP JUNCTION STRUCTURE AND COMPOSITION

In thin section electron microscopy, gap junctions are detected as regions of closely apposed plasma membranes separated by a 2-3 nm intercellular gap. In tangential view, they appear as macular regions of subunits, each termed a 'connexon' (13), 8-9 nm in diameter with a central 2 nm electron dense core, slightly larger in some arthropod junctions. This is especially evident when the junctions have been isolated and negatively stained (Fig. 1.).

In freeze-fracture replicas, gap junctions exist as specialized regions of the plasma membrane with plaques of intramembrane particles (or complementary pits) corresponding to the connexons seen in thin section or negative stain studies (Fig. 2.). The partitioning and packing of the particles varies with tissue type and the preparative procedures used, but in most cases, they occur on the P fracture face (internal membrane half) and are

337

338   C.R. GREEN AND N.B. GILULA

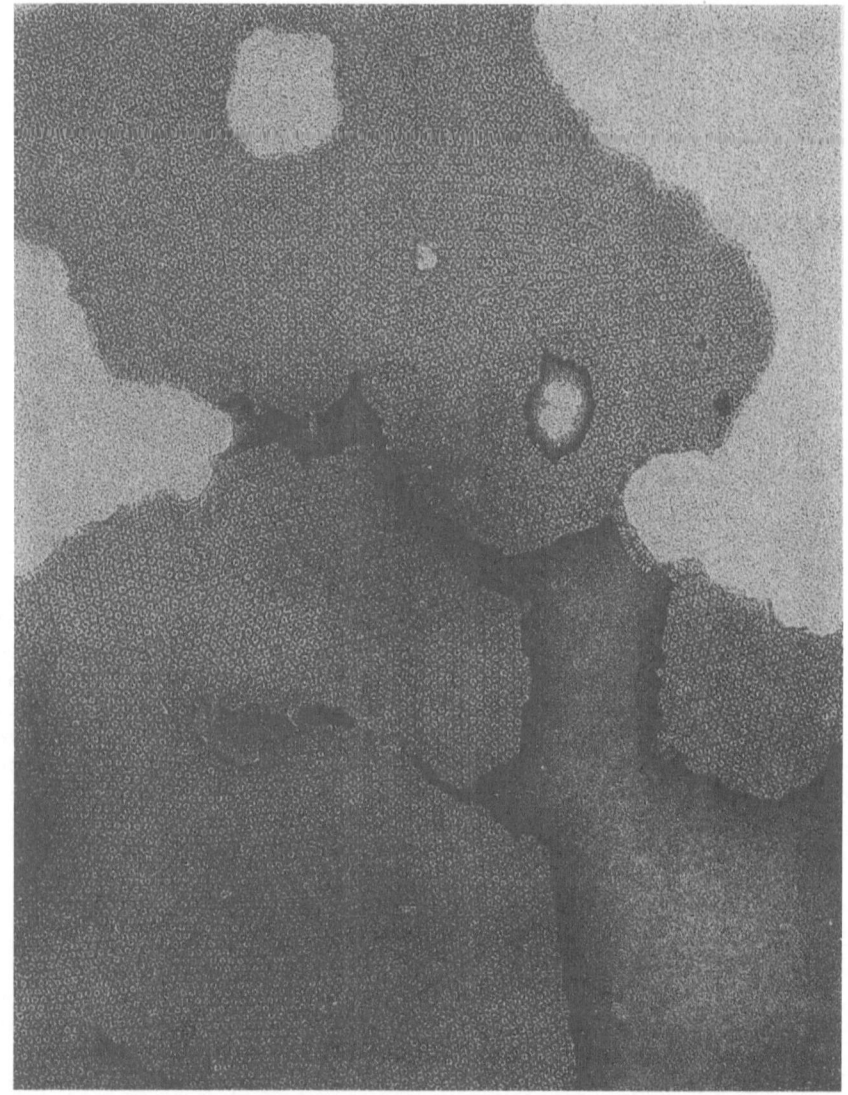

Fig. 1. Negative stained gap junctions isolated from rat
         liver. The stain outlines the connexons and fills
         their 2 nm diameter central core. Each connexon has
         a total diameter of 8 nm. X200,000.

7-8 nm in diameter. Again, the arthropods are a notable exception
with particles not only on the opposite fracture face (the E face),
but also larger and heterogeneous in size. For some time, it was
thought that the arrangement of the connexons reflected the

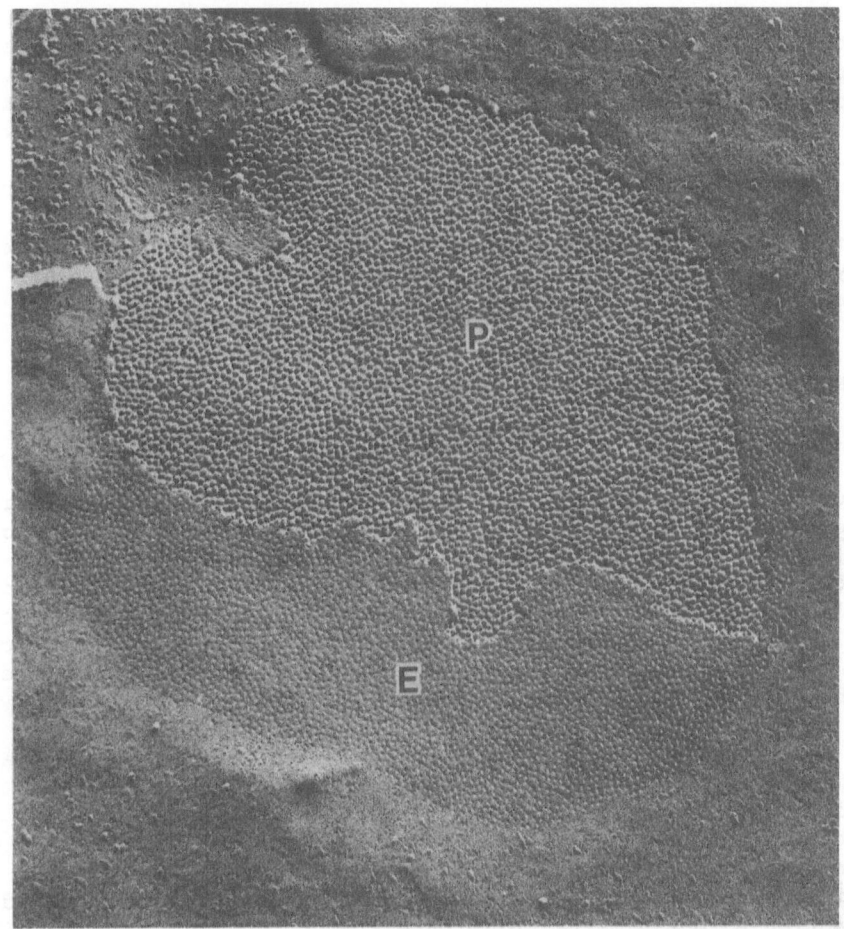

Fig. 2.   A freeze-fracture replica of a gap junction in rat
           pancreatic   tissue.   The junctional particles (each
           corresponding to a connexon) are on the P face (P)
           with pits visible where portions of the E face (E)
           have been fractured. X100,000.

functional state of the junctions ; randomly arrayed particles
indicating coupled junctions (at the time of freezing), and packed
hexagonal arrays reflecting the uncoupled state. This does not,
however, appear to be a consistent feature (14) and any structural
changes related to junctional permeability are more likely to exist
at the level of individual connexons (15, 16).

     Several laboratories have now isolated gap junctions in
sufficient quantities to obtain some chemical information and to

permit the production of antibodies directed against the major junctional polypeptide (7, 17-20). In most cases, the major polypeptide of the mammalian liver junction is described as having an apparent molecular weight of 26-28,000, although not without exception (17). Using immunoblot analysis and immunolocalization techniques, it is now evident that the junctional protein is highly conserved in many tissues and across several animal phyla. Although some amino acid sequence data is available (21), the major gap junction protein (as distinct from the eye lens junction protein) has yet to be fully sequenced as a protein or via recombinant DNA approaches.

INTERCELLULAR COMMUNICATION VIA GAP JUNCTIONS

The presence of gap junctions, observed cytologically, has been correlated with ionic coupling and the ability of cells to communicate by the transfer of low molecular weight molecules in a large number of systems (e.g., ref. 31). Three primary techniques have been used to demonstrate the concommitant existence of communication pathways : electrophysiological measurement of ionic coupling, dye transfer in which low molecular weight fluorescent dyes are injected into cells and cell-cell dye spread monitored, and autoradiography in which the transfer of small radiolabeled metabolites between cells is assayed.

The initial physiological description of low resistance pathways between cells was in invertebrate nervous tissue (1). The rapid transfer of current pulses between cells via electrical or electrotonic synapses, with very little voltage attenuation, has advantages over chemical synapses when rapid responses are required. In myocardial tissue, gap junctions allow the spread of the action potential from cell-to-cell synchronizing the heart into a functional syncytium (22, 23). The synchronization of myocardial cells in tissue culture is also contact dependent with the cytological appearance of gap junctions occurring concommitantly (24). While these are more obvious examples of ionic coupling, non-excitable tissues also have similar pathways (25-27). Since there is no apparent requirement for electrical synchronization in these tissues, it is presumed that cell-cell communication can be involved in other phenomena such as growth control, development and differentiation (32). Electrophysiological studies have demonstrated that the junctional pathways have resistance properties similar to cytoplasm (10-100 Ω cm^2) as opposed to the high resistance non-junctional membrane (10^6-10^8 Ω cm^2). The ions responsible for the observed current transfer are most probably small and inorganic (Na$^+$, K$^+$, or Cl$^-$).

Metabolic coupling and the transfer of molecular components between cells is also well documented in several tissues. Cyclic

AMP, for example, which mediates the effects of some hormones, is apparently transferred readily from cell-to-cell via the junctions (24, 28, 29). Similarly, the tranfer of metabolites is contact dependent (30, 31). In such co-culture experiments, a donor cell type can 'rescue' recipient cells (metabolically deficient) by transferring labeled metabolites, presumably small nucleotides, necessary for their survival. Conversely, a mouse cell line, the A9 derivative of the L cell, is unable to form gap junctions under the experimental conditions used and is therefore unable to be 'rescued' (31).

GAP JUNCTION PERMEABILITY

The size of the junction channels is finite ; large poly-nucleotides or polypeptides are unable to be transferred from cell-to-cell (33). Therefore, although the cytoplasms of adjacent cells are in close contact and the transfer of molecules up to approximately 1500 daltons is possible, each cell retains its own identity. The actual size of the junctional channels has been probed using fluorescein isothiocyanate (FITC) covalently attached to neutral linear oligosaccharides, neutral branched glycopeptides and charged linear peptides. From the molecular dimensions of molecules transferred between cells in culture, it has been possible to deduce that the channel diameter is 1.6-2 nm in mammalian cells and 2-3 nm in insect cells (from insect salivary gland) (34). These figures are not inconsistent with those obtained from structural studies (16) or electrophysiological studies (35). There is a possibility that cells can regulate these channel diameters, since a channel size increase allowing synchronization of a tissue response has been reported following the application of hormones (35).

Several factors have been implicated in regulating or gating the gap junction channels, implying that the junctional channels between cells are not continuously open. In some systems, junctional conductance can be altered by applied voltage potentials. Amphibian embryos, for example, can be dissected to leave coupled cell pairs in which the application of a current pulse to one of the cells causes a voltage increase within in, but a decrease in the neighboring cell (12). Since this effect is symmetrical (current of either sign can be applied to either cell with analogous results), the active element is assumed to be the junctional membrane. It has been suggested that the junction voltage-dependence in the amphibian blastomeres may provide a mechanism for gating intercellular communication during tissue differentiation ; a boundary between developing regions being formed if permeability changes, electrogenic pumping or other factors shift gap junctions past a threshold to an uncoupled state (12). A similar voltage-dependent gating has been observed in the

Limulus ommatidium (36). However, gap junction conductivity in several other systems (both vertebrate and invertebrate) seems independent of transjunctional potential changes (reviewed in 12).

Increase in cytoplasmic acidity can also decrease junctional conductivity. In fact, intracellular pH dependent gating appears to be common to many tissues (12) with only slight cytoplasmic acidification resulting in a substantial decrease in junctional conductivity. Two good examples of this effect come from experiments carried out in the late 1970's. In 32 cell stage Xenopus embryos, the lowering of pH from about 7.4 to 6.8 by elevating carbon dioxide levels in the surrounding medium causes a rapid but reversible uncoupling of intercellular communication, measured electrophysiologically (37). Similarly, cardiac Purkinje fibers show electrical uncoupling with lowered pH levels (38). In this latter experiment, pH was lowered by the intracellular injection of hydrogen ions.

The third major candidate proposed for direct gap junction regulation is calcium, an increase in cytoplasmic calcium ion concentration leading to a coupling decrease in a variety of systems. This is apparently true for both excitable tissues, such as electrotonic synapses in the buccal ganglion of Navanax (39), cardiac cells (40), and non-excitable tissues (Chironomus salivary gland (41)). Calmodulin, a calcium binding protein which regulates a number of calcium dependent processes (42), has subsequently been demonstrated to bind to purified gap junction polypeptides in vitro (43). Furthermore, drugs which block calmodulin can inhibit coupling in an insect epidermis (44), further indicating a potential reglatory role for calcium.

However, while voltage gating and pH effects appear at present to be separate phenomena (12), the separation of calcium ion and pH effects is less evident. Turin and Warner (37) could detect no increase in intracellular calcium levels (using ion selective electrodes) during their pH regulation experiments in Xenopus, but Rose and Rick (41), using combined pH electrode and aequorin studies on Chironomus salivary glands, did find a variable increase in intracellular calcium concentration with pH induced uncoupling. Conversely, a calcium concentration increase may act independently of pH (41), or result in lowered intracellular pH levels (37, 12) indicating that Ca^{++} induced uncoupling could operate via a pH change. Therefore, the problem remains in separating all these gating factors from one another, and from other perturbations they may be causing within a cell (resulting in a junctional conductivity change which is a secondary rather than direct effect).

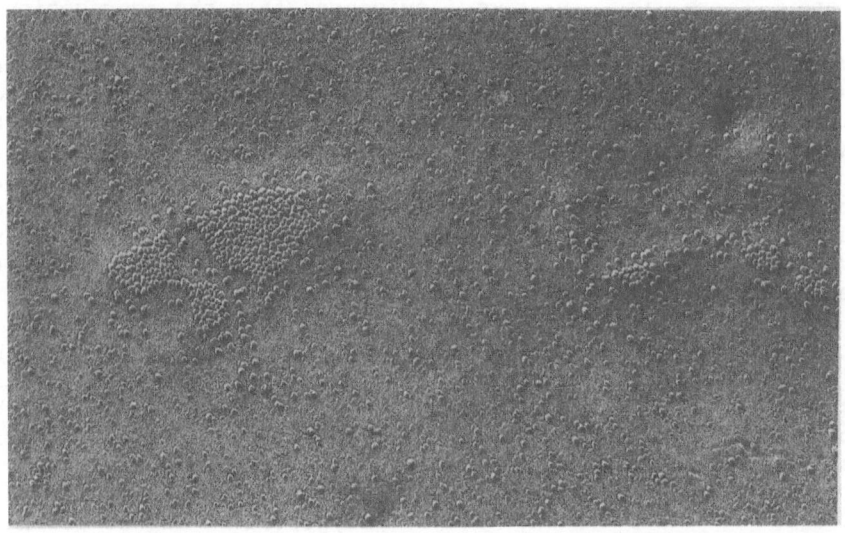

Fig. 3. Gap junction formation in the mammalian ovary. This
         freeze-fracture replica shows small aggregates of
         junctional particles at sites where junctions are
         presumably forming. These will subsequently enlarge
         by the accretion of other connexons migrating to the
         "formation" sites. X65,000.

GAP JUNCTION SYNTHESIS AND TURNOVER

     In addition to the gating of existing channels, communication
between cells, especially in developing tissues, may be modulated
by the rate of junction formation and subsequent removal and
degradation. Unfortunately, despite the advances made in the
structural, physiological, and biochemical characterization of gap
junctions, relatively little information is available on their
synthesis and turnover.

     Freeze-fracture studies have indicated that junction formation
tends to follow a set sequence. Namely, the appearance of a
flattened   formation   plaque   into   which small aggregates of
junctional particles gather, then an enlargement of these
aggregates by the accretion of additional particles (Fig. 3. ;
refs. 45-47). The onset of junctional communication can occur very
rapidly once adjacent cells come into contact, within minutes (and
possibly even seconds) in culture (48, 49), or may appear over
several hours (50). This means that in some cases at least, the
formation and assembly of gap junctions must virtually be
independent of transcriptional and translational events, being
brought about by utilizing pools of pre-existing material in the

membrane. There is, in fact, evidence that the aggregation of junctional particles may be under cytoskeletal control (51, 52). Junction breakdown or removal can be by a reversal of this process with particles dispersing away from the junctional plaques under conditions which result in uncoupling (53, 54), but in other instances, 'internalization' of entire junctions by endocytosis occurs (e.g., ref. 55). Even under relatively stable conditions, junctions are turned over ; the half-life of the mouse liver gap junctional polypeptide is only 5 hours (56). Hence, minor alterations to formation or turnover rates may greatly affect the degree of intercellular communication.

On the other hand, several studies have demonstrated that partial hepatectomy results in an initial loss of gap junction structures (up to 35 hours after surgery) with a reappearance within 48 hours (see ref. 57). Traub et al. (58) have used liver junction antiserum to follow the expression of junctional protein during this cycle of events. Their antibody blotting procedure showed a substantial reduction of the polypeptide in 36-45 hours post-operative samples with a return to normal levels by 48 hours. These results have been interpreted to represent degradation and resynthesis of junctional protein after surgery, rather than a dispersal and reassembly of junctional plaques.

Control of communication may, therefore, be occurring at several levels. Theoretically, cell shape changes, biosynthesis of junctional components (protein and lipid), the assembly of channels, the aggregation of channels at intercellular contact sites, and subsequent channel removal could all play a role. The influence of many of these processes could be regulated, in turn, by post-translational modifications to the junctional protein. It is also conceivable that other cellular factors such as metabolism, pH and calcium ion concentration will effect the rate at which these processes occur, in addition to their gating effects on intact channels. It is only with the recent production of specific molecular probes that some progress is now being made in resolving issues of this type (58-60).

GAP JUNCTION COMMUNICATION DURING DEVELOPMENT AND DIFFERENTIATION

Gap junctions can be located very early on during embryogenesis and are believed to play a vital role in growth control, pattern formation and differentiation in developing systems (2). With the exception of one recent study (20), the principal evidence fot this remains correlative, but is none the less substantial. Junctions can be located in regions where cell-cell interactions might be expected during embryogenesis and are absent or show reduced coupling between areas which form separate developmental compartments (reviewed in 2).

Gap junctions at the segmental border in insect epidermis for example, have different permeability properties from those between cells within the same segment (61, 62). Warner and Lawrence (62) examined cell-cell communication between epidermal cells of milkweed bug (Oncopeltus) and blowfly (Calliphora) larvae using ionic coupling measurements and the transfer of injected molecules to map the pattern of intercellular communication. They reported that while all cells were ionically coupled, spread of the dye lucifer yellow (MW 450) at the segmental borders was restricted. In both species, dye was readily transferred between cells within a segment, but not across the compartmental border (in all cases in Calliphora and in 90% of the Oncopeltus preparations examined). On the other hand, the slightly smaller and more compact anion, lead-EDTA (MW 374), passed freely across the segmental boundary. Similar results were reported by Blennerhassett and Caveney (61), but in addition, their study of Oncopeltus indicated the existence of a specific cell type with reduced junctional permeability at the boundary. In both studies, the authors proposed that a modulation of junction permeability, rather than a reduction in channel numbers, was the controlling factor.

Another arthropod study which demonstrates the existence of communication distinct developmental compartments is that of Weir and Lo (63). Ionophoretic injection of small fluorescent molecules (sodium fluorescein, MW 330 ; sodium carboxyfluorescein, MW 376 ; dilithium lucifer yellow CH, MW 450) into Drosophila wing imaginal discs indicated that dye spread was not uniform, the movement of dye from cell-to-cell being partially restricted at several defined boundaries. The 'communication compartments' mapped appeared to match those identified by cell lineage studies. While the authors were unable to discern between the possibility of there being fewer junctions between cells at the boundaries as opposed to channels with reduced permeability, their patterns of communication do reflect a probable role for gap junctions in regulating developmental processes within the imaginal disc. This role is, in fact, probably assumed much earlier in Drosophila development. Even though nuclear divisions in the embryo are initially acellular, gap junctions appear prior to cellularization of the blastoderm (in preparation ; see also ref. 64).

Experiments of this type also indicate the presence of developmental compartments in the mollusc Patella. De Laat et al. (65) have used lucifer yellow CH transfer to demonstrate that regional and time-specific intercellular coupling in this species can be correlated with determinative events, in particular, dorso-ventral polarity.

Similarly, within the vertebrates, intercellular communication may provide a mechanism for regulating developmental events. In the

mammalian ovarian follicle, a complete cycle of development and
differentiation involving gap junction interactions is represented.
Using electrophysiological and morphological techniques, Gilula et
al. (66) demonstrated that the granulosa cells of the cumulus
oophorus communicate extensively with the oocyte at different
stages of follicular development. The ionic coupling however,
decreases and is lost by the time of ovulation, indicating that
communication regulates the maturation of the oocyte during
follicular development prior to ovulation. Upon disruption of
coupling, the oocyte is released from meiotic arrest (67). Gap
junctions are also present in the early mouse embryo from the 8
cell stage onwards and intercellular coupling coincides with
blastomere compaction (68). Following implantation, at least two
communication compartments can be detected in vitro; the inner
cell mass destined to be the embryo proper, and the trophoblast
region which forms part of the placenta. As with arthropod
compartmentalization, the reduction in coupling is not complete.

     The major problem which has faced investigators has been the
lack of specific probes for selectively disrupting junctional
communication without eliciting secondary, metabolic effects. With
the recent advances in gap junction isolation techniques allowing
the production of junctional peptide antisera (and hopefully, other
molecular probes in the near future), it is now possible to study
the influence of intercellular communication more directly.
Already, researchers in two laboratories have demonstrated that
specific antibodies can be used to uncouple junctions (20, 69).
Warner et al. (20) used them to demonstrate, for the first time,
the direct effect of a lack of communication during vital stages of
amphibian development. In this experiment, antibodies to the major
protein of rat liver gap junctions (MW 27,000) were microinjected
into one identified cell of an 8 cell stage Xenopus embryo. The
selection of the cell injected was based on earlier work by Guthrie
(70) in which she demonstrated that cell-cell transfer of lucifer
yellow at the 32 cell stage was not uniform within the animal pole,
transfer being maximal near the dorsal side and minimal ventrally.
Following injection of the antibody into a cell which would
normally give rise to four well coupled progeny at the 32 cell
stage, it was found that intercellular coupling had been reduced or
abolished, communication being monitored by dye transfer and
electrophysiological techniques. This antibody-induced disruption
of communication provides direct evidence that the protein
(antigen) obtained from isolated rat liver gap junctions is
directly involved in the physiological coupling pathway.
Furthermore, embryos left to develop to stage 36 (Nieuwkoop and
Faber staging - see ref. 20) produced characteristic defects, the
main features being the absence of the eye and lack of brain
development on the side previously injected with antibody. A
blockage of intercellular communication, at least during the early
part of development, can therefore result in a failure of cells to

differentiate, presumably because vital 'signals' are not transmitted from cell-to-cell.

CONCLUSION

It is apparent that the modulation of gap junction channel numbers and/or the transfer of molecules through the junctions provides a mechanism for regulating growth, development and differentiation. To date, there is little knowledge of the composition of specific signals or metabolites crossing the junctional membrane, the exception being in Hydra where morphogens with a molecular weight range of 500-1100 daltons have been identified (71). These at least fall within the known permeability limits of the junctions. However, it should soon be possible to identify determinants which control the functional properties of gap junctions, to identify the molecules which pass through them to direct development and differentiation, and certainly, to further analyze the consequences of blocking intercellular communication during developmental processes.

ACKNOWLEDGEMENTS

The research carried out in this laboratory has been supported by grants from the National Institutes of Health (General Medical Sciences and Heart-Lung) and the Welch Foundation. The authors sincerely appreciate the excellent secretarial assistance from Ms. Suzanne Saltalamacchia.

REFERENCES

1. E.J. Furshpan and D.D. Potter, J. Physiol., 145 : 289-325 (1959).
2. S. Caveney, Ann. Rev. Physiol., 47 : 319-335 (1985).
3. M.E. Finbow, In: "The Functional Integration of Cells in Animal Tissues", eds. Pitts J.D. and Finbow M.E., Cambridge University Press, Cambridge, pp. 1-37 (1982).
4. N.B. Gilula and E.L. Hertzberg, In: "The Liver : Biology and Pathobiology", eds. Arias I., Popper H., Schaehter D. and Shafritz D.A., Raven Press, New York, pp. 615-623 (1982).
5. E.L. Hertzberg, T.S. Lawrence and N.B. Gilula, Ann. Rev. Physiol., 43 : 479-491 (1981).
6. E.L. Hertzberg, Ann. Rev. Physiol., 47 : 305-318 (1985).
7. J. Hope, A. Zervos and W.H. Evans, In: Matrices and Cell Differentiation", eds. Kemp R.B. and Hinchliffe J.R., Alan R. Liss, Inc., New York, pp. 261-274 (1984).
8. J.P. Revel, B.J. Nicholson and S.B. Yancey, Ann. Rev. Physiol., 47 : 263-279 (1985).

9. R.M. Schultz, Biol. Repro., 32 : 27-42 (1985).
10. J.D. Sheridan and M.M. Atkinson, Ann. Rev. Physiol., 47 : 337-353 (1985).
11. D.C. Spray and M.V.L. Bennet, Ann. Rev. Physiol., 47 : 281-303 (1985).
12. D.C. Spray, R.L. White, A.C. Campos de Carvalho, A.L. Harris and M.V.L. Bennet, Biophys. J., 45 : 219-230 (1984).
13. D.A. Goodenough, J. Cell Biol., 68 : 220-231 (1976).
14. C.R. Green and N.J. Severs, J. Cell Biol., 99 : 453-463 (1984).
15. P.N.T. Unwin and P.D. Ennis, J. Cell Biol., 97 : 1459-1466 (1983).
16. P.N.T. Unwin and G. Zampighi, Nature, 283 : 545-549 (1980).
17. M.E. Finbow, J. Shuttleworth, A.E. Hamilton and J.D. Pitts, The EMBO J., 2 : 1479-1486 (1983).
18. R. Dermietzel, A. Leibstein, U. Frixen, U. Janssen-Timmen, O. Traub and K. Willecke, The EMBO J., 3 : 2261-2270 (1984).
19. E.L. Hertzberg and R.V. Skibbens, Cell, 39 : 61-69 (1984).
20. A.E. Warner, S.C. Guthrie and N.B. Gilula, Nature, 311 : 127-131 (1984).
21. B.J. Nicholson, M.W. Hunkapiller, I.B. Grim, L.E. Hood and J.P. Revel, Proc. Natl. Acad. Sci. USA, 78 : 7594-7598 (1981).
22. S. Weidmann, J. Physiol., 118 : 348-360 (1952).
23. N.J. Severs, In: "Advances in Myocardiology", eds. Harris P. and Poole-Wilson P.A., Plenum, London, Volume 5, pp. 223-242 (1985).
24. T.S. Lawrence, W.H. Beers and N.B. Gilula, Nature, 272 : 501-506 (1978).
25. E.J. Furshpan and D.D. Potter, Curr. Top. Devel. Biol., 3 : 95-127 (1978).
26. R.G. Johnson and J.D. Sheridan, Science, 174 : 717-719 (1971).
27. W.R. Loewenstein, S.J. Socolar, S. Higashino, Y. Kanno and N. Davidson, Science, 149 : 295-298 (1965).
28. R.W. Tsien and R. Weingart, J. Physiol., 260 : 117-141 (1976).
29. S.A. Murray and W.H. Fletcher, J. Cell Biol., 98 : 1710-1719 (1984).
30. R.P. Cox, M.R. Kraus, M.E. Balis and J. Dancis, In: "Cell Communication",ed. R.P. Cox, Wiley, New York, pp. 67-96 (1974).
31. N.B. Gilula, O.R. Reeves and A. Steinbach, Nature, 235 : 262-265 (1972).
32. L. Wolpert, In: "Intercellular Junctions and Synapses", eds. Feldman J., N.B. Gilula and J.D. Pitts, Chapman and Hall, London, pp. 83-94 (1978).
33. J.D. Pitts and J.W. Simms, Exp. Cell Res., 104 : 153-165 (1977).
34. G. Schwarzmann, H. Weigandt, B. Rose, A. Zimmerman, D. Ben-Haim and W.R. Loewenstein, Science, 213 : 551-553 (1981).
35. R.C. Berdan and S. Caveney, Cell Tissue Res., 239 : 111-122 (1985).

36. T.G. Smith and F. Baumann, Prog. Br. Res., 31 : 313-349 (1969).
37. L. Turin and A.E. Warner, J. Physiol., 300 : 489-504 (1980).
38. W.C. DeMello, Cell Biol. Int. Rep., 4 : 51-58 (1980).
39. G. Baux, M. Simonneau, L. Tauc and J.P. Segundo, Pro. Natl. Acad. Sci. USA, 75 : 4577-4581 (1978).
40. W.C. DeMello, J. Physiol., 250 : 231-245 (1975).
41. B. Rose and R. Rick, J. Membr. Biol., 44 : 377-415 (1976).
42. L.J. VanEldik, J.G. Zendegui, D.R. Marshak and D.M. Watterson, Intl. Rev. Cytol., 77 : 1-61 (1982).
43. L.J. VanEldik, E.L. Hertzberg, R.C. Berdan and N.B. Gilula, Biochem. Biophys. Res. Commun., 126 : 825-832 (1985).
44. J.P. Lees-Miller and S. Caveney, J. Membr. Biol., 69 : 233-245 (1982).
45. R.S. Decker, J. Cell Biol., 69 : 685-699 (1976).
46. R. Johnson, M. Hammer, J. Sheridan and J.P. Revel, Proc. Natl. Acad. Sci. USA, 71 : 4536-4540 (1974).
47. J.D. Sheridan, In: "Intercellular Junctions and Synapses", eds. Feldman J., Gilula N.B. and Pitts J.D., Chapman and Hall, London, pp. 37-60 (1978).
48. M.L. Epstein and N.B. Gilula, J. Cell Biol., 75 : 769-787 (1977).
49. W. Michalke and W.R. Loewenstein, Nature, 232 : 121-122 (1971).
50. J. Flagg-Newton and W.R. Loewenstein, J. Membr. Biol., 50 : 65-100 (1979).
51. G. Tadvalkar and P. Pinto da Silva, J. Cell Biol., 96 : 1279-1287 (1983).
52. C.R. Green and N.J. Severs, Cell Tissue Res., 237 : 185-186 (1984).
53. N.J. Lane and L.S. Swales, Cell, 19 : 579-586 (1980).
54. W.M. Lee, D.G. Cran and N.J. Lane, J. Cell Sci., 57 : 215-218 (1982).
55. R.D. Ginzberg, and N.B. Gilula, Develop. Biol., 68 : 110-129 (1979).
56. R.F. Fallon and D.A. Goodenough, J. Cell Biol., 90 : 521-526 (1981).
57. D.J. Meyer, S.B. Yancey and J.P. Revel, J. Cell Biol., 91 : 505-523 (1981).
58. O. Traub, P.M. Druge and K. Willecke, Proc. Natl. Acad. Sci. USA, 80 : 755-759 (1983).
59. D.L. Paul and D.A. Goodenough, J. Cell Biol., 96 : 636-638 (1983).
60. E.H. Williams, N.M. Kumar and N.B. Gilula, J. Cell Biol., 95 : 386 a (1982).
61. M.G. Blennerhassett and S. Caveney, Nature, 309 : 361-364 (1984).
62. A.E. Warner and P.A. Lawrence, Cell, 28 : 243-252 (1982).
63. M.P. Weir and C.W. Lo, Develop. Biol., 102 : 130-146 (1984).
64. S. Eichenberger-Glinz, Roux's Arch. Dev. Biol., 186 : 333-349 (1979).

65. S.W. de Laat, L.G.J. Tertoolen, A.W.C. Dorresteijn and J.A.M. van den Biggelaar, Nature, 287 : 546-548 (1980).
66. N.B. Gilula, M.L. Epstein and W.H. Beers, J. Cell Biol., 78 : 58-75 (1978).
67. N. Dekel, T.S. Lawrence, N.B. Gilula and W.H. Beers, Develop. Biol., 86 : 356-362 (1981).
68. C.W. Lo and N.B. Gilula, Cell, 18 : 399-409 and 411-422 (1979).
69. E.L. Hertzberg, D.C. Spray and M.V.L. Bennett, Proc. Natl. Acad. Sci. USA, 82 : 2412-2416 (1985).
70. S.C. Guthrie, Nature, 311 : 149-151 (1984).
71. H.C. Schaller and H. Bodenmuller, Proc. Natl. Acad. Sci. USA, 78 : 7000-7004 (1981).

RELATION BETWEEN STRUCTURE AND FUNCTION OF GAP JUNCTIONS

Françoise Mazet

Laboratoire de physiologie comparée et de
physiologie cellulaire associée au CNRS
Université Paris XI
91405 Orsay, France

It has become clear over the past fifteen years that the cells
in organized tissues commonly form interconnected systems. The
cell-to-cell interaction arises when the plasma membrane of
individual cells contact each other. At the level of these contacts
a specialization of the cell surface takes place, which exhibits
various structural differentiations depending on the cell-to-cell
interaction. The name gap junction is now widely accepted for the
cell-to-cell interaction where the two adjacent cells can exchange
part of their molecular content (metabolic coupling) or different
ions (ionic coupling). Several reviews have been published on the
structural and functional aspects of the gap junction (1-2-3).

The first indication of an electrical coupling between cells
came from work with electrical excitable tissues. In 1952 Weidmann
(4) working on heart muscle found that a voltage shift in one cell
caused a parallel shift in a contigous cell, as if they were linked
by a conductive pathway. Similar results have been found in
different tissues later (conf. 3). This transmission probably
occurs at the level of close apposition between adjacent cell
membranes. At that time, in thin sections of fixed and embedded
material, electron microscopic (E.M.) observations revealed areas
of close apposition of various extensions of adjoining plasma
membranes (Fig. 1). These junctional regions appear as a
pentalaminar profile due to the tight apposition of the two
opposite outer layers of each trilaminar junctional plasma
membrane.

After the discovery by Kanno and Loewenstein (5) that not only
small ions but large hydrophilic molecules (ranging from 300 to

351

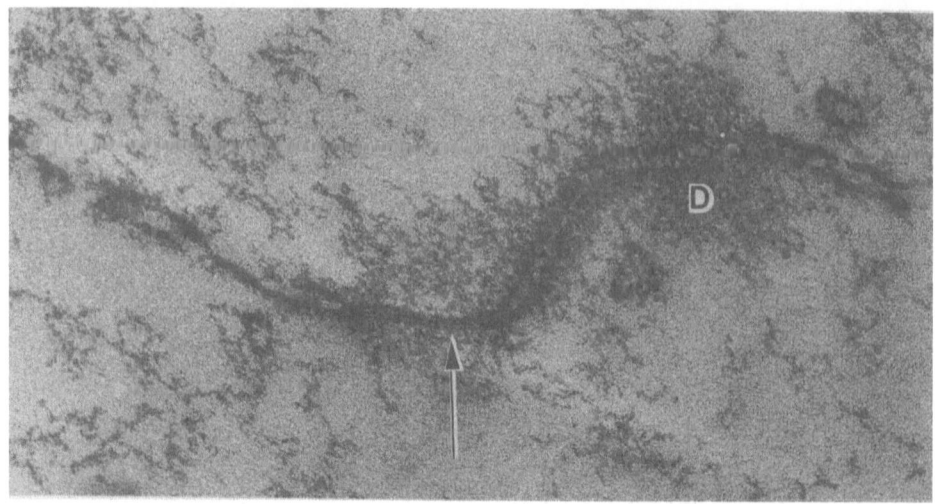

Fig. 1.   Thin section of embryonic chicken cardiac cells sho-
          wing the close apposition of the two opposite outer
          layers of each trilaminar junctional plasma membrane
          (arrow). Desmosome : D. (X 120 000).

1000 daltons) can flow through the junctional pathway it became
necessary to think of them as specialized channels linking the
cells (conf. 6).

     In 1966 Loewenstein (6) proposed a model for the junctional
membrane channel. According to this hypothesis each channel unit is
composed of a pair of permable elements, one from each opposing
cell membrane, and tightly joined.

     At the same time, the morphological data, using colloïdal
lanthanum elctron opaque tracers (7) as a "marker" of the
intercellular space, demonstrated a pattern of interlocking
entities between the junctional membrane. The isolation of gap
junctional membranes and their observation by E.M. after negative
staining clearly permits detection of a network of thin
anastomosing entities (9 to 10 nm in diameter) which bridge the
intercellular space (8).

     The application of freeze-fracture techniques has brought new
information showing that the lattice of subunits penetrates the
lipid hydrophobic core of both junctional membranes. On the
protoplasmic freeze-fracture face (P.F.) the junctional area is
characterized by a polygonal lattice of 9 nm particulate entities
at an average center to center distance of 10 nm. The extracellular
fracture face (E.F.) displays a complementary arrangement of pits
or depressions (Fig. 2). Typical gap junctions are usually found as

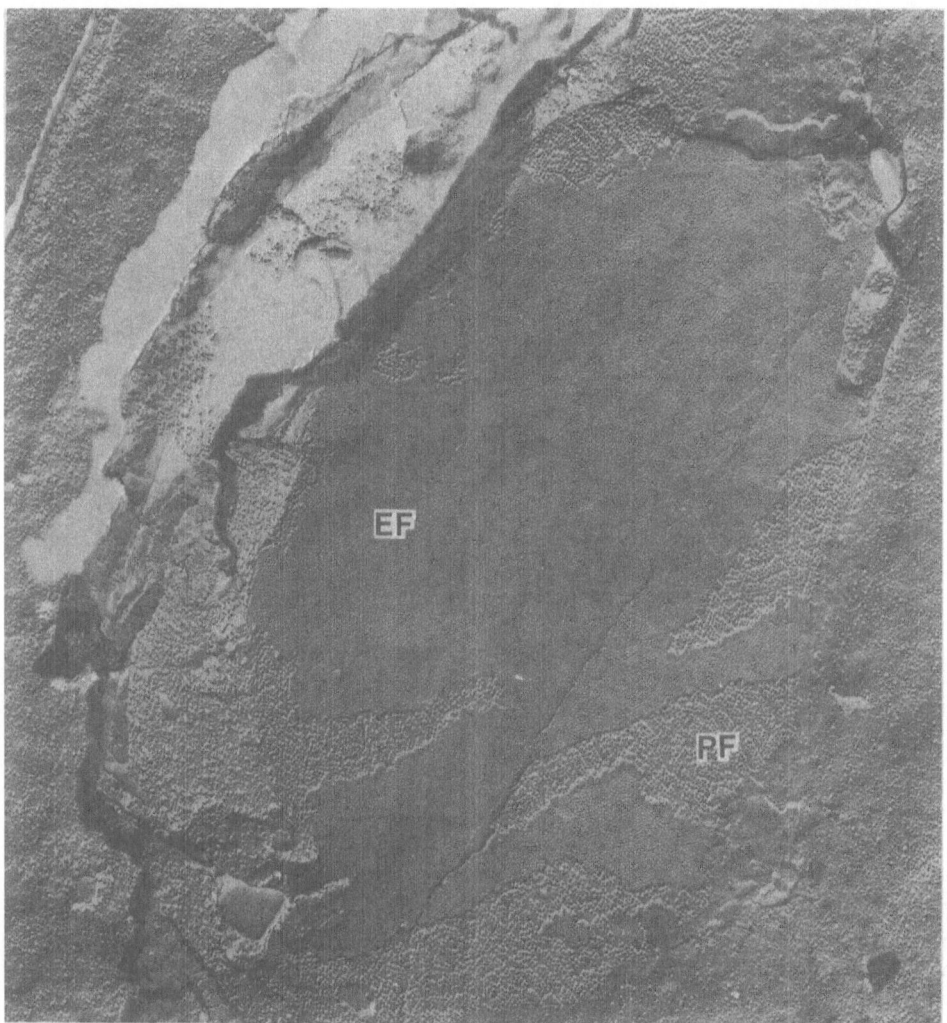

Fig. 2.   Adult rat   cardiac cell. P and E fracture   face of a
          large   gap   junction formed mainly by closely packed
          particles   on   the   P-face   with   complementary   de-
          pressions on the E face. (X 48 000).

round   membrane   patches composed of particles that are homogeneous
in shape and size arranged in more or less   hexagonal   arrays.   The
size   of   individual   gap   junctions as well as the total areas per
cell-to-cell interface may   vary   considerably.   For   instance,   in
mammalian   heart the cell surface area occupied by gap junctions is
approximately 3.7 %, in liver : 1.5 %, in   brown   fat   :   1   %,   in
fibroblasts : 0.5 % and in smooth muscle : 0.2 % (9-10).

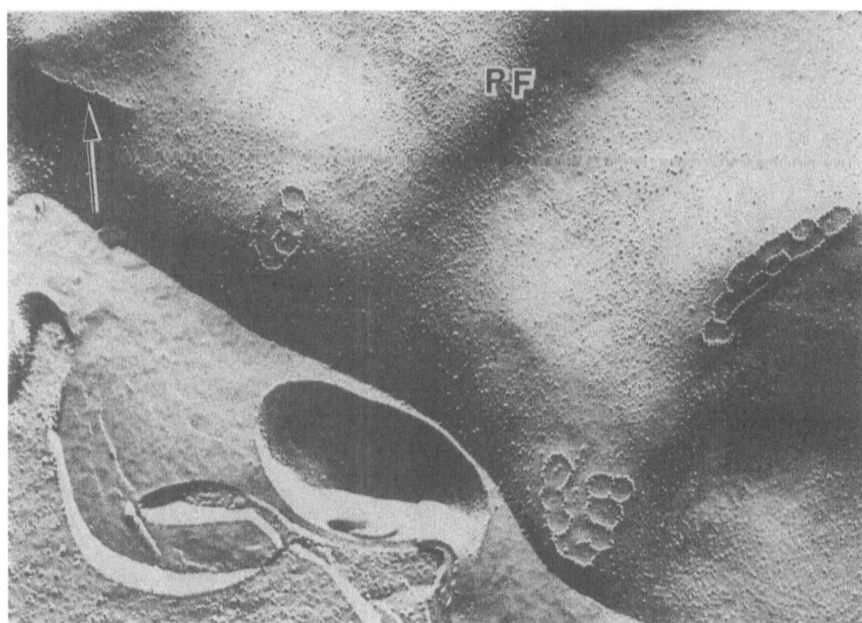

Fig. 3. P fracture face (PF) of frog myocardial cell showing
         the polymorphism of the junctional particle
         arrangements : a single row (arrow) and anastomosed
         circular profiles. (X 60 000). (From : Mazet and
         Cartaud, 1976).

     However in some excitable, and conductive tissues the question
remained open up to 1976 (11). As in frog cardiac tissues, the
freeze-fracture technique was the only one which permits
unambiguous identification of gap junction structures. The frog
myocardial cells are connected by gap junctions exhibiting an
unusual pattern : the junctional particles are gathered in small
assemblies of lines, circles and associations of circles (Fig. 3)
(ref. 11-12). The junctional nature of the arrays of particles
observed was demonstrated by the fact that they were found in
regions in which the intercellular space between adjacent plasma
membranes was absent, and that on the extracellular fracture face
the complementary pits were visible. The main structural feature
was the presence of anastomosed circular profiles of particles
circumscribing smooth areas (conf. 11). Similar arrangements have
been found in different tissues : photoreceptor cells (13),
reptilian cardiac tissues (14), and embryonic mammalian cardiac
tissues (15-2).

     In one amphibian cardiac muscle : Xenopus laevis, the only gap
junction feature observed is a linear or circular row of particles

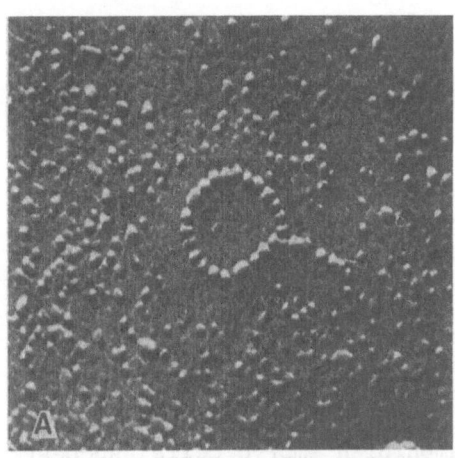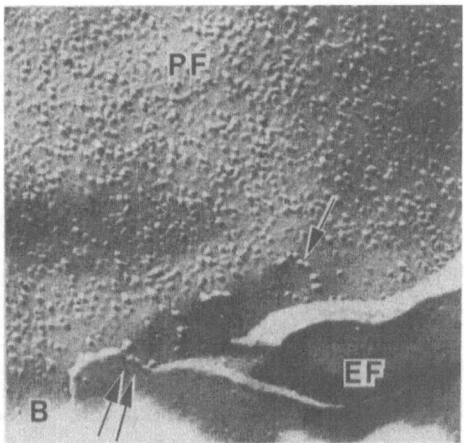

Fig. 4. A) Freeze fracture replica of Xenopus adult ventricular myocardial cells. On the PF a circular row of
particles associated with a short linear row is
observed (X 170 000).
B) Freeze fracture of young Rana tadpole myocardial
cells. The two fracture faces are in close
apposition at the level of small clusters of
particles (arrow) set in a field which is relatively
devoid of other membrane particles. These junctional
particles are assembled on the P face (PF) and
corresponding depressions are visible (double arrow)
on the matching EF. It is believed that such
clusters may represent the beginning of new
communicating junctions forming between myocardial
cells (X 80 000). (A and B, from Mazet, 1977).

circumscribing a smooth area (Fig. 4 A). The most complex
arrangement of the junctional particles is found in this adult
conductive tissue (16). The synchronous heart beat implicates the
propagation of electrical signals from cell-to-cell which implies
the existence of low resistance pathways between them. These
observations show that one cannot attribute electrical conduction
exclusively to the large patches of junctional particles. Indeed
linear arrays of intramembranous particles may also permit
electrical conduction.

The junctions existing between myocardial cells in young Rana
tadpoles (8 days after hatching) have been observed ; this suggests
that the linear array correspond to a step in gap junction
formation. In this myocardium the freeze-fracture replicas of
membrane fusion revealed small clusters (two or three particles) of
structurally identical particles in a smooth field of membrane

leaflets which are relatively devoid of particles (Fig. 4 B). These
formations of small matching clusters of intramembranous particles
interlock the plasma membrane of adjacent cells and obliterate the
intercellular space as indicated by the arrow (Fig. 4 b). The small
clusters (2 or 3 particles) seem to be the candidate to ensure the
conduction between the myocardial cells (with synchronous beats).
This result reinforced the idea that the most important is the
existence of particles making contact between the two cytoplasms
whatever be the arrangement of the particles in the plane of the
membrane.

The question that arises now is : how does this unit work ?

        The experiment of Loewenstein et al. (1978, ref. 17) showing
that junctional conductance develops by quantum steps during
junction formation between pair of amphibian embryo cells, suggests
that the opening of each channel is not progressive but corresponds
to a sudden change of conformation. This experiment suggests that
each particle is an independant functional unit.

        In parallel morphological studies done by Unwin and Zampighi,
1980 (18) and Unwin and Ennis, 1984 (19) two defined configuration
of the gap junction particle have been demonstrated. By tilting
negatively stained specimens at various angles to the incident beam
and by analysing the electron microscope pictures by Fourier and
image reconstruction technique, these authors have obtained an 18
Å resolution three-dimensional map of isolated rat liver gap
junctions protein. According to these authors the junctional unit
can be depicted as an annular oligomer composed of six protein
subunits which span the membrane and protrude from either side. The
subunits are roughly rod shaped, about 25 Å in diameter and 75 Å
long. The path taken by the subunits is also inclined tangentially
with respect to the six-fold axis of the junctional unit, giving to
the whole assembly a left hand character (Fig. 5). Depending on the
isolation process Unwin and Ennis (19) described two possible
states for the junctional unit. These state were interpreted in
terms of two kinds of displacements of the subunits, one radial and
the other tangential, about the 6-fold axis through the center of
the connexon. The transition between the two forms was produced by
radial inward motion of the subunits near their cytoplasmic
extremities and a reduction of their inclination tangential to the
6-fold axis. A similar model was also proposed by Makowsky et al.
1982 (20). If this model really represents a gap junction particle,
it provides a mechanism to explain the opening and closing of a
channel. As this model was done on isolated junctions and using
several detergents the mechanism is strictly presumptive, however
plausible, and the identification of a particular functional state
remains an open question.

        Also it is not yet possible to relate the conducting state of

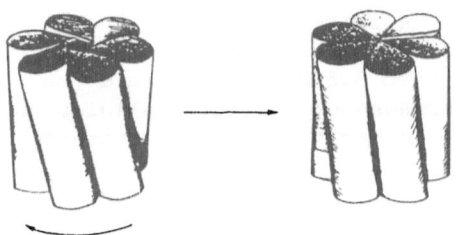

Fig. 5.   Model of the connexon, depicting the transition   from
          the 'open' to the 'closed' configuration suggested by
          X-ray experiments and the   three-dimensional   map   of
          isolated   rat   liver   gap   junctions. (From Unwin and
          Zampighi, 1980 ; ref. 18).

a gap junction to the appearance of an individual particle, several
studies   have   tried to relate the conductance state to the overall
features of the particle assemblies. Therefore, it was   interesting
to   determine   if   it   is   possible   in   uncoupled   conditions,   to
demonstrate ultrastructural changes   in   the   organisation   of   the
particles or in the diameter of each particles as has been shown in
mammalian cardiac tissues (21-22). It has been found in non-cardiac
tissues   that   the   uncoupling   is associated with a highly regular
hexagonal packing and a closely spaced junctional particle   pattern
in   the   plane of the junctional membrane (23-24-25). This view has
been challenged by the observation of dispersion of the   junctional
particle patches associated with electrical uncoupling (26-27).

     Therefore   it   was   of   interest   to   explore   the   structural
correlates of gap junctions in frog myocardium which   exhibit   an
atypical   configuration   of   the   gap junction as it has been shown
previously (Fig. 3). The fate of gap   junctions   has   been   studied
after   different   exposures   of   the   frog auricle to CO_2-saturated
Ringer (30-31), since CO_2 has been reported to induce uncoupling in
the early amphibian embryo by intercellular acidification (28).

     In   parallel   to   the morphological study we have investigated
the junctional conductance changes in   frog   atrial   fibers   during
exposure   to CO_2-saturated Ringer,   using   the double sucrose gap
technique (29). This technique allows us   to   monitor   the   voltage
drop   along the whole length of the preparation when current pulses
of constant amplitude was passed from one end to the other. Due   to
the   sucrose   gap   arrangement   the   potential   variations   reflect
changes   in   internal   resistance. On   the   same   preparation   in
alternation   with the longitudinal resistance, the action potential
was recorded also.

     The Ringer's solution bubbled with 100 % CO_2 abbreviated   the
action   potential   and   increase   its   threshold : no or very small

responses could be elicited after 8-10 mn . Meanwhile the longitudinal resistance increased up to three times its original value (0.8 - 1.5 M up to 3.5 M) which could be attributed to a large significant decrease of the intercellular conductance. It was partially reversed by going back to the reference perfusion.

Before fixation of isolated frog auricles for subsequent freeze-fracture analysis, propagation of electrical activity elicited by point stimulation was checked by intercellular microelectrodes. When compared to the reference conditions, several modifications occured in the CO_2-enriched medium : a marked decrease in the number of particles per junctional assembly and a dispersion of these assemblies on the membrane freeze-fracture faces. This indicates a lowering of the organization of the gap junctions. In parallel loose clusters of large particles became visible on the P-fracture face of these cardiac cells with no complementary pits on the E-fracture face. On return to the reference Ringer perfusion, the number of gap junctions on the membrane fracture faces increased to a value close to the control value (31).

So in frog myocardial cells in uncoupled conditions the main effect observed was a lowering of the organization of the gap junctions and a dispersion of the assemblies over the cell membrane. The appearance of loose clusters could represent hemi-channels undergoing an enlargement and a flattening following their disconnection.

In conclusion

We believe that several structural modifications to gap junctional transmembrane particles could be associated with the electrical uncoupling.

Firstly, the channel would be closed by a simple change of conformation without any modification of the particle arrangement. This modification should be difficult to detect in a non cristalline gap junction organization such as the frog myocardium. Secondly, the closure would be followed by a loosening of particle arrangement and finally the disconnection of the two opposing half gap junctions could generate in the plane of the junctional membrane an internal rearrangement of the oligomer resulting in the enlargement and flattening of the freeze-fracture junctional particles.

REFERENCES

1.   C. Peracchia, Inter. Rev. Cytol., 66 : 81-146 (1980).
2.   F. Mazet and J.C. Ehrhart, J. Physiol. Paris, 76 : 529-549 (1980).

3. W.R. Loewenstein, Physiological Review, 61 : 829-913 (1981).

4. S. Weidmann, J. Physiol. London, 118 : 348-360 (1952).

5. Y. Kanno and W.R. Loewenstein, Science, 143 : 959-960 (1964).

6. W.R. Loewenstein, Ann. N.Y. Acad. Sci., 137 : 441-472 (1966).

7. J.P. Revel and M.J. Karnovsky, J. Cell Biol., 33 C : 7-18 (1967).

8. E.L. Benedetti and P. Emmelot, J. Cell Biol., 38 : 15-28 (1968).

9. J.D. Sheridan, M. Hammer-Wilson, D. Preus and R.G. Jonhson, J. Cell Biol., 76 : 532-544 (1978).

10. S. Gabella and D. Blundell, J. Cell Biol., 82 : 239-247 (1979).

11. F. Mazet and J. Cartaud, J. Cell Sci., 22 : 427-434 (1976).

12. R.W. Kensler, P. Brink and M.M. Dewey, J. Cell Biol., 73 : 768-781 (1977).

13. E. Raviola and N.B. Gilula, Proc. Natl. Acad. Sci. USA, 70 : 1677-1681 (1973).

14. Y. Shibata and T. Yamamoto, J. Ultrastruc. Res., 67 : 79-88 (1979).

15. D. Gros, J.P. Mocquard, C.E. Challice and J. Schrevel, J. Cell Sci., 30 : 45-61 (1978).

16. F. Mazet, Develop. Biol., 60 : 139-152 (1977).

17. W.R. Loewenstein, Y. Kanno and S.J. Socolar, Nature London, 274 : 133-136 (1978).

18. P.N.T. Unwin and G. Zampighi, Nature London, 283 : 545-549 (1980).

19. P.N.T. Unwin and P.D. Ennis, Nature London, 307 : 609-613 (1984).

20. L. Makowski, D.L.D. Caspar, D.A. Goodenough and W.C. Phillips, Biophys. J., 37 : 189-191 (1982).

21. D. Dahl and G. Isenberg, J. Membrane Biol., 53 : 63-75 (1980).

22. J. Delèze and J.C. Hervé, J. Membrane Biol., 74 : 203-215 (1983).

23. C. Peracchia, J. Cell Biol., 72 : 628-641 (1977).

24. C. Peracchia and L.L. Peracchia, J. Cell Biol., 87 : 708-718 (1980).

25. C. Peracchia and L.L. Peracchia, J. Cell Biol., 87 : 719-727 (1980).

26. K.L. Campbell and D.F. Albertini, Tissue and Cell, 13 : 651-668 (1981).

27. W.M. Lee, D.C. Cran and N.J. Lane, J. Cell Sci., 57 : 215-228 (1982).

28. L. Turin and A.E. Warner, J. Physiol. London, 300 : 489-504 (1980).

29. O. Rougier, G. Vassort and R. Stämpfli, Pflügers Arch. Physiol., 301 : 91-108 (1968).

30. I. Dunia, F. Mazet, J.L. Mazet and G. Vassort, J. Physiol. London, 349 : 46 pp.

31. F. Mazet, I. Dunia, G. Vassort and J.L. Mazet, J. Cell Sci., (in press).

(1) Gratecos, D. (2) Salbas, A. (3) Falugi, C. (4) Segmueller, M.
(5) Salbas, K. (6) Hauch, A. (7) Green, C.R. (8) Burger, P.M. (9)
Bloch, R. (10) Levin, L. (11) Clara, N. (12) Pugh, D.M. (13) Hazan,
R. (14) Oezer, N. (15) Dieterlen, F. (16) Borges, M.T.C. (17)
Smith, O. (18) Acan, N.L. (19) Longo, A. (20) Pinto, R. (21)
Raineri, M. (22) Broadley, K.N. (23) Burger, M.M. (24)
Augusti-Tocco, G. (25) Nghiêm, H.O. (26) Erkman, L. (27) Hennequin,
E. (28) Gualandris, L. (29) Morin, L.D.R. (30) Kiortsis, V. (31)
Sharma, K.K. (32) Pippia, P. (33) Streit, A. (34) Pugh, D.M. (35)
Marthy, H.J. (36) Alder, H. (37) Allen, F.L. (38) Vermeulen, J.
(39) Stewart, W.W. (40) Guthrie, S.C. (41) Baroffio, A. (42)
Curtis, A.S.G. (43) Padmanabhan, R. (44) Bernheim, L. (45)
Johnston, D.A. (46) Anastasio, G. (47) Kilinc, K. (48) Jonas, J.
(49) Dean, M.F. (50) Caruso, R.L. (51) Goudou, D. (52) Mazet, F.
(53) Olive, J. (54) Kordeli, C. (55) Schmid, V. (56) Garrod, D.R.
(57) Cartaud, J. (58) Poiana, G. (59) Dertinger, H. (60) Kinn, S.R.
(61) Bee, J.A. (62) Van den Hoef, M.H.F. (63) Fischer, E.G. (64)
Reggio, H. (65) Measures, H. (66) Falke, N.

PARTICIPANTS

Acan, N.L.	Hacettepe University, Faculty of Medicine, Biochemistry Department, Ankara, Turkey
Alder, H.	Zoologisches Institut, Universität Basel, 4051 Basel, Switzerland
Allen, F.L.	University College London, Department of Anatomy, Gower Street, London WCIE 6BT, U.K.
Anastasio, G.	Instituto di Istologia ed Embriologia, Facolta di Scienza, Via Mezzocannone 8, 80134 Napoli, Italia
Augusti-Tocco, G.	Universita di Roma La Sapienza, Dipartimento di Biologia Cellulare e dello Sviluppo Piazzale A. Moro I-00185 Roma, Italia
Baroffio, A.	Institut de Physiologie, Université de Lausanne, rue de Bugnon 7, 1001 Lausanne, Switzerland
Bee, J.A.	Department of Anatomy, The Royal Veterinary College, University of London, Royal College Street, London NWL OTU, U.K.
Bernheim, L.	Department de Physiologie, Centre médical Universitaire, 1 rue Michel Servet, 1211 Genève 4, Switzerland
Bloch, R.	University of Maryland, Department of Physiology, 660 West Redwood Street, Baltimore, Maryland 21201, U.S.A.
Borges, M.T.C.	Faculdade de Ciencias, Departamento de Zoologia, Rua da Escola Politécnica, 1200 Lisboa, Portugal
Broadley, K.N.	Department of Cell Biology, The University of Glasgow, Glasgow G12 8QQ, Scotland, U.K.
Burger, M.M.	Department of Biochemistry, Biocenter of the University of Basel, Klingelbergstrasse 70, CH-4056 Basel, Switzerland
Burger, P.M.	Universität zu Köln, Zoologisches Institut, Weyertal 119, 5000 Köln 41, West Germany

Cartaud, J.   Institut Jacques Monod, Laboratoire de
                               Microscopie Electronique, Université
                               Paris VII, 2 Place Jussieu, 75251 Paris
                               cedex 05, France
Caruso, R.L.   Department of Pediatrics and Human
                               Development, College of Human Medicine,
                               Michigan State University, B-240 Life
                               Science Building, East Lansing, Mi 48864,
                               USA
Curtis, A.S.G.   Department of Cell Biology, The University
                               of Glasgow, Glasgow C12 8QQ, Scotland,
                               U.K.
Dean, M.F.   Kennedy Institute of Rheumatology, 6 Bute
                               Gardens, Hammersmith, London W6 7DW,
                               U.K.
Dertinger, H.   Kernforschungszentrum Karlsruhe GmbH.,
                               Institut für Genetik, Postfach 3640, 7500
                               Karlsruhe, West Germany
Dieterlen-Lièvre, F   Institut d'Embryologie du CNRS et du
                               Collège de France, 49 bis, avenue de la
                               Belle-Gabrielle, 94130 Nogent-sur-Marne,
                               France
Duband, J.-L.   Institut d'Embryologie du CNRS et du
                               Collège de France, 49 bis Avenue de la
                               Belle-Gabrielle, 94130 Nogent-sur-Marne,
                               France
Erkman, L.   Université de Genève, C.M.U., Department de
                               Pharmacologie, 9 avenue Champel / 21 rue
                               Lombard, 1211 Genève, Switzerland
Falke, N.   Lindenstrasse 31, 7900 Ulm, West Germany
Falugi, C.   Istituto di Anatomia Comparata, Università
                               degli Studi, Via Balbi 5, 16126 Genova,
                               Italy
Fischer, E.G.   Wagnerstrasse 59, 7900 Ulm, West Germany
Franke, W.W.   German Cancer Research Center, P.O. Box
                               101949, 6900 Heidelberg 1, West Germany
Garrod, D.R.   University of Southampton, Faculty of
                               Medicine, Centre Block CF 99, Southampton
                               General Hospital, Southampton SO9 4XY,
                               U.K.
Gilula, N.B.   Baylor College of Medicine, Department of
                               Cell Biology, Texas Medical Center,
                               Houston, Texas 77030, USA
Goudou, D.   U 153 ISERM, 17 rue du Fer-à-Moulin, 75005
                               Paris, France
Gratecos, D.   L.G.B.C.-C.N.R.S., Faculté de Luminy, 70
                               route de Léon Lachamp, 13009 Marseille,
                               France

Green, C.R.	Baylor College of Medicine, Department of Cell Biology, Texas Medical Center, Houston, Texas 77030, USA
Gualandris, L.	Laboratoire de Biologie Générale, Université Paul Sabatier, 118 route de Narbonne, 31062 Toulouse, France
Guthrie, S.C.	Department of Anatomy, University College, Gower Street, London WC1 E 6BT, U.K.
Hauch, A.	Zoologisches Institut, Im Neuenheimerfeld 230, 69 Heidelberg, West Germany
Hazan, R.	German Cancer Research Center, P.O. Box 1°949, 6900 Heidelberg 1, West Germany
Hennequin, E.	Faculté des Sciences et Techniques de Rouen, Laboratoire Echanges Cellulaires, 76130 Mont Saint-Aignan, France
Imhof, B.A.	Institut d'Embryologie du CNRS et du Collège de France, 49 bis, avenue de la Belle-Gabrielle, 94130 Nogent-sur-Marne, France
Johnston, D.A.	C.R.C. Medical Oncology University, Southampton General Hospital, Tremorna Road, Southampton S09 4XY, U.K.
Jonas, J.	Zoologisches Institut der Universität, Abteilung Experimentelle Morphologie, Weyertal 119, 5000 Köln 41, West Germany
Kilinc, K.	Hacettepe University, Faculty of Medicine / Department of Biochemistry, Hacettepe, Ankara, Turkey
Kinn, S.R.	Department of Cell Biology, University of Glasgow, Glasgow 12 8QQ, Scotland, U.K.
Kiortsis, V.	Zoological Laboratory, University of Athens, Panepistimiopolis, 15771 Athens, Greece
Kordeli, C.	Laboratoire de Microscopie Electronique, Institut Jacques Monod, Université Paris VII, 2 Place Jussieu, 75221 Paris, France
Levin, L.	Harvard Medical School, Department of Neurobiology, 25 Shattuck Street, Boston Mass. 02115, USA
Longo, A.	Istituto di Istologia ed Embriologia, Facoltà di Scienze, Via Mezzocannone 8, 80134 Napoli, Italy
Mc Carthy, R.	Biozentrum der Universität, Klingelbergstrasse 70, 4056 Basel, Switzerland
Marthy, H.-J.	Laboratoire Arago, Université Pierre et Marie Curie (Paris VI) / C.N.R.S. (U.A. 117), 66650 Banyuls-sur-mer, France

Mazet, F.   Laboratoire de Physiologie Comparée,
                           Université d'Orsay, Bâtiment 443, 91405
                           Orsay, France

Measures, H.R.   Medical Oncology Unit, Level F, General
                           Hospital CF 99, Tremorna Road,
                           Southampton SO9 4XY, U.K.

Misevic, G.   Department of Biochemistry, Biocenter of
                           the University of Basel,
                           Klingelbergstrasse 70, CH-4056 Basel,
                           Switzerland

Morin, L.D.R.   Université Nationale de Rwanda, Butare,
                           Rwanda

Nghiêm, H.O.   Institut Pasteur I.B.M., 25 rue du Dr Roux,
                           75724 Paris, France

Nowakowski, R.S.   Department of Anatomy, The University of
                           Mississipi / Medical Center, 2500 North
                           State Street, Jackson, Miss. 39216, USA

Oezer, N.   Hacettepe University, Department of
                           Biochemistry, Hacettepe, Ankara, Turkey

Olive, J.   Laboratoire de Microscopie Electronique,
                           Institut Jacques Monod, Université Paris
                           VII, 2 Place Jussieu, 75221 Paris,
                           France

Padmanabhan, R.   Kuwait University, Faculty of Medicine /
                           Department of Anatomy P.O. Box 24923,
                           Kuwait, Arabian Gulf

Pinto, M.R.   Stazione Zoologica, Villa Communale, 80121
                           Napoli, Italy

Pippia, P.   Istituto di Fisiologia Generale e Chimica
                           Biologica, Via Muroni 25, Sassari, Italy

Pitts, J.D.   The Beatson Institute for Cancer Research,
                           Wolfson Laboratory for Molecular
                           Pathology, Garscube Estate, Switchback
                           Road, Bearsden, Glasgow G61 1BD, U.K.

Poiana, G.   Università di Roma La Sapienza,
                           Dipartimento di Biologia Cellulare e
                           dello Sviluppo Piazzale A. Moro I-00185
                           Roma, Italy

Porter, T.L.   National Science Foundation, Washington
                           D.C. 20550, USA

Pugh, D.M.   Department of Veterinary Medicine, Faculty
                           of Veterinary Medicine UCD, Veterinary
                           College Dublin, Ballabridge, Dublin 4,
                           Ireland, U.K.

Raineri, M.   Istituto di Anatomia Comparata, Università
                           degli studi, Via Balbi 5, 16126 Genova,
                           Italy

Reggio, H.   Institut de Cytologie et de Biologie
                            Cellulaire, Université d'Aix-Marseille,
                            Centre de Marseille-Luminy, 70 route Léon
                            Lachamp, 13288 Marseille, France
Salbas, K.   Biophysics Research Laboratory, Faculty of
                            Ankara University, Ankara, Turkey
Schmid, V.   Zoologisches Institut, Universität Basel,
                            Rheinsprung 9, 4051 Basel, Switzerland
Segmuller, M.   Zoologisches Institut, Universität Basel,
                            Rheinsprung 9, 4051 Basel, Switzerland
Sharma, K.K.   Zoologisches Institut der Universität,
                            Isotopen Laboratorium Weyertal 119, 5000
                            Köln, Lindenthal, West Germany
Smith, J.   Institut d'Embryologie, 49 bis avenue de la
                            Belle-Gabrielle, 94130 Nogent-sur-Marne,
                            France
Smith, O.   Department of Biochemistry, Royal Free
                            Hospital School of Medicine, Rowland Hill
                            Street, London NW3 2FF, U.K.
Stewart, W.W.   Department of Health and Human Services,
                            Room 339, Building 4, National Institutes
                            of Health, Bethesda, Maryland 20205, USA
Streit, A.   Rathenauplatz 17, 5000 Köln, West Germany
Thiery, J.P.   Institut d'Embryologie, 49 bis avenue de la
                            Belle-Gabrielle, 94130 Nogent-sur-Marne,
                            France
Turin, L.   Station Zoologique, Université Pierre et
                            Marie Curie, 06230 Villefranche-sur-mer,
                            France
Vaheri, A.   Department of Virology, University of
                            Helsinki, Haartmaninkatu 3, 00290
                            Helsinki, Finland
Van den Hoef, M.H.F.   Kariboestraat 257, 3584 Utrecht, The
                            Netherlands
Vermeulen, J.   Hubrecht Laboratorium, Uppsalalaan 8, 3584
                            CT Utrecht, The Netherlands

AUTHOR INDEX

Augusti-Tocco, G.   Kam, E.
Baetscher, M.   Kapprel, H.P.
Bee, J.A.   Kordeli, K.
Behrens, J.   Krieg, J.
Birchmeier, W.   Le Douarin, N.M.
Bloch, R.J.   Louvard, D.
Burger, M.M.   Marthy, H.-J.
Buultjens, J.   Mazet, F.
Cartaud, J.   Misevic, G.
Coudrier, E.   Nghiêm, H.-O.
Cowin, P.   Nowakowski, R.S.
Curtis, A.S.G.   Pitts, J.D.
Dieterlen-Lièvre, F.   Pumplin, D.W.
Duband, J.-L.   Reggio, H.
Eldridge, T.   Shuttleworth, J.
Feder, N.   Smith, J.
Franke, W.W.   Stewart, W.W.
Finbow, M.E.   Thiery, J.P.
Garrod, D.R.   Turin, L.
Gilula, N.B.   Vaheri, A.
Green, C.R.   Von der Mark, K.
Imhof, B.A.   Vollmers, H.P.